MARINE DIESEL ENGINES

THIRD EDITION

MARINE DIESEL ENGINES

MAINTENANCE, TROUBLESHOOTING, AND REPAIR

THIRD EDITION

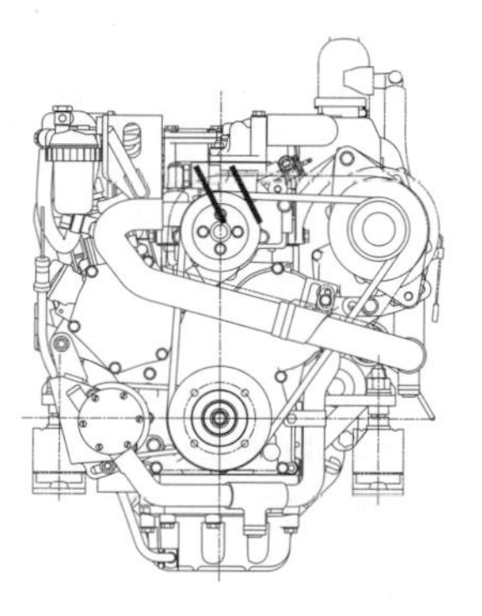

NIGEL CALDER

International Marine/McGraw-Hill
Camden, Maine • New York • Chicago • San Francisco • Lisbon
London • Madrid • Mexico City • Milan • New Delhi • San Juan
Seoul • Singapore • Sydney • Toronto

Other Books by Nigel Calder

Boatowner's Mechanical and Electrical Manual: How to Maintain, Repair and Improve Your Boat's Essential Systems, Third Edition

The Cruising Guide to the Northwest Caribbean: The Yucatan Coast of Mexico, Belize, Guatemala, Honduras, and the Bay Islands of Honduras

Cuba: A Cruising Guide

How to Read a Nautical Chart: A Complete Guide to the Symbols, Abbreviations, and Data Displayed on Nautical Charts

Nigel Calder's Cruising Handbook: A Compendium for Coastal and Offshore Sailors

Refrigeration for Pleasureboats: Installation, Maintenance, and Repair

Repairs at Sea

The **McGraw·Hill** *Companies*

16 17 18 19 20 LCR 21 20 19 18

Library of Congress Cataloging-in-Publication Data
Calder, Nigel.
Marine diesel engines : maintenance, troubleshooting, and repair / Nigel Calder.—3rd ed.
p. cm.
Includes index.
ISBN 0-07-147535-4 (hardcover : alk. paper)
1. Marine diesel motors—Maintenance and repair. I. Title.
VM770.C25 2006
623.87'2368—dc22 2006008729

ISBN-13: 978-0-07147535-8
ISBN-10: 0-07-147535-4

Questions regarding the content of this book should be addressed to

International Marine
P.O. Box 220
Camden, ME 04843
www.internationalmarine.com

Questions regarding the ordering of this book should be addressed to

The McGraw-Hill Companies
Customer Service Department
P.O. Box 547
Blacklick, OH 43004
Retail customers: 1-800-262-4729
Bookstores: 1-800-722-4726

All photos by the author unless otherwise noted.
Illustration page iii courtesy Yanmar.

To Terrie,
who never minds getting grease under her fingernails

CONTENTS

Appendices

LIST OF TROUBLESHOOTING CHARTS

An untrained observer will see only physical labor, and often gets the idea that physical labor is what the mechanic does. Actually, the physical labor is the smallest and easiest part of what the mechanic does. By far the greatest part of his [or her] work is careful observation and precise thinking.

Robert M. Pirsig
Zen & the Art of Motorcycle Maintenance

PREFACE TO THE THIRD EDITION

More than ten years have gone by since I wrote the second edition of this book. In this time, there have been significant changes in the diesel engine world. Whereas in the past such changes have occurred largely as a result of economic and competitive pressures, in recent years the motivation for change has come from what is known as "technology forcing" legislation, primarily in the form of ever-tightening emissions standards (see the Technology Forcing Legislation sidebar next page). When such legislation is first introduced, many in the industry argue that the new standards will be impossible to meet, but in fact, as each successive deadline has approached, manufacturers have invariably succeeded in exceeding the new requirements. Some will admit off the record that the legislative pressure has been good for the industry.

When first proposed, most of the tightened standards were not applicable to marine engines. But because of the relatively small size of the marine marketplace (approximately 50,000 diesel engines up to 800 horsepower worldwide each year, as opposed to millions of engines in the automotive and trucking industries), many marine diesel engines have always been adapted from other applications, and to the degree that the new standards applied to these applications, the technology found its way onto boats. From about 2004 onward, marine engines have been specifically included in both international and U.S. EPA regulations, with increasingly stringent emissions requirements being phased in over the five-year period from 2004 to 2009. This has resulted in numerous technological changes, most of which are invisible to boatowners, consisting of refinements in materials and design elements that have little impact on operating and maintenance practices. As such the changes have had, and continue to have, little practical impact on most boatowners. The two notable exceptions are electronic engine controls and common rail fuel injection.

Electronic engine controls and common rail fuel injection are different, with considerable practical implications, so I have worked them both into this new edition. Even so, it is worth noting that these technologies have typically not yet filtered down to marine diesel engines below 100 hp (76 kW), and in terms of the major players in this marketplace—Volvo Penta and Yanmar—are not likely to make their way into this horsepower range anytime soon. Thus, they can be ignored by the owners of most small auxiliary diesel engines.

The net result of this picture is that despite numerous modifications to diesel engines, as far as most boatowners with an inboard diesel engine are concerned, there has been little change over the past ten years at the propulsion end of things. On the transmission side, things are a little different. We have seen major inroads into the sailboat market by saildrives, an innovation that replaces the conventional propeller shaft and shaft seal, and which consists of an inboard diesel engine connected to a drive leg that passes through the bottom of the boat. I look at saildrives in Chapter 8, Marine Transmissions.

Readers of earlier editions of this book will notice that many of the illustrations have been updated, and I have added a considerable number of new ones. I would particularly like to thank Jan Dahlsten at Volvo Penta and Greg Eck at Yanmar for helping me assemble this art. Scattered throughout the text are also numerous small pearls of wisdom and practical tips supplied by my readers over the years—my thanks to all of them.

Once again, it has been a pleasure to work with the crew at International Marine, especially Molly Mulhern, Janet Robbins, and Margaret Cook.

Nigel Calder
Cruising on *Nada* in Northern Europe,
Summer 2006

Technology Forcing Legislation

The U.S. Environmental Protection Agency (EPA) began working on emissions standards for cars and trucks soon after the passage of the 1970 Clean Air Act, but those did not apply to marine engines. The first major standards regulating marine diesel engines were adopted in 1973 as Annex VI to the International Convention on Prevention of Pollution from Ships. This was subsequently modified by the Protocol of 1978, known as MARPOL 73/78.

The Annex VI standards themselves only apply to engines over 130 kW (170 hp) installed on vessels constructed after January 2000 and were not enforceable until May 2005. In the meantime, as a consequence of the passage of amendments to the Clean Air Act in 1990, the EPA adopted these standards as Tier 1 standards, first as voluntary and then as mandatory starting in January 2004 for all *commercial* marine engines with cylinder sizes from 2.5 liters (that's a fairly large engine) to 30 liters (that's a huge engine). Next, the EPA developed much tougher standards, known as Tier 2 standards, for all *commercial* marine diesel engines over 37 kW (48 hp) and up to 30 liters per cylinder, with phased implementation (depending on engine size) between 2004 and 2007 (the smaller engines have to comply first). Marine diesel engines below 37 kW are included in tough emissions standards that govern off-road diesel engines.

There is a separate category in the regulations for *recreational* marine diesel engines. These must be installed on vessels used for recreational purposes with a power output above 37 kW (48 hp) and a cylinder size less than 5 liters. They are subject to the Tier 2 standards also; the only difference is that the implementation phase extends to 2009, once again based on engine size. However, since the compliance deadline for all but the largest recreational marine diesel engines is January 2007, to all intents and purposes there is no difference between commercial and recreational marine diesels in the engine sizes covered by this book.

The Tier 2 standards progressively enforce radical reductions in certain exhaust emissions. Given the current state of diesel engine technology, large engines (above 2.5 liters per cylinder) can only meet the Tier 2 standards through the medium of electronic engine controls (see pages 28–33), hence the rapid spread of this technology on high-powered engines. Smaller marine diesel engines can meet the Tier 2 standards with more traditional technology, hence the survival—for the time being—of these engines in the marine world, although their days are almost certainly numbered.

The Tier 2 standards presuppose, and can only be met through, the use of ultra-low-sulfur diesel (ULSD) fuel (see the EPA Sulfur Rules sidebar on page 51). Although the sulfur limits became mandatory in the United States in 2006, diesel fuel with a much higher sulfur content is still common in many parts of the world. It is not at all clear to me what impact, if any, higher-sulfur fuel will have on the operation of a diesel engine designed to run on ULSD fuel.

PREFACE TO THE SECOND EDITION

It is five years now since I went to work on the first edition of this book. Since then it has been a best-seller, which is very gratifying. In the intervening years, there have been no startling or revolutionary technological breakthroughs in the world of marine diesel engines, but as always there has been a steady and continuing process of improvement and change. This second edition has enabled me to catch up with some of these developments.

Of more significance than changes in the diesel world is the expansion and refinement of my own ideas on a number of the topics covered in this book. The new edition reflects this. In particular, I pay more attention to the electrical side of marine diesel engines—a subject that was given pretty short shrift in the first edition—and I have expanded the section on engine overhauls to include many new illustrations, for most of which I am greatly indebted to William Gardner Ph.D., author and lecturer in the diesel engine field, and the Caterpillar Tractor Co. The coverage of Detroit Diesel two-cycle engines, marine transmissions, and engine installations is also more thorough.

I have extensively rewritten and reorganized the introductory chapters, utilizing material from a series of articles I wrote for *National Fisherman* in 1987; my thanks to them, too. In spite of all these changes, the purpose of the book is unchanged: to provide the basic information necessary to select, install, maintain, and carry out repairs on a marine diesel engine. As such, the book remains neither a simple how-to book nor a technical manual on the thermodynamics of internal-combustion engines. Rather, it falls somewhere between the two.

As I stated in the first edition, this reflects my own experience as a self-taught mechanic with some twenty-five years of experience on a variety of engines from 10 to 2,000 hp. With specific enough instructions it is possible to dismantle an engine and put it back together again without having any understanding of how it works, but troubleshooting that engine is not possible without a basic grasp of its operating principles.

In order to grasp these operating principles, you need to know a little of the most basic theory behind internal-combustion engines. This I try to present, but only insofar as it is necessary to provide a good understanding of the practical side of diesel engine maintenance. My objective is to help turn out competent amateur mechanics, not automotive engineers.

This particular blend of theory and practical mechanics represents a mix that has worked well for me over the years. It should help a boatowner to see trouble coming and to nip it in the bud before an engine breaks down.

Although this book has been written with engines from 10 to 100 hp in mind, the principles are virtually the same as those associated with engines of hundreds or even thousands of horsepower. The information in this book applies to just about all diesels.

Sources of data and drawings are indicated throughout the book, but nevertheless my thanks to all those who have helped me, in particular Paul Landry, Bill Osterholt, and Dennis Caprio, my editor, who made a mass of detailed suggestions that improved the book greatly. The following companies provided drawings and other help: the AC Spark Plug Division of General Motors, Allcraft Corporation, Aquadrive, BorgWarner Automotive, Caterpillar Tractor Co., Deep Sea Seals, Detroit Diesel, Garrett Automotive Products Co., Halyard Marine, Hurth, Hart Systems Inc., Holset Engineering Co. Ltd., ITT/Jabsco, Kohler Generators, Lucas/CAV Ltd., Morse Controls, Paragon Gears, PCM, Perkins Engines Ltd., PYI, Racor, Sabb Motor A.S., Shaft Lok, United Technologies Diesel Systems, VDO, Volvo Penta, Wilcox Crittenden, and Yanmar Diesel Engines. Many of the line drawings were done by Jim Sollers, a wonderful illustrator.

For this second edition, Caterpillar Tractor Co. once again dug deep into its incomparable educational material. Don Allen of Allen Yacht Services (St. Thomas, USVI), Russell Dickinson, Joseph Joyce of Westerbeke, and Steve Cantrell of Volvo Penta all critiqued the text and made many valuable comments.

I extend my thanks to Dodd, Mead for permission to use the material from Francis S. Kinney's *Skene's Elements of Yacht Design*, and to Reston Publishing for information contained in Robert N. Brady's *Diesel Fuel Systems*. My publisher, International Marine, gave me permission to incorporate one or two relevant sections from my *Boatowner's Mechanical and Electrical Manual* in this book, and information from Dave Gerr's *Propeller Handbook* (quite the best book available on this subject). Sections of the revised chapter on engine installations first appeared in an article I wrote for *WoodenBoat* magazine.

Jonathan Eaton, Tom McCarthy, Janet Robbins, and the crew at International Marine have been as helpful and encouraging as ever. Any errors remaining are solely mine.

Nigel Calder
Montana, November 1990

INTRODUCTION

For very good reasons, the diesel engine is now the overwhelming choice for sailboat auxiliaries, and it is becoming more popular in sportfishing boats, high-performance cruisers, and large sportboats. Diesels have an unrivaled record of reliability in the marine environment; they have better fuel economy than gasoline engines; they are more efficient at light and full loads; they emit fewer harmful exhaust pollutants; they last longer; and they are inherently safer because diesel fuel is less volatile than gasoline.

Despite its popularity, the diesel engine is still something of a mystery, in large part because of the differences that distinguish it from the more familiar gasoline engine. Boatowners are frequently nervous about carrying out maintenance and repairs on a diesel. This is surprising because in many respects diesels are easier to understand and maintain than gasoline engines.

The first objective of this book, then, is to explain how a diesel engine works, to define new terms, and to lift this veil of mystery.

If the owner of a diesel engine has a thorough understanding of how it works, the necessity for certain crucial aspects of routine maintenance and the expensive consequences of habitual neglect will be fully appreciated. Properly maintained, most diesel engines will run for years without trouble, which leads me to my second objective—to drive home the key areas of routine maintenance.

If and when a problem arises, it normally falls into one or two easily identifiable categories. The ability to visualize what is going on inside an engine frequently enables you to go straight to the heart of a problem and to rapidly find a solution, without making blind stabs at it. My third objective is to outline troubleshooting techniques that promote a logical, clear-headed approach to problem solving.

The fourth objective of the book is to outline various maintenance, overhaul, and repair procedures that can be reasonably undertaken by an amateur mechanic. It also describes one or two that should not be attempted but which might become necessary in a dire emergency. (Cruising sailors, in particular, sometimes must tackle repairs that they would never even consider when they are ashore. As a fellow sailor, I have at all times tried to keep the rather special needs of cruisers in the forefront of my mind when writing and revising this book.) Major mechanical breakdowns and overhauls are not included. This kind of work can only be carried out by a trained mechanic.

The book ends with a consideration of some criteria to assist you in the selection of a new engine for any given boat and of correct installation procedures. Much of the last chapter may throw some light on problems with an engine already installed.

There is no reason for you as a boatowner not to have a long and trouble-free relationship with a diesel engine. All you need is to set up the engine correctly in the first place, pay attention to routine maintenance, learn to spot the early warning signs of impending trouble, and know how to correct small problems before they become large ones.

CHAPTER 1

PRINCIPLES OF OPERATION

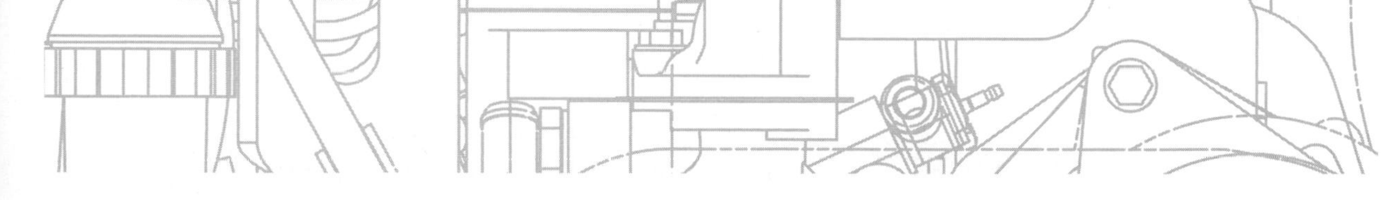

In the technical literature, diesel power plants are known as compression ignition (CI) engines. Their gasoline counterparts are of the spark ignition (SI) variety. This idea of compression ignition is central to understanding a diesel engine.

When a given amount of any gas is compressed into a smaller volume, its pressure and temperature rise. The increase in temperature is in direct relation to the rise in pressure, which is directly related to the degree of compression. That is, the rise in temperature is the result of compressing the existing gas into a smaller space, rather than by the addition of extra heat.

For a better understanding, imagine two heaters with exactly the same output. Each has been placed in a separate room. To begin with, both rooms are the exact same temperature, but one is twice the size of the other. Both heaters are turned on. The small room will heat up faster than the large one, even though the output of the heaters is the same. In other words, although the same quantity of heat is being added to both rooms, the temperature of the smaller one rises faster because the heat is concentrated into a smaller space.

This is crudely analogous to what happens when a gas is compressed. At the outset, it has a given volume and contains a certain amount of heat. As the gas is compressed, this quantity of heat is concentrated into a smaller space and the temperature rises, even though no more heat is being added to the gas.

Compression Ignition

All internal-combustion engines consist of one or more cylinders that are closed off at one end and have a piston driving up the other. In a diesel engine, air enters the cylinder, then the piston is forced up into it, compressing the air.

As the air is compressed, the heat contained in it is concentrated into a smaller and smaller space. The pressure and temperature rise steadily. In a compression ignition engine, this process continues until the air is extremely hot, say around 1,000°F (538°C). This temperature has been attained purely and simply by compression (see Figure 1-1).

Diesel fuel ignites at around 750°F (399°C); therefore, any fuel sprayed into a cylinder filled with air superheated to 1,000°F is going to catch fire. This is exactly what happens: at a precisely controlled moment, fuel is injected into the cylinder and immediately starts to burn. No other form of ignition is needed.

To attain high enough temperatures to ignite diesel fuel, air generally has to be compressed into a space no larger than 1/14 the original size of the cylinder. This is known as a compression ratio of 14:1. The compression ratio is the volume of the cylinder when the piston is at the bottom of its stroke relative to the volume of the cylinder when the piston is at the top of its stroke (see Figure 1-2).

Most diesel engines have compression ratios ranging from 16:1 to 23:1. This is much higher than

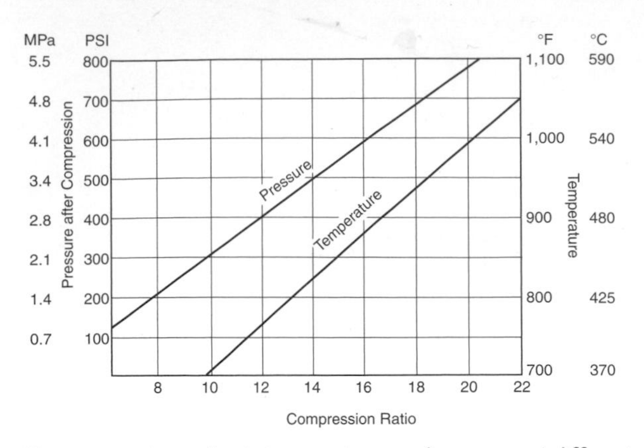

FIGURE 1-1. *Approximate temperatures and pressures at different compression ratios. At "typical" cranking speeds, most healthy diesel engines develop cylinder pressures of 400 psi or higher. (Note: To convert MPa to psi, multiply by 145.)*

the average gasoline engine's compression ratio of 7:1 to 9:1. The lower compression ratios of gasoline engines produce lower cylinder pressures and temperatures. As a consequence, the ignition temperature of gasoline is not reached through compression, and the gas-air mixture has to be ignited by an independent source—a spark (hence the designation *spark ignition* for gasoline engines).

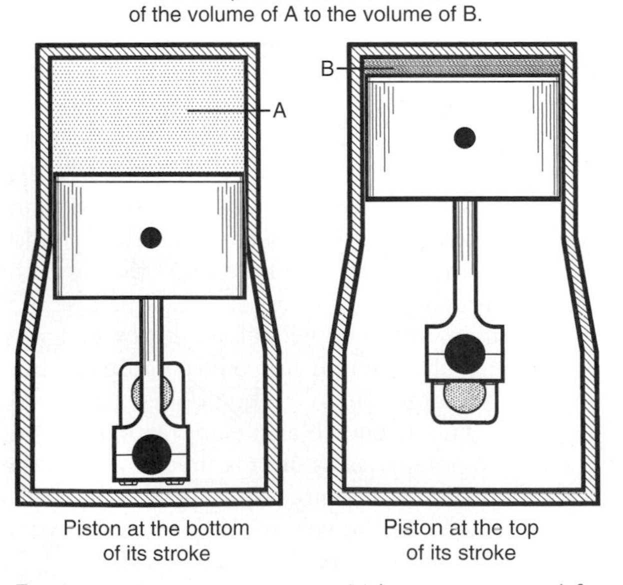

FIGURE 1-2. *Compression ratios. (Valves, etc., omitted for simplicity.) (Fritz Seegers)*

CONVERTING HEAT TO POWER

I've established that when a gas is compressed its temperature rises. It is also true that when a gas is heated in a sealed chamber its pressure rises. In the first instance, no heat is added—the existing gas is merely concentrated into a smaller space, thereby raising its temperature. In the second instance, heat is added to a gas that is trapped in a closed vessel, and this causes the pressure to rise.

This is what happens during ignition in an internal-combustion engine: A body of air is trapped in a cylinder by a piston and compressed. The temperature rises. Fuel is introduced by some means and ignited. The burning fuel raises the temperature in the cylinder even higher, and this raises the pressure of the trapped gases. The increased pressure is used to drive the piston back down the cylinder, resulting in what is termed the piston's *power stroke*. The engine has converted the heat produced by the burning fuel into usable mechanical power. For this reason, internal-combustion power plants are sometimes known as *heat engines*.

It is possible to calculate the heat content of the fuel by measuring how many Btu (or joules, in the metric system) are produced by burning a given amount (e.g., 1 gallon or 1 liter). An engine's horsepower (hp) rating (or kilowatt, in the metric system) can also be converted into Btu or joules; e.g., 1 hp = 2,544 Btu. In this way, the heat energy going into an engine (the number of gallons or liters burned per hour × the Btu or joule content of the fuel) can be compared with the power being produced by it. This enables the thermal efficiency of the engine to be determined; i.e., how much of the fuel's heat energy is being converted to usable power.

The average diesel engine has a thermal efficiency of 30% to 40%. In other words, only about one-third of the heat energy contained in the fuel is being converted to usable power. Roughly half of the remaining two-thirds is lost through the exhaust system in the form of hot gases. The rest is dissipated into the atmosphere through the cooling system and by contact with hot engine surfaces (see Figure 1-3). As wasteful as this sounds, diesels are still considerably more efficient than gasoline engines, which have a thermal efficiency of 25% to 35%.

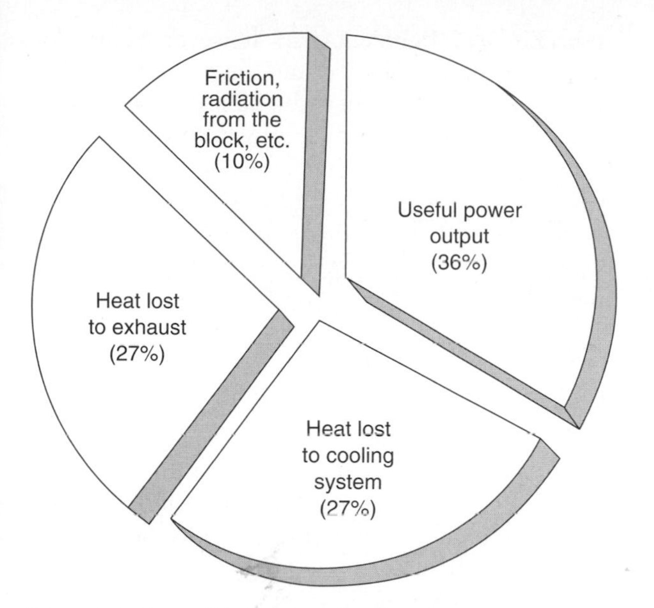

FIGURE 1-3. *Heat utilization in a diesel engine.*

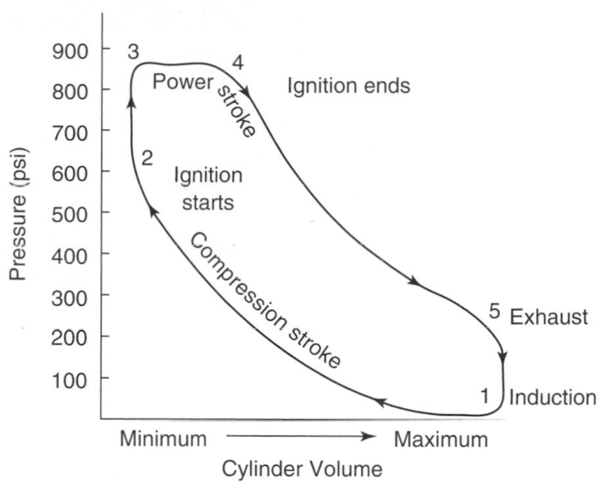

FIGURE 1-4. *Pressure-volume curve for a diesel engine. From position 1 (P1) to position 2 (P2), the pressure steadily rises to around 600 psi as cylinder volume decreases. At around P2, injection occurs and ignition commences. The rapidly rising temperature sharply boosts pressure to about 850 psi (P3). The piston is now on its way back down the cylinder in its power stroke, increasing the volume and decreasing pressure. However, fuel is still burning, and the increasing temperature temporarily counteracts the effect of increasing cylinder volume, which causes a relatively constant pressure from P3 to P4. Ignition is now tailing off, and the cylinder volume continues to increase. This results in a steady decline in pressure and temperature while the piston is still on its power stroke from P4 to P5. From P5 to P1 the engine expels the remaining gases of combustion as exhaust and draws in a fresh charge of air. The cycle starts over.*

EXPANSION AND COOLING

Just as compressing a gas raises its temperature, so too can reducing the pressure lower the temperature. This is due purely to the expansion of the gas into a larger space, or volume, not to any loss of heat. The greater the reduction in pressure, the lower the resulting temperature of the gas.

As a piston moves down on its power stroke, the volume inside its cylinder increases, causing a fall in pressure and consequently a fall in the temperature of the gases in the cylinder. These declining temperatures reflect the heat of combustion being converted into mechanical power, i.e., the movement of the piston (see Figure 1-4).

The higher the compression ratio of an engine, the greater the expansion of gases on the power stroke. In an engine with a compression ratio of 22:1, for example, the gases will expand into a volume twenty-two times the size of the compression chamber. In an engine with a compression ratio of 7:1, the degree of expansion will only be seven times greater. Since the temperature drops as a gas expands, diesel engines, because of their higher compression ratios, are able to convert more of the heat of combustion into mechanical power than their gasoline counterparts. Hence diesels are more thermally efficient than gasoline engines.

GASOLINE ENGINES

You might well ask, why not increase the compression ratio on the gasoline engine and thereby improve its efficiency?

A gasoline engine draws in fuel with its air supply before compression, either through a carburetor or through fuel injection into the inlet manifold (not the cylinder). A diesel, on the other hand, has the fuel injected after the air has been compressed. Increasing the compression ratio for a gasoline engine would raise its compression temperature beyond the ignition point of gasoline, which would lead to premature combustion of the fuel-air mixture. This would rapidly wreck the engine. To avoid this, the compression ratio on a gasoline engine must be kept low, and the fuel-air mixture must then be set off by a spark at the appropriate moment. This accounts for the need for an electrical ignition system on these engines.

On occasion, gasoline engines can become sufficiently overheated to cause the fuel-air mixture to ignite before it should. This is known as *auto-ignition*, or *pre-ignition*, and most often occurs when the overheated engine is turned off but refuses to quit, even though the ignition has been turned off.

You might next ask, why not jack up the compression ratios on gasoline engines and use fuel injection directly into the cylinders to prevent premature ignition, just as is done on a diesel? Apart from the obvious fact that you now have a diesel engine to all intents and purposes and might as well use the often cheaper diesel fuel, you run up against the nature of gasoline itself, which is far more volatile than diesel.

Even though a diesel engine may be turning over at 3,000 rpm (revolutions per minute), with the power stroke of any one piston lasting no more than 1/100 of a second, the injected diesel fuel burns at a controlled rate, rather than exploding. Indeed, if it fails to burn at the correct rate, ignition problems result and engine damage is likely.

Because of its greater volatility, the same degree of control cannot be maintained over gasoline. Explosive combustion would occur, destroying the high-compression power plant. The gasoline engine, at the current level of technology, is therefore locked into lower compression ratios and decreased thermal efficiency.

Cost and Power-to-Weight

Because diesels feature higher compression ratios than gasoline engines, they are subjected to greater stresses and must be built more ruggedly. To sustain these higher compression ratios and loads, diesels generally have to be machined to closer tolerances. The heavier construction and closer machining tolerances account for the increase in weight and price of diesel engines over gasoline engines of the same power output. In recent years, however, tremendous advances in metallurgy and engine design have achieved drastic weight reductions on many diesels, considerably narrowing this power-to-weight gap.

Types of Diesels

Diesel engines can be four-cycle or two-cycle. The differences will become clear in a moment. Let us first look at a four-cycle engine.

A Four-Cycle Diesel

1. Imagine the piston is at the top of its cylinder. The inlet valve opens as the piston descends to the bottom of its cylinder. The descending piston draws air into the cylinder. When the piston reaches the bottom of its cylinder, the inlet valve closes, trapping the air inside the cylinder (see Figure 1-5). This movement of the piston from the top to the bottom of its cylinder is known as a *stroke*. It also constitutes one of the four cycles of a four-cycle engine—in this case the *suction* (induction), or *inlet*, cycle.
2. The piston now travels up the cylinder, compressing the trapped air. The pressure

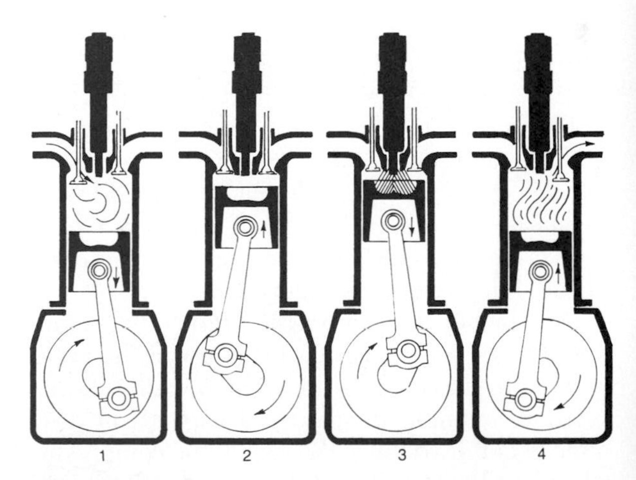

Figure 1-5. *The four cycles of a four-stroke diesel engine. 1. Inlet stroke—the piston draws air into the cylinder via the inlet valve. 2. Compression—the piston compresses the air, which also heats it. 3. Injection—fuel is sprayed into the hot air, ignites, and burns in a controlled manner, resulting in the power stroke. 4. Exhaust—the piston forces burned gases out the open exhaust valve. (Courtesy Lucas/CAV)*

rises to between 450 and 700 pounds per square inch (psi; as compared to 80 to 150 psi in a gasoline engine) and the temperature to 1,000°F (538°C) or more. This is the *compression* cycle.

3. Somewhere near the top of the compression stroke, fuel enters the cylinder via the fuel injector and starts to burn. The temperature climbs rapidly—from 2,000°F (1,093°C) to 5,000°F (2,760°C). This increase in temperature causes a rise in pressure to around 850 to 1,000 psi, pushing the piston back down its cylinder. As the piston descends, the cylinder volume increases rapidly, leading to a sharp reduction in the pressure and temperature. This is the third cycle and is known as the *power stroke*.
4. When the piston nears the bottom of the power stroke, the exhaust valve opens. The cylinder still contains a considerable amount of residual heat and pressure, and most of the gases rush out. The piston then travels back up the cylinder, forcing the rest of the burned gases out of the exhaust valve. This is the fourth, or *exhaust*, cycle. At the top of the exhaust stroke, the exhaust valve closes and the inlet valve opens, ready to admit a fresh charge of air when the piston descends the cylinder once again. This brings the engine back to the starting point of the four cycles.

A Two-Cycle Diesel

Note: The following is a description of the operation of a Detroit Diesel two-cycle engine, the most common and widely known type. There are engines with other forms of two-cycle diesel operation, but such engines are unlikely to be encountered in small-boat applications.

A Detroit Diesel two-cycle engine operates in essentially the same manner as a four-cycle engine, but condenses the four strokes of the piston into two—once up the cylinder and once down. Here's how:

1. We start with the piston at the top of its cylinder on its compression stroke. The cylinder is filled with pressurized, superheated air. Diesel is injected and ignites. The piston starts down the cylinder on its power stroke. As it descends, the cylinder pressure and temperature fall. When the piston nears the bottom of its power stroke, the exhaust valve opens and most of the burned gases rush out of the cylinder (see Figure 1-6). So far all is the same as for a four-cycle diesel. Now as the piston continues to descend the cylinder, it uncovers a series of holes, or ports, in the cylinder wall. A supercharger or turbocharger blows pressurized air through these ports, pushing the rest of the burned gases out of the cylinder and refilling it with a fresh air charge. The piston has

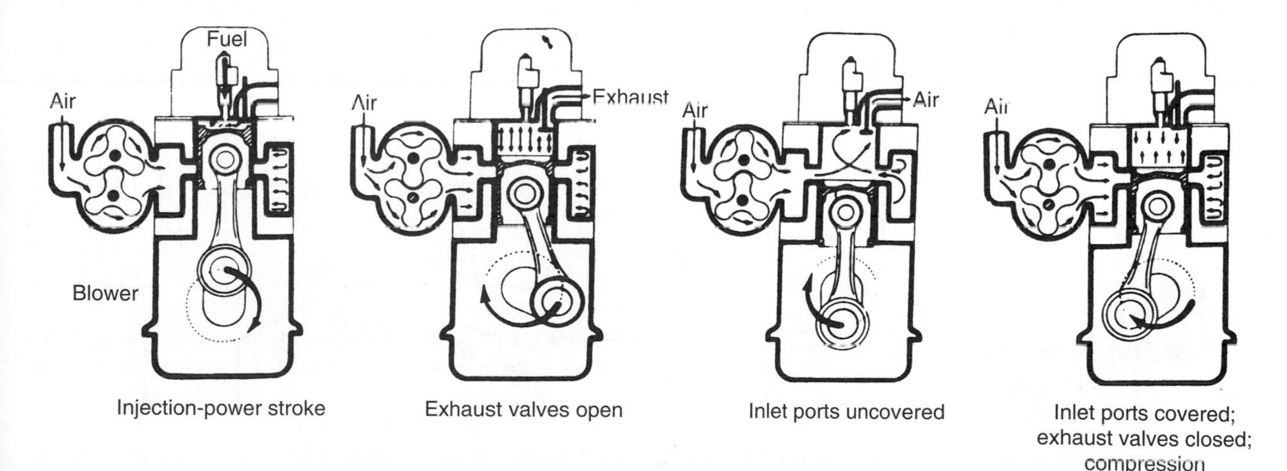

Figure 1-6. *Operation of a two-cycle Detroit Diesel. (Courtesy Detroit Diesel)*

only now reached the bottom of its cylinder and is starting back up again. The exhaust valve closes.

2. As the piston moves back up, it blocks off the inlet ports, trapping the charge of fresh air in the cylinder. Although the piston has only covered a little over one stroke, it has already completed its power stroke, the exhaust process, and the inlet cycle. As the piston comes back up the cylinder on its second stroke, it compresses the fresh air. When it reaches the top of the cylinder, injection and combustion take place. The cycle starts over. The engine has done in two strokes what a four-cycle diesel does in four.

A two-cycle engine has two power strokes for every one power stroke of a four-cycle engine. For a given engine size, a two-cycle engine develops considerably more power than a four-cycle. This leads to lower costs per horsepower and improved power-to-weight ratios.

A two-cycle diesel, however, is less thermally efficient than a four-cycle and has a higher fuel consumption, resulting in higher levels of exhaust pollution. The life of a two-cycle diesel tends to be shorter than that of a four-cycle model because of the higher loads placed on the engine. What's more, for reasons that will become clear later, two-cycle diesels tend to be far noisier in operation than four-cycles, which makes them unsuitable for a wide range of pleasure boat applications.

In recent years, tightening emissions regulations have pretty much put an end to the manufacture of two-cycle diesels.

Principal Engine Components

The Crankshaft

So far I have talked in a purely abstract fashion of a piston moving up and down its cylinder. To harness a piston to the rest of the engine, and to utilize the mechanical energy developed by its power stroke, this reciprocal motion must be converted to rotary motion. This is done by a crankshaft and connecting rod.

A *crankshaft* is a sturdy bar set in bearings in the base of an engine. Beneath each cylinder, it has an offset pin forming a *crank*. The *connecting rod* ties the piston to this crank. A bearing at each end of the connecting rod allows the crank to rotate within the connecting rod's lower end, while the piston, mounted on a *piston pin* or *wrist pin*, oscillates around its upper end. As the piston moves up and down, the crankshaft turns (see Figure 1-7).

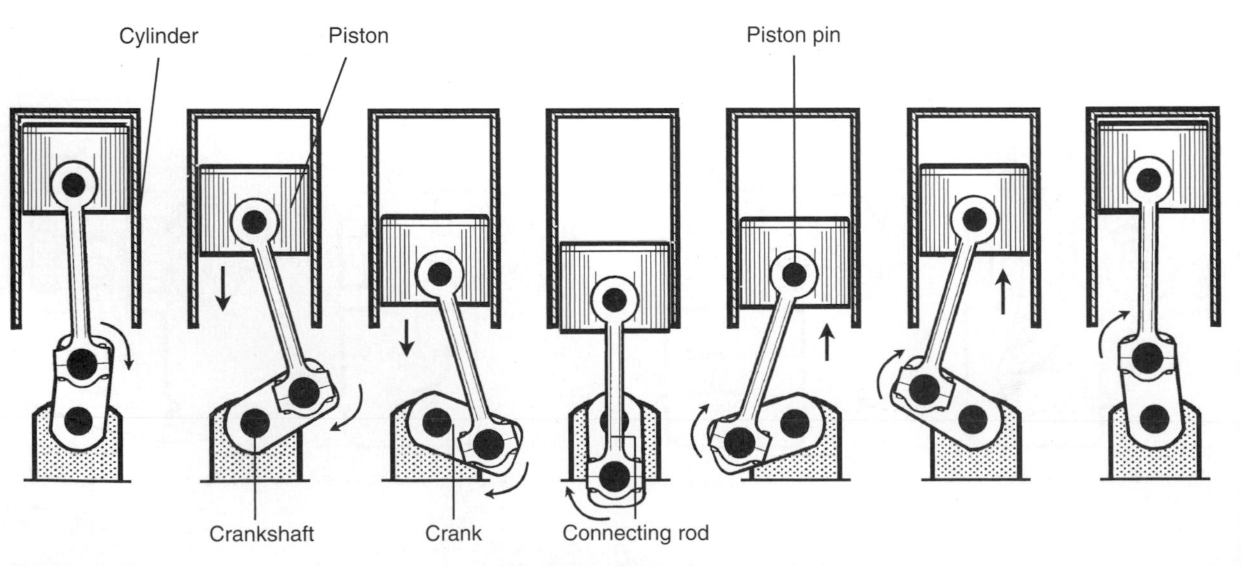

Figure 1-7. *Converting reciprocal motion to rotary motion. (Fritz Seegers)*

VALVES AND TIMING

The effective operation of four-cycle and two-cycle engines requires the precise coordination of piston movement with valve opening and closing times, as well as with the moment of fuel injection. This is known as valve and fuel injection timing.

Valves are set in cylinder heads and held in a closed position by a valve spring. A lever known as a *rocker arm* opens the valve. The rocker arm, moved up and down directly or indirectly by a *camshaft*, pivots in the top of the cylinder head (see Figures 1-8 and 1-9).

A camshaft has a series of elliptical protrusions, or cams (one for each valve in the engine), along its length. As the camshaft turns, these protrusions push the rocker arms up and down. Some camshafts are set in cylinder heads with the cams in direct contact with the rocker arms—these are overhead camshafts. Others are placed in the engine block and actuate the rocker arms indirectly via *push rods*.

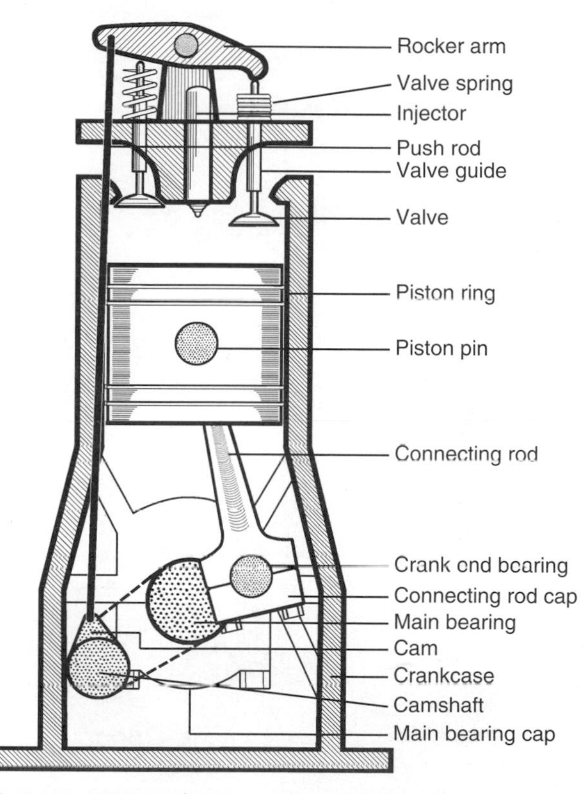

FIGURE 1-8. *Principal components in a traditional diesel engine using push rods. Many modern engines have an overhead camshaft and no push rods (see Figure 1-9). (Fritz Seegers)*

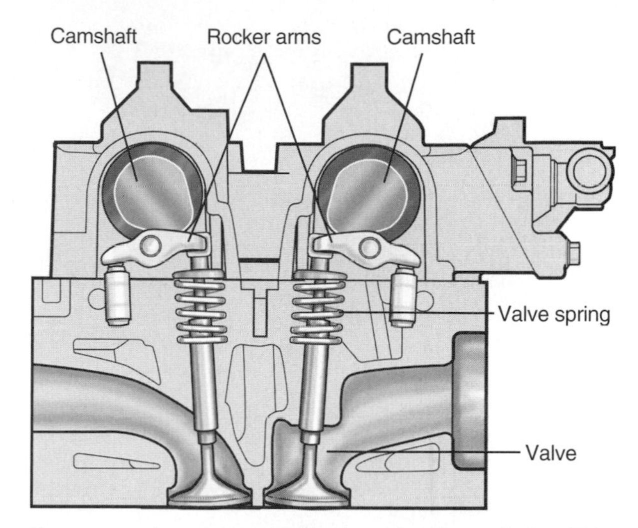

FIGURE 1-9. *An engine with dual overhead camshafts. (Fritz Seegers)*

A gear keyed (locked) to the end of the crankshaft rotates with it. Another gear keyed to the end of the camshaft rotates with it. The valve timing, or valve opening and closing times, is coordinated with the movement of the pistons by linking the two gears with an intermediate gear, a belt, or a chain so that they rotate together (see Figure 1-10).

On a four-cycle engine, the inlet and exhaust valves open and close on every other stroke of the engine. As a consequence, the gear on the camshaft is twice the size of that on the crankshaft, causing the former to rotate at half the speed of the latter and the valves to open and close on every other engine revolution.

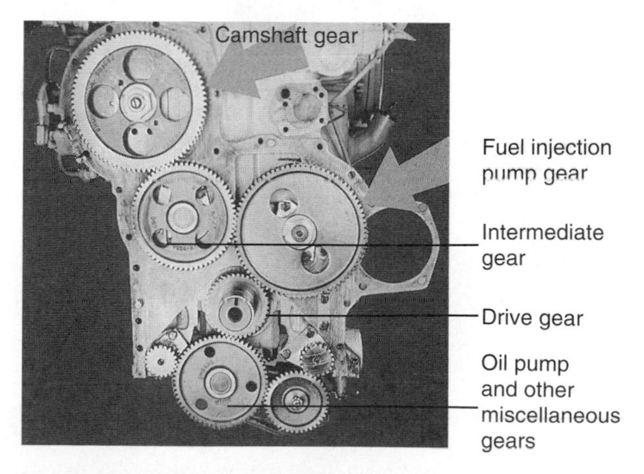

FIGURE 1-10. *Engine timing gears. (Courtesy Caterpillar)*

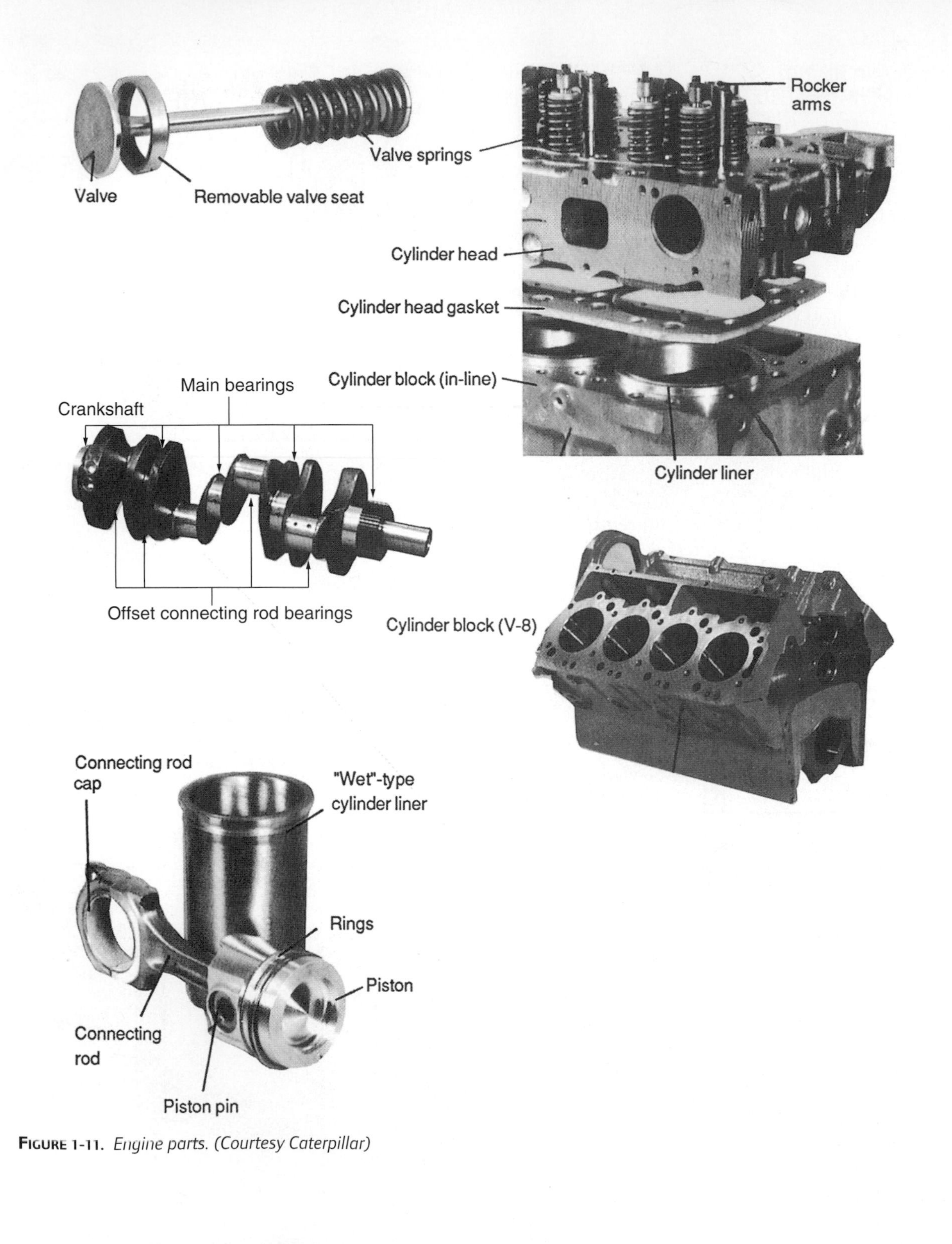

FIGURE 1-11. *Engine parts. (Courtesy Caterpillar)*

Detroit Diesel two-cycle engines have only exhaust valves. The exhaust valves open on every downward stroke of the piston; therefore, the camshaft gear is the same size as that on the crankshaft. The two shafts rotate at the same speed, and the valves open and close on every engine revolution.

By setting the camshaft and crankshaft gears in different relationships with one another, the valves can be made to open and close at any point of piston travel. Timing engine valves consists of placing the gears in the precise relationship needed for overall maximum engine performance.

Conventional mechanical fuel injection timing is set the same way (electronic may be different). A gear fitted to the end of the fuel injection pump driveshaft is driven by the crankshaft via an intermediate gear, belt, or chain. Altering the relationship of the gears allows the moment of fuel injection to be set at any point of piston travel. Since injection occurs on every other revolution on a four-cycle engine, the pump drive gear is twice the size of the crankshaft gear, causing the pump to rotate at half the engine's speed. On two-cycle engines, the gears are the same size, producing injection at every engine revolution.

Cylinders and Other Parts

The central frame to which the rest of an engine is assembled is known as the *block*, or *crankcase* (see Figure 1-11). On all except air-cooled engines, which are rarely found on boats, this is a casting containing numerous passages for air, cooling water, oil, and other engine parts, such as the crankshaft and camshaft.

There are two types of cylinders, or liners, available for diesel engines—dry liners and wet liners. In a dry-liner diesel, the cylinders and block are in continuous metal-to-metal contact. In a wet-liner arrangement, the block comes into contact with the cylinders only at their tops and bottoms, and engine cooling water circulates directly around the liners (see Figures 1-12 and 1-13).

Wet liners have the advantage of being relatively easy to replace in the field when a major engine overhaul is necessary, whereas to change dry liners, the whole block must generally be taken to a machine shop.

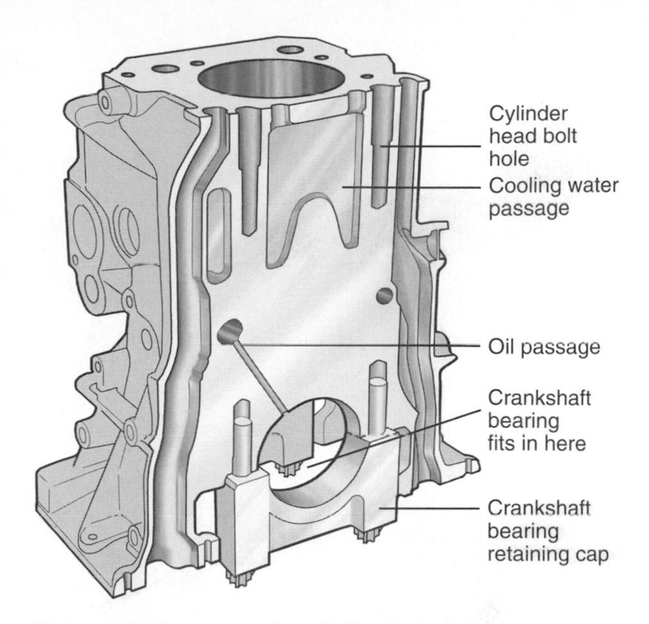

Figure 1-12. *A cutaway of a cylinder block. (Fritz Seegers)*

Pistons are fitted with a number of piston rings set in grooves cut into the outside circumference of the piston. The piston rings press out against the walls of the cylinder to make a gas-tight seal. The top end of a cylinder is sealed with a casting known

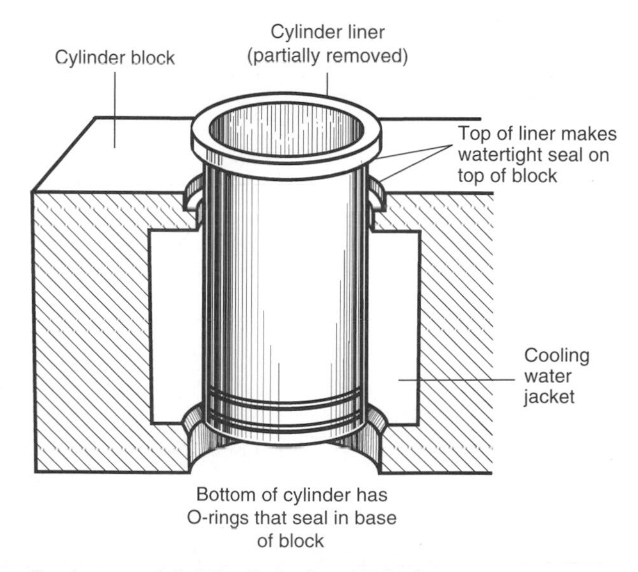

Figure 1-13. *A "wet" cylinder liner. (Fritz Seegers)*

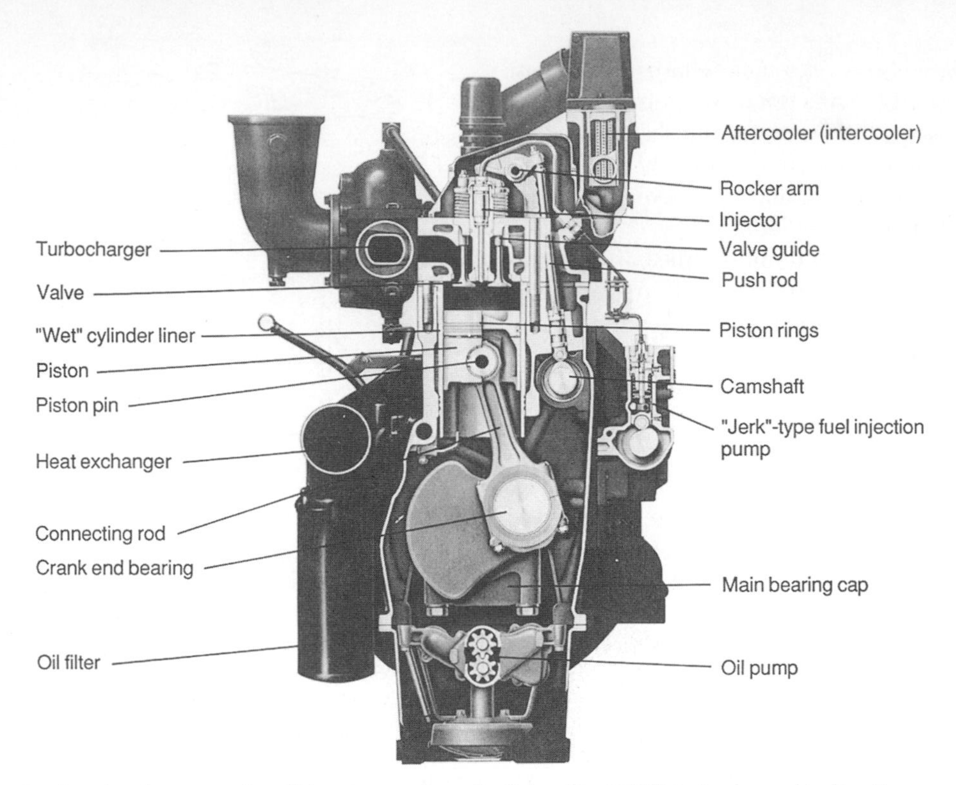

FIGURE 1-14. *Putting the pieces together: This cutaway view of a Caterpillar 3406B turbocharged in-line 6 represents typical modern diesels. (Courtesy Caterpillar)*

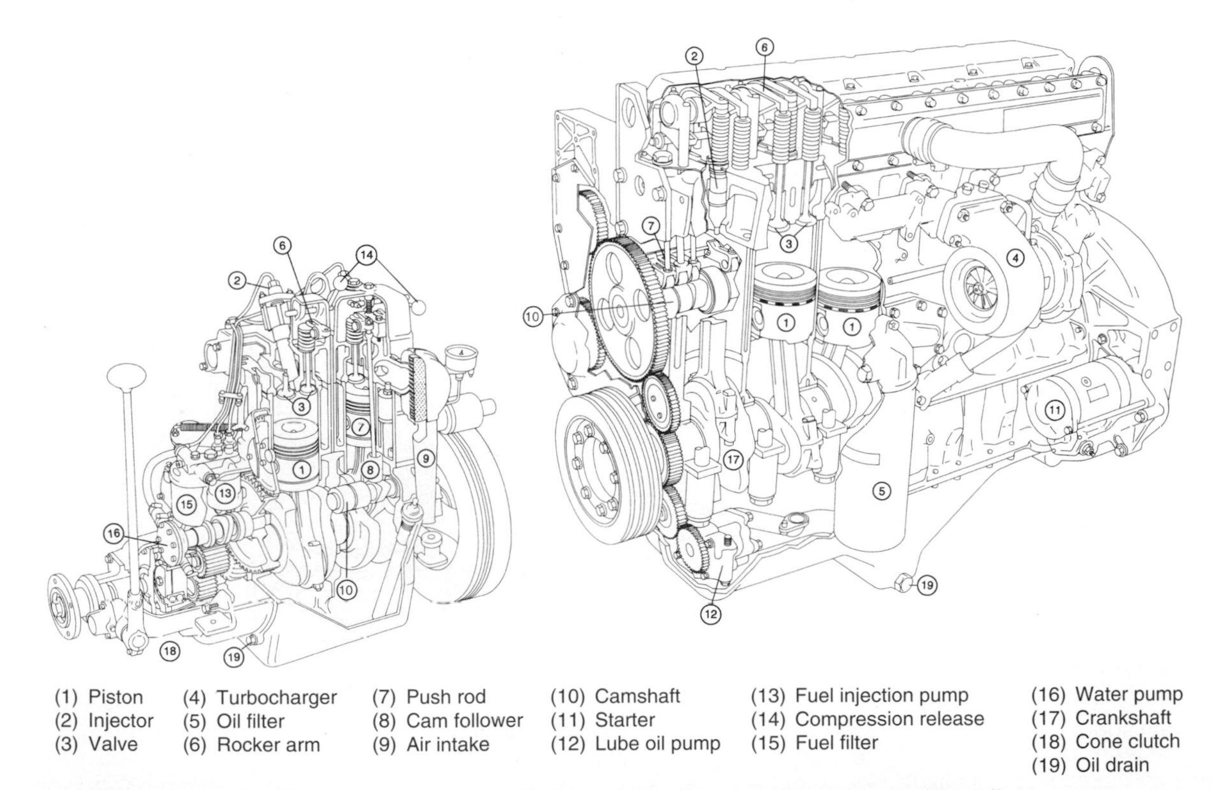

FIGURE 1-15. *Putting the pieces together: Side views of four-cylinder, four-cycle diesel engines. (Jim Sollers)*

as a *cylinder head*, which contains the injectors; valves; and frequently combustion chambers, water passages, and so on. The valves are set in guides, which are replaceable sleeves pressed into the cylinder head.

When an engine begins to show appreciable wear, these valve guides can be pressed out and replaced. Better-quality engines also have replaceable *valve seats*, the area that a valve contacts to seal the combustion chamber. Remachining valve seats, replacing valve guides, and installing new valves renew the cylinder head, giving this expensive part an almost indefinite life (see Figures 1-14 and 1-15).

CHAPTER 2

DETAILS OF OPERATION

A diesel engine creates power by burning fuel. The more fuel that an engine of any given size can burn, the more heat it will generate and the greater will be its potential power output. Increasing the power-to-weight ratio reduces the cost per horsepower produced. Manufacturers are therefore continually striving to increase the amount of fuel a given engine can burn.

Getting more diesel into a cylinder is no problem—all it takes is a bigger fuel injection pump and injector. Getting it to burn is another matter. Incomplete combustion lowers fuel economy and causes excessive exhaust pollutants. For effective combustion, three interdependent factors must be present:

1. An adequate supply of oxygen.
2. Effective atomization of the injected diesel.
3. Thorough mixing of the atomized diesel with the oxygen in the cylinder.

Let's look first at the oxygen supply. Atomization of the fuel and proper mixing with the oxygen are covered in the next section, Combustion.

SECTION ONE: THE AIR SUPPLY

What actually takes place when diesel fuel burns is a reaction between oxygen in the air and hydrogen and carbon in the diesel fuel. Starting this reaction takes a temperature of around 750°F (399°C). Once it begins, the oxygen combines with the hydrogen to form water and with the carbon to form carbon dioxide (and occasionally carbon monoxide when combustion is incomplete). In the process of these chemical reactions, considerable quantities of heat and light are released—the fuel burns.

Air is only about 23% oxygen by weight (21% by volume); the rest is principally nitrogen, plus one or two trace gases, and plays no part in the combustion process. This idea of air having weight may be a little confusing, so let's take a moment to look at this.

You are probably familiar with the concept of atmospheric pressure, the notion that the weight of the atmosphere exerts pressure on the surface of the earth. As you climb to higher elevations, the atmosphere becomes thinner; i.e., it exerts less pressure. At the top of Mount Everest, for example, you need an oxygen mask to breathe. Deep space has no atmosphere—it is a vacuum, with no pressure whatsoever.

At sea level, 1 cubic foot of air weighs approximately 0.076 pounds at 60°F (15.6°C). At higher altitudes it weighs less. It also weighs less at higher temperatures, because as the temperature rises, air expands so that each cubic foot contains less of it.

To completely burn 1 pound of diesel fuel requires approximately $3^1/_3$ pounds of oxygen. Diesel weighs around 7.2 pounds per gallon. Doing a little arithmetic, and keeping in mind that air is only 23% oxygen, we find that burning 1 gallon of diesel fuel at sea level requires almost 1,500 cubic feet of 60°F air. This is as much air as is contained in a good-size room! At higher elevations and temperatures, it can take quite a lot more. The ability to get sufficient air into an engine for complete combustion of the fuel becomes the limiting factor on how much fuel the engine can burn.

Volumetric Efficiency

Engineers must do everything possible to avoid restricting the flow of air to an engine and the flow of exhaust gases from it. Air filters are made as large as possible, and manufacturers recommend changing them frequently. Inlet manifolds are made as smooth and direct as possible to reduce friction from the incoming air. Valves are made as large as can be accommodated in the cylinder head (on two-cycle diesels, the inlet ports are given as large an area as possible). On many modern engines, there are four valves per cylinder—two inlet and two exhaust—to improve the airflow (see Figure 2-1). The unavoidable inefficiencies created by the remaining friction in the air passages (including the exhaust—more on this later) are referred to as *pumping losses*.

The degree to which an engine succeeds in completely refilling its cylinders with fresh air is known as its *volumetric efficiency*. From the bottom of its stroke to the top, a piston occupies, or displaces, a certain volume. This is known as its *swept volume*. If an engine were able to draw in enough air on its inlet stroke to completely fill this swept volume at atmospheric pressure, it would have a volumetric efficiency of 100%. Volumetric efficiency, then, is the proportion of the volume of air drawn in, relative to the swept volume, at atmospheric pressure.

Naturally Aspirated Engines

An engine that draws in its air charge through the action of its pistons is known as a *naturally aspirated* engine. Here's how it works.

On a standard four-cycle engine, the downward movement of the piston on the inlet cycle reduces the pressure in the cylinder and pulls air into the

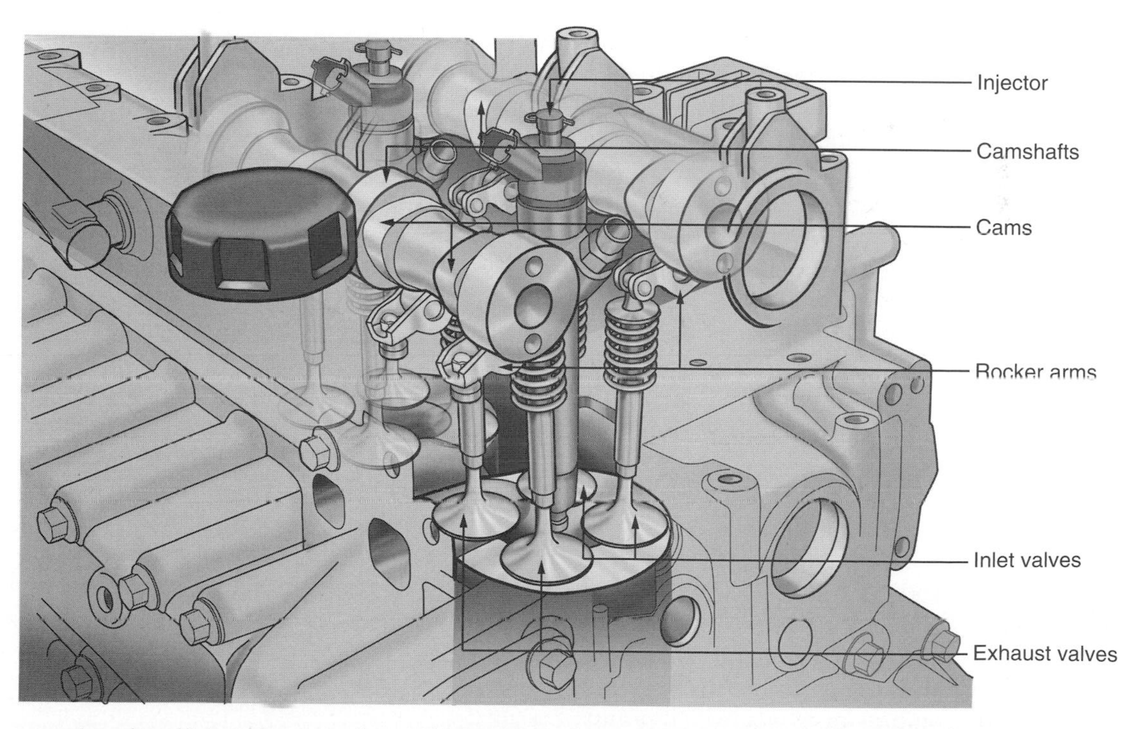

Figure 2-1. *A modern, high-performance, four-cycle diesel engine with dual overhead camshafts and four valves per cylinder (two inlet and two exhaust). (Fritz Seegers)*

cylinder. (Strictly speaking, reduced pressure in the cylinder causes the higher atmospheric pressure on the outside to push air into the engine, but it is easier to visualize it as the piston sucking in air.)

Friction caused by the air filter and air-inlet piping (manifold) partially obstructs the flow of air into the engine. As a consequence, when the piston reaches the bottom of its stroke, the pressure inside the cylinder is still marginally below that of the atmosphere; i.e., there is a slight vacuum. This means that the cylinder has not been completely refilled with air under atmospheric pressure.

As the air rushes into a cylinder, however, it gains momentum. When a piston of a four-cycle engine reaches the bottom of its stroke on the inlet cycle and starts back up on the compression stroke, air continues to enter the cylinder but only for a moment. To take advantage of this, the inlet valve is set to close a little after the piston has started its compression stroke.

The same valve is set to open a short time before the piston has reached the top of its stroke on the exhaust cycle and before the exhaust valve is fully closed. This is known as valve overlap, and the two valves are said to be *rocking* at this point (see Figure 2-2). This ensures that the inlet valve is wide open by the time the piston begins its induction stroke, which enables the piston to start drawing in air immediately. These techniques can usually bring up volumetric efficiency to 80% to 90%.

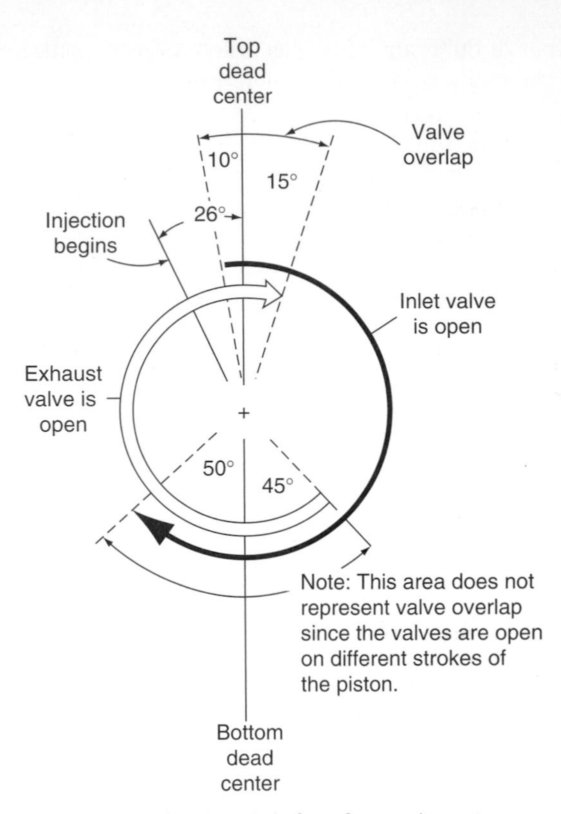

FIGURE 2-2. *Typical timing circle for a four-cycle engine.*

SUPERCHARGERS AND TURBOCHARGERS

The volumetric efficiency, and therefore the power output, of a naturally aspirated diesel can be increased dramatically by forcing more air into it under pressure. This is the principle of *supercharging* and *turbocharging*.

A supercharger pumps air into the inlet manifold by means of a fan, or blower, that is mechanically driven off the engine via a belt, chain, or gear. A turbocharger consists of a turbine installed in the exhaust manifold of an engine and connected to a compressor wheel in the inlet manifold. As the exhaust gases rush out, they spin the turbine. The turbine spins the compressor wheel, which pumps air into the inlet manifold. A turbocharger has no mechanical drive (see Figure 2-3).

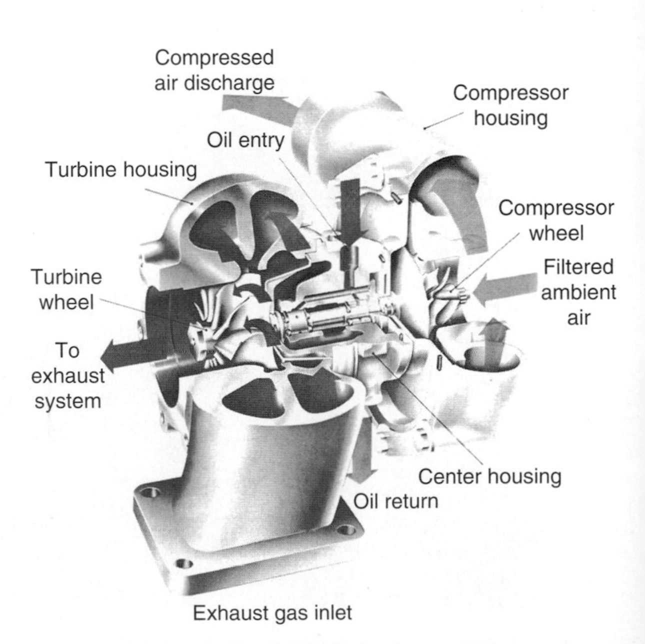

FIGURE 2-3. *A cutaway view of a turbocharger. (Courtesy Garrett Automotive Products Co.)*

When the load increases on an engine, more fuel is injected, leading to a rise in the volume of the exhaust gases. This spins the turbocharger's exhaust turbine and compressor wheel faster, forcing more air into the engine. A turbocharger is extremely responsive to changes in load, driving up the power output of an engine just when it is needed most.

A turbocharger won't work on a two-cycle diesel, at least not unless it's used in conjunction with a supercharger, because when the inlet ports are uncovered by the descending pistons, a two-cycle diesel depends on a supply of pressurized air to blow the exhaust gases out of its cylinders and refill them with fresh air. This process is called *scavenging*. Since during start-up the engine doesn't have any exhaust gases to spin a turbocharger, it needs a mechanically driven blower (a supercharger) to pump air into the cylinders and get everything moving (see Figure 2-4).

The efficiency with which fresh air is introduced into the cylinders in a two-cycle diesel is known as *scavenging efficiency*, which is much the same concept as volumetric efficiency on four-cycle engines. If the inlet air drives all the exhaust gases out of a cylinder and completely refills it with fresh air at atmospheric pressure, the engine has 100% scavenging efficiency.

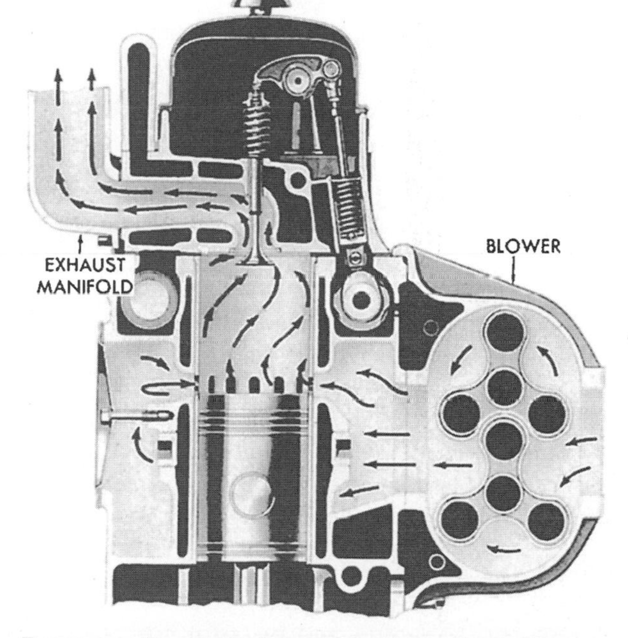

FIGURE 2-4. *"Scavenging" on a two-stroke diesel. (Courtesy Detroit Diesel)*

INTERCOOLERS AND AFTERCOOLERS

Air compressed by a turbocharger or supercharger heats up. Because hot air at a given pressure weighs less than cold, it contains less oxygen per cubic foot. To counteract the turbocharger's and supercharger's loss of efficiency caused by this rise in temperature, the air must be cooled. Most supercharged or turbocharged engines accomplish this with a *heat exchanger*, known as an *intercooler* or *aftercooler*, fitted in the inlet manifold between the supercharger or turbocharger and the engine block. (An intercooler is fitted between a turbocharger and a supercharger; an aftercooler is fitted between a supercharger or turbocharger and the engine block. The two terms tend to get used interchangeably, so to avoid cumbersome dual references, I will use aftercooler.) Cooling water is circulated through these units, much like the radiator on your automobile, lowering the temperature of the air as it passes into the engine (see Figure 2-5).

Some aftercoolers are plumbed into the engine cooling circuit, receiving water that has already been warmed by circulation through the engine. Others are plumbed directly to a separate raw-water supply (see Section Six: Keeping Things Cool). This latter arrangement produces the maximum possible drop in the temperature of the inlet air, thereby enabling the greatest possible amount of air to be pumped into the cylinders. Manufacturers will sometimes list three separate horsepower ratings for the same

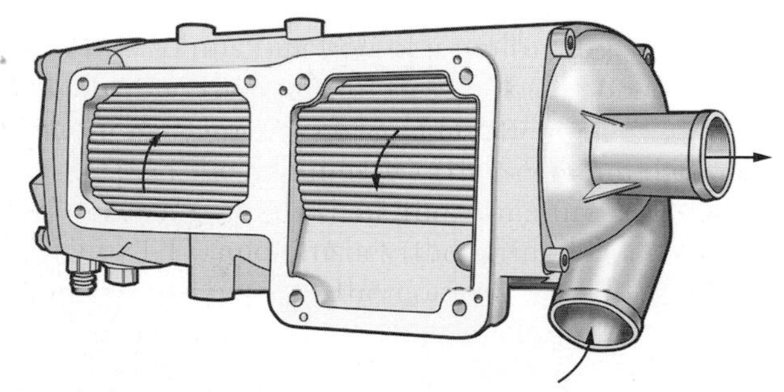

FIGURE 2-5. *A cutaway view of an aftercooler. Water is circulated through the copper tubes while the inlet air passes between them. The water cools the air. (Fritz Seegers)*

engine to reflect these differing arrangements: (1) naturally aspirated, (2) turbocharged and aftercooled using the engine water circuit, and (3) turbocharged and aftercooled using a separate water circuit.

Superchargers and turbochargers frequently will raise volumetric efficiency to 150% or more, which is to say that the air in the cylinder at the end of the induction stroke is well above atmospheric pressure. Such an engine commonly develops 50% more power, and sometimes almost 100% more power, than the same-sized naturally aspirated diesel.

Of course, the price paid for this improved efficiency is more complex and expensive engines (although cost per horsepower is frequently less). Also, turbocharging and supercharging accelerate engine wear and increase the costs of servicing. But the considerable improvement in power-to-weight ratios is a major benefit to many weight-conscious boatowners.

SECTION TWO: COMBUSTION

When diesel fuel is injected into a cylinder containing high-pressure, superheated air, it does not explode—it burns. The relatively slow burn rate of diesel fuel produces a more even rise in cylinder temperatures and pressures than does gasoline, exerting a more gradual force on the piston over the whole length of its power stroke. This is one of diesel's advantages over gasoline, and as a result, diesel engines have far more constant torque (the turning force exerted by the crankshaft), especially at low speeds.

THE IMPORTANCE OF TURBULENCE

At the moment of injection, pressure in a cylinder can be as high as 700 psi. The temperature can be more than 1,000°F (538°C). Into this dense mass of superheated air, an injector sprays fuel in the form of one or more streams of minute particles. Only a little more than 20% of the air is oxygen. When a particle of diesel encounters an oxygen molecule, it begins to burn, consuming the oxygen in the process. Complete combustion requires the fuel to come into contact with fresh oxygen, but as the burning progresses, fewer and fewer oxygen molecules remain in the cylinder. Those that do remain are not always in the right place.

Failure to engage a molecule of oxygen causes the half-burned diesel to be blown out of the exhaust pipe as black smoke, lowering fuel economy, increasing exhaust pollutants, and robbing power from the engine.

Injectors can spray fuel only in straight lines. When the particles of diesel first emerge from the tip of an injector, they are fairly densely packed and encounter most of the oxygen molecules. As they move farther from the injector, they spread out, and fewer of the oxygen molecules lie directly in their path (see Figure 2-6). Because of this, mixing the fuel and air is imperative.

This business of mixing fuel and air is highly scientific. It has been the subject of much investigation and experimentation over the years, and in modern engines several approaches are taken. The principal variations lie in the nature of the pattern in which the diesel fuel is injected into the cylinders, and in the techniques used for creating turbulence inside the cylinders so that the air and diesel become thoroughly mixed.

INJECTOR SPRAY PATTERNS

The fuel injection spray pattern is determined by the size and shape of the openings in the injector nozzle, of which there are two basic styles:

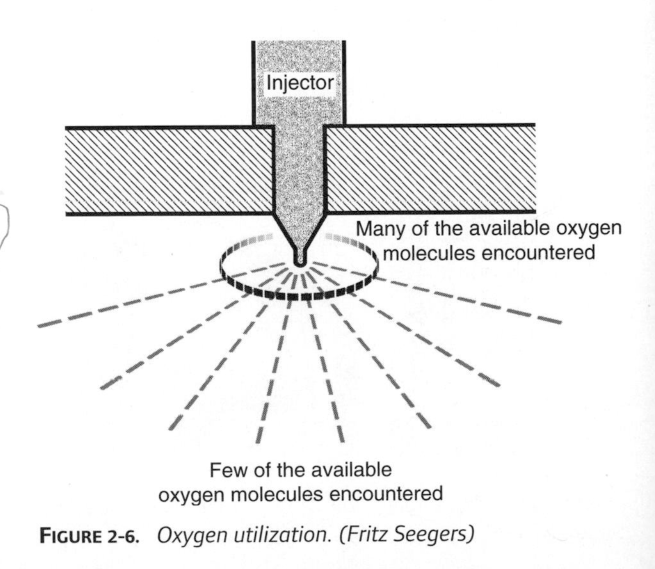

FIGURE 2-6. *Oxygen utilization. (Fritz Seegers)*

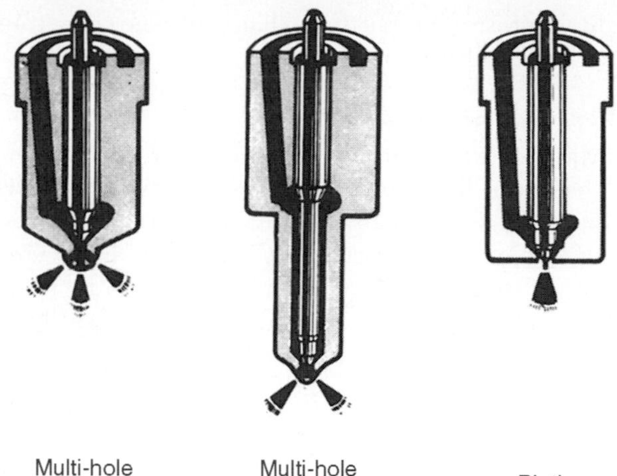

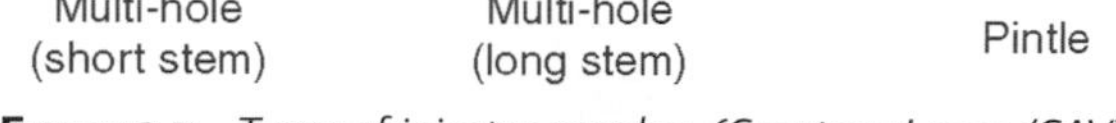

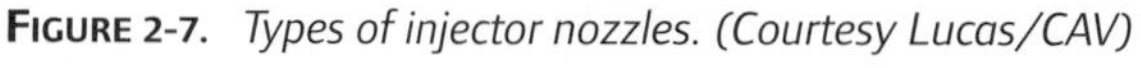

FIGURE 2-7. *Types of injector nozzles. (Courtesy Lucas/CAV)*

- *Hole-type nozzles* force the fuel through one or more tiny orifices. Varying the size of the hole(s) atomizes (breaks up) the fuel to a greater or lesser extent. Changing the number and angles of the holes projects the fuel to different parts of the combustion chamber (see Figure 2-7).
- *Pintle (pin) nozzles* inject a conical pattern of fuel out of a central hole down the sides of a pin (pintle). This type cannot atomize the fuel to the same degree as a hole-type nozzle. Varying the angle of the side of the pintle makes the cone of fuel larger or smaller.

There are also various hybrid injectors, such as the Lucas/CAV Pintaux, which consists of a pintle nozzle with an auxiliary hole.

Pintle nozzles have a major advantage over hole nozzles—the action of the fuel scrubbing down the sides of the pintle helps keep it clean, whereas the tiny orifices in hole-type nozzles can be blocked by even the smallest piece of debris.

Techniques for Creating Turbulence

In order to mix the injected fuel particles with the oxygen in the cylinders, manufacturers design pistons and combustion chambers in ways that impart a high degree of turbulence to the air in the cylinders. Almost all modern diesels use one of the following designs.

Direct Combustion Chambers

A direct, or open, combustion chamber is really no more than a space left at the top of a cylinder when the piston is at the top of its stroke. The space also may be hollowed out of the piston crown or cylinder head (see Figure 2-8). This is the simplest kind of combustion chamber and has a number of advantages.

Its surface area, relative to the volume of the combustion chamber, is less than in any other type of chamber. This means that less heat is lost to engine surfaces. As a result, the thermal efficiency is higher. It also makes starting easier, because less of the heat of compression dissipates to the cold

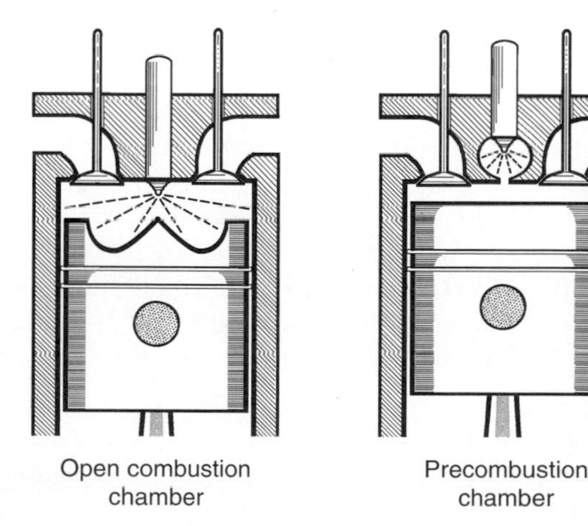

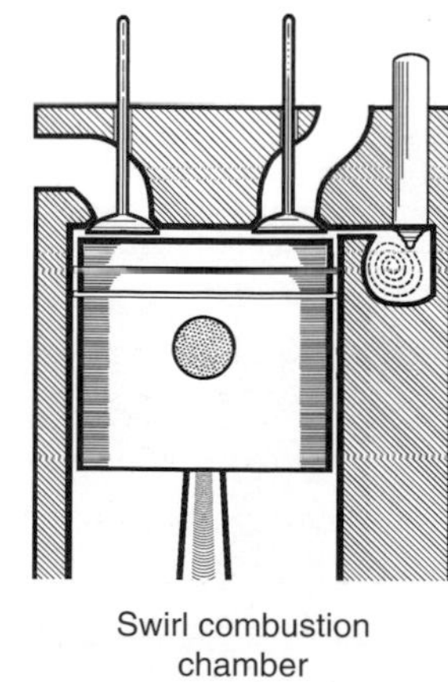

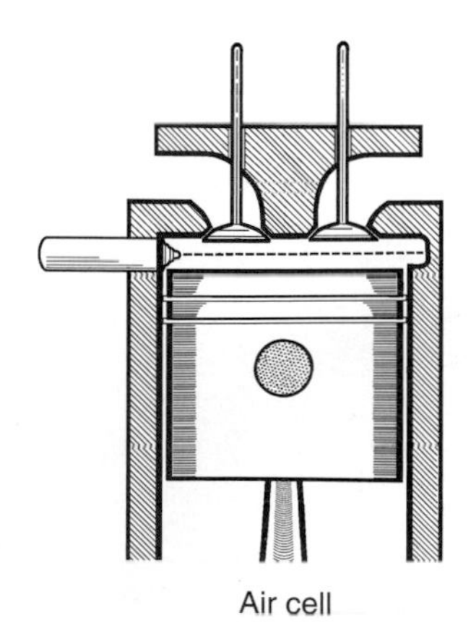

FIGURE 2-8. *Types of combustion chambers. (Fritz Seegers)*

engine. This feature allows lower compression ratios than those required with other types of chambers (often 16:1 as opposed to 20:1 or higher), leading to less stress on the engine and longer life.

In other types of combustion chambers, a portion of the air charge is forced in and out through small openings. This is often referred to as "work being done on the air." The process generates a good deal of friction. It also consumes power and contributes to pumping losses. Of all the different designs, direct chambers do the least amount of work on the air, but they have drawbacks.

A direct combustion chamber creates less turbulence than any other type; therefore, it uses the least amount of oxygen in the cylinder. For a given cylinder size, direct combustion chambers generate less power than the others. In an attempt to offset this, inlet valves and seats are shaped and positioned in ways that impart a swirling motion to the air charge as it enters the cylinder. Further, the piston crowns on these engines are frequently given a convoluted shape (known as a *toroidal crown*—see Figure 2-9), which has the same effect. In recent years, a great deal of research has been done on improving the efficiency of direct combustion chambers, with many manufacturers making a move back to this type of chamber.

Engines with direct combustion chambers almost always use hole-type injectors, which break fuel into smaller particles than pintle nozzles, thus aiding combustion.

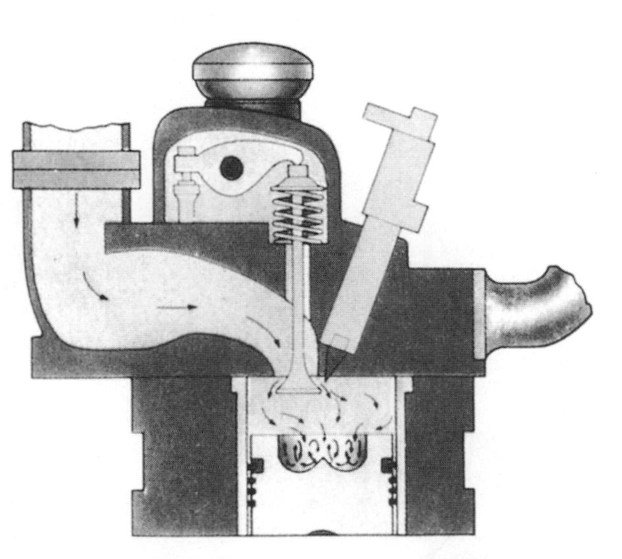

FIGURE 2-9. *Direct injection with a toroidal crown piston. (Courtesy Volvo Penta)*

PRECOMBUSTION CHAMBERS

Manufacturers often cast separate precombustion chambers into a cylinder head. These chambers range in size from 25% to 40% of the total compression volume of the cylinder. When fuel is injected into a precombustion chamber, it starts to burn, causing the temperature and pressure to rise above those in the main combustion chamber. This forces the unburned balance of the fuel and air mixture to rush through the precombustion chamber's relatively small opening into the main chamber, resulting in a high degree of turbulence and a thorough mixing of the fuel and air.

This type of engine generally uses pintle injectors because their conical pattern distributes the diesel throughout the precombustion chamber. The extreme turbulence set up in the main combustion chamber offsets the pintle injector's reduced degree of atomization, compared with a hole-type nozzle.

Precombustion chambers make better use of the oxygen than direct chambers, resulting in more power from a given cylinder size. More work is done on the air, however, and the increased surface area of the two combustion chambers reduces thermal efficiency. Engine starting is harder due to the greater heat losses. For this reason, compression ratios are generally higher (from 20:1 to 23:1), and glow plugs (see Chapter 4) are invariably installed in the precombustion chambers to assist in cold starting (see Figure 2-10).

SWIRL CHAMBERS

Swirl chambers are similar to precombustion chambers, but their volume is almost equal to that of the main chamber. A very high degree of turbulence is imparted to the air charge as it enters the swirl chamber. Pintle nozzles inject the fuel into this swirling mass. Air use is high, but so too is the amount of work done on the air. Thermal efficiency suffers, so compression ratios must be high. Glow plugs are needed for cold starting.

OTHER VARIATIONS

These three chamber types account for most injector nozzle–combustion chamber combinations found in marine diesels, but there are other variations. For

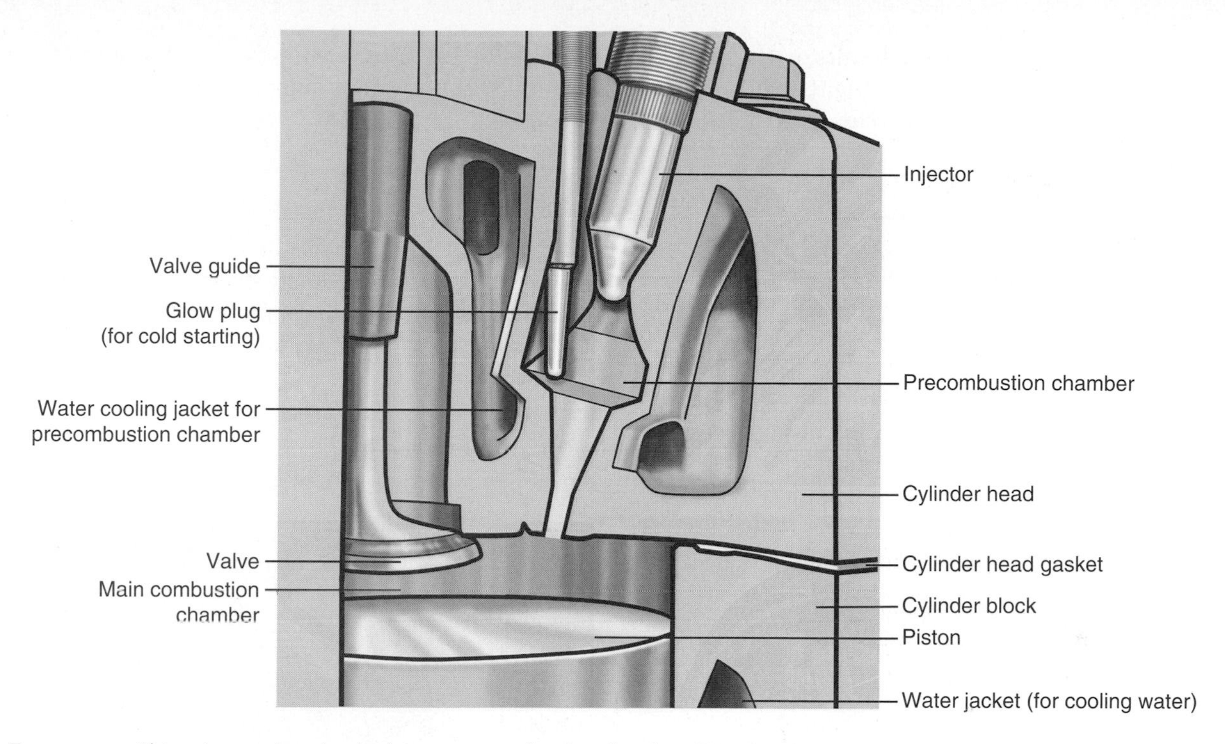

FIGURE 2-10. *This cutaway view clearly shows a precombustion chamber. (Fritz Seegers)*

example, air cells are sometimes used. In this arrangement, an open chamber is set opposite the injector, the fuel is sprayed across the piston top into the air cell, and combustion takes place throughout.

Despite the variety of combustion chambers and injectors employed, fuel and air never mix or burn 100%. For this reason, diesel engines are always designed to draw in more air than is strictly required to burn the fuel charge so that full combustion of the diesel is assured. The more complete the combustion, the more fuel efficient and the less polluting the engine. Mixing fuel and air as thoroughly as possible reduces the amount of excess air needed for a clean burn, resulting in greater power from any cylinder of a given size.

WHY BOTHER?

At this point you may ask, "Why do I need to know these things?" For one thing, the next time you see a glossy engine brochure advertising the merits of a "high-swirl combustion engine," you will know just what is being described. More important, the basis for effective troubleshooting is an understanding of what is going on inside your engine. For example, say your engine has been increasingly difficult to start of late, has been emitting black exhaust smoke, and is now beginning to overheat and seize up while in use. You've determined that the cooling system is working fine, that the crankcase has plenty of oil, and that the oil pressure is normal (or perhaps marginally on the low side due to the overheating). You need to ask yourself a couple of questions: Does this engine have direct chambers, precombustion chambers, or swirl chambers? Does it have hole or pintle injectors?

If the engine has direct combustion chambers with hole type injectors, one or more injectors may be malfunctioning, leading to poor atomization of the fuel and possibly injector dribble. This could be the cause of the difficult starting and the black smoke from unburned fuel. The liquid fuel in the cylinder is now washing away the film of lubricating oil on part of a cylinder wall, creating excessive friction with the piston. The piston and cylinder are heating up, and a partial seizure is underway. Engines with precombustion and swirl chambers are less likely to suffer these symptoms.

This is only one possibility (more troubleshooting follows later). But this example does illustrate that you can never have too much information about an engine when diagnosing problems, and even the facts that seem most obscure can come in handy.

SECTION THREE: FUEL INJECTION

The first two sections of this chapter described what must happen inside a cylinder if the injected fuel is to burn effectively. This section takes a look at the injection system itself, which is surely one of the miracles of modern technology. Today's breed of small, high-speed, lightweight, and powerful diesels are possible in large part because of dramatic improvements in fuel injection technology.

Consider a four-cylinder, four-cycle engine running at 3,000 rpm, and burning 2 gallons of diesel per hour. On every compression stroke, the fuel system injects 0.0000055 gallon of fuel (5.5 millionths of a gallon!). Depending on the type of injector, injection pressures vary from 1,500 psi to over 20,000 psi, so the fuel must be raised to this pressure as well.

Each stroke of the piston of an engine running at this speed takes only 1/100 of a second. In less than an instant, the injection system must initiate injection, continue it at a steady rate, then cut it off cleanly. The rate of injection must be precisely controlled. If it is too fast, combustion accelerates, leading to high temperatures and pressures in the cylinder and to engine knocks (for more on this see the Knocks section on pages 118–20). If it is too slow, combustion retards, leading to a loss of power and a smoky exhaust. The fuel, as we have seen, must be properly atomized, and there must be no dribble from the tip of the injector, either before or after the injection pulse.

The actual beginning point of injection must be timed to an accuracy of better than 0.00006 second! Finally, every cylinder must receive exactly the same amount of fuel. It also must be constant from revolution to revolution to avoid vibration and uneven cylinder loading, which could lead to localized overheating and piston seizure. These facts and figures serve only to illustrate that a diesel engine fuel injection system is an incredibly precise piece of engineering that needs to be treated with a great deal of respect.

The overwhelming majority of diesel engines use one of the following approaches to fuel injection: a jerk pump (or in-line pump), a distributor pump (or rotary pump), a unit injector, or a common rail system.

JERK (IN-LINE) PUMPS

A schematic of this type of fuel injection system is shown in Figure 2-11. A *lift*, or *feed*, pump draws the fuel from the tank through a primary fuel filter. It then pushes fuel at low pressure through a secondary filter to the jerk injection pump.

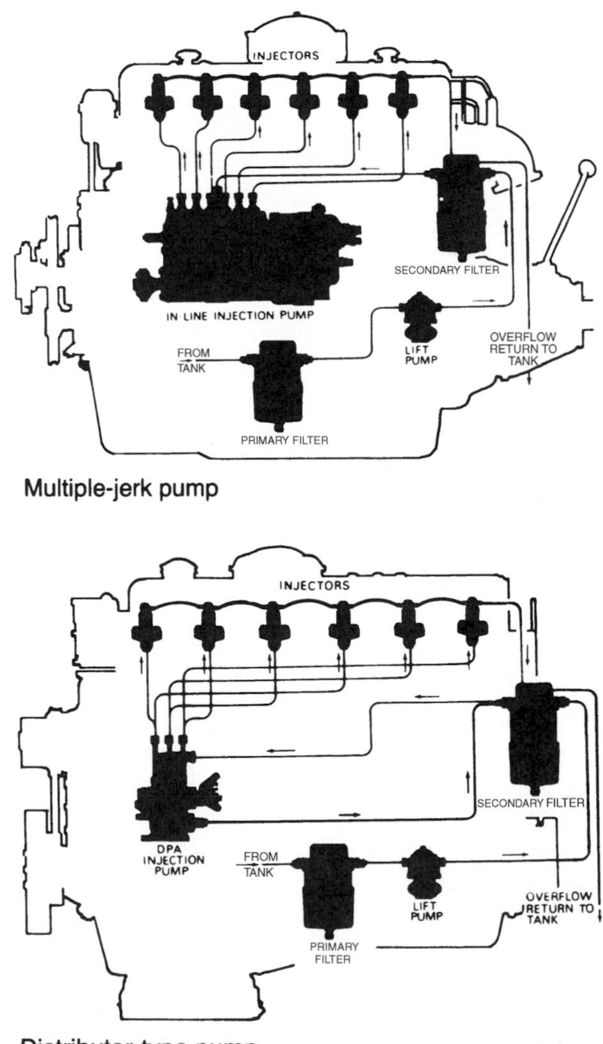

FIGURE 2-11. *Schematics of jerk pump (top) and distributor pump (bottom) fuel injection systems. (Courtesy Lucas/CAV)*

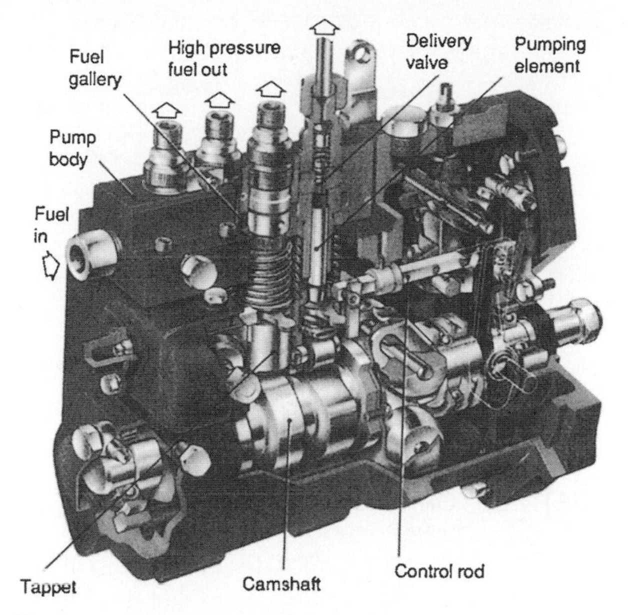

FIGURE 2-12. *Arrangement of plungers and barrels of an in-line jerk pump. (Courtesy Lucas/CAV)*

Jerk pumps consist of a plunger driven up and down a barrel by a camshaft. At the bottom of the plunger stroke, fuel enters the barrel. As the plunger moves upward, it forces this fuel out of the barrel through a check, or delivery, valve to the injector via a delivery pipe (see Figure 2-12). The pressure generated by the pump forces open another valve in the injector, allowing injection to take place.

The speed of a diesel engine is regulated by controlling the amount of fuel injected into the cylinders. To make this possible, the jerk pump's plunger has a curved spill groove machined down its edge. A hole drilled through the top of the plunger connects with this groove so that the fuel in the barrel can flow to it (see Figure 2-13). Another hole, a bleed-off port, is also drilled into the pump barrel.

Anytime the spill groove lines up with the bleed-off port, the fuel in the barrel runs out, pressure falls, and injection ceases. By rotating either the barrel or the plunger, the groove and port can line up at a variety of points on the plunger's stroke, thereby varying the amount of fuel injected before the moment of bleed-off.

A gear fitted to the plunger or the barrel is turned via a geared rod, which is known as the fuel rack. The throttle is connected to this rack. Varying the position of the throttle regulates the flow of diesel to the injectors, which controls engine output.

Many turbocharged engines include a smoke limiter, which consists of a diaphragm plumbed to the inlet manifold on one side and the fuel rack on the other. When the inlet air pressure is low (the turbocharger is not up to speed), the maximum travel of the fuel rack is limited. This eliminates the characteristic cloud of black smoke that older

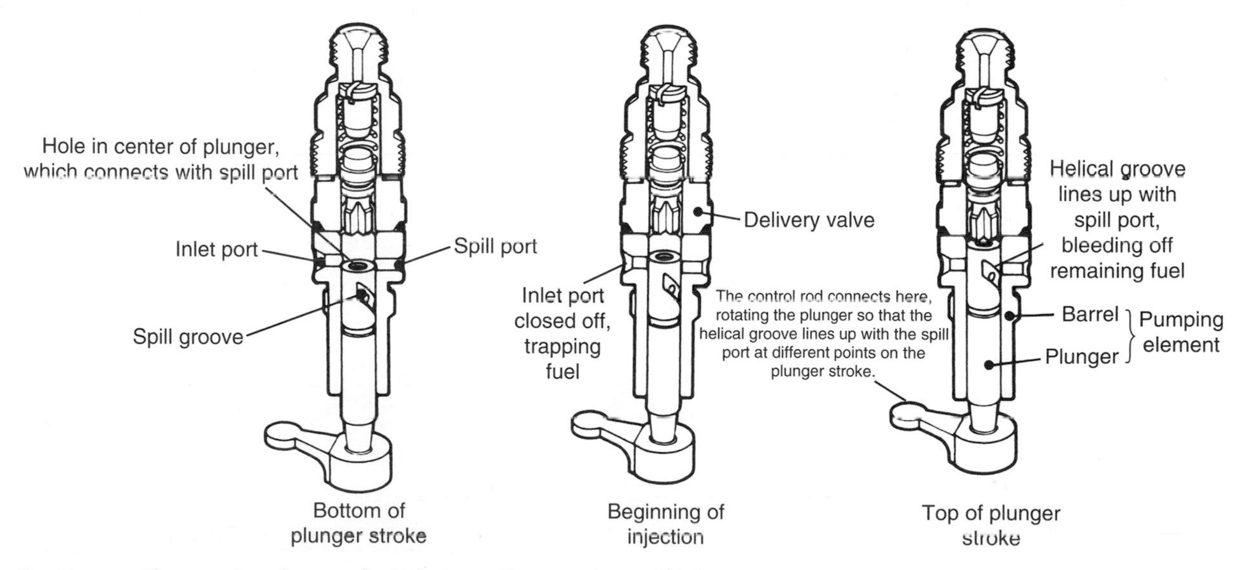

FIGURE 2-13. *The pumping element of a jerk pump. (Courtesy Lucas/CAV)*

engines released when they accelerated, and it also prevents smoking at higher throttle settings.

On some engines, an electronic device connected to the engine's computer operates the rack, as opposed to the traditional mechanical linkage. Typically, the computer monitors the inlet air pressure and temperature, and from these readings, it calculates the quantity of air available for combustion, adjusting the maximum fuel setting accordingly. Once again, this process eliminates the cloud of black smoke when an engine accelerates and stops smoking at higher throttle settings.

Each cylinder requires its own jerk pump, but all of them are normally housed in a common block and driven by a common camshaft (with a separate cam for each pump). These are known as in-line pumps (see Figure 2-14).

Smooth engine operation from an in-line system demands that each pump put out exactly the same quantity of fuel to within millionths of a gallon. A jerk pump plunger has no piston rings to seal it in its barrel—it relies solely on the accuracy of the fit of the plunger and barrel. Nowadays, the two are machined to within 0.0004 inch of each other, and plungers and barrels must be as smooth as glass or they will seize up. Because of this degree of precision, amateurs should never tamper with fuel injection pumps. The only possible result is expensive damage!

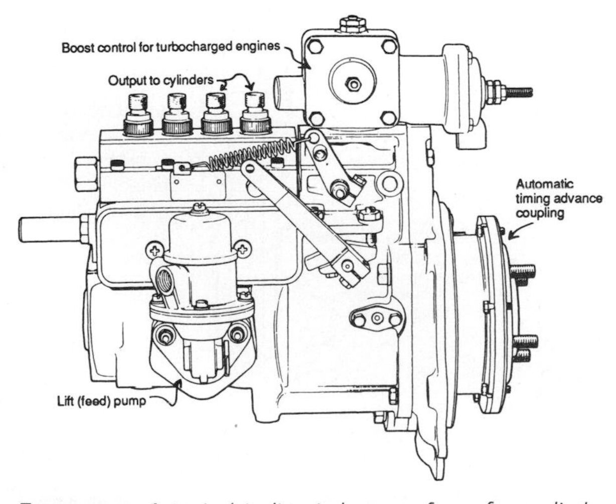

FIGURE 2-14. *A typical in-line jerk pump for a four-cylinder engine. (Courtesy Lucas/CAV)*

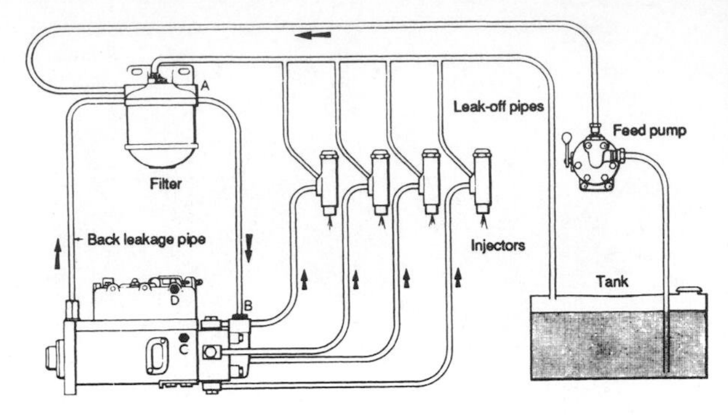

FIGURE 2-15. *A typical fuel system with a distributor pump. (Courtesy Lucas/CAV)*

DISTRIBUTOR (ROTARY) PUMPS

Figures 2-11, 2-15, and 2-16 show a distributor pump fuel injection system. Note that it is broadly similar to the jerk pump system. The two systems operate in the same way, use the same injectors, etc. The only difference between the two lies in the injection pumps themselves.

Whereas a jerk pump has a separate pump for each engine cylinder, a distributor pump uses one central pumping element and a rotating head that sends fuel to each cylinder in turn. This is done in much the same way that a gasoline engine's distributor used to send a spark to each spark plug in turn (as in the days before electronic ignition systems), hence the name distributor pump.

Because the same pump feeds all the cylinders, every injector gets an equal amount of fuel, ensuring even engine loading and smooth running at idle

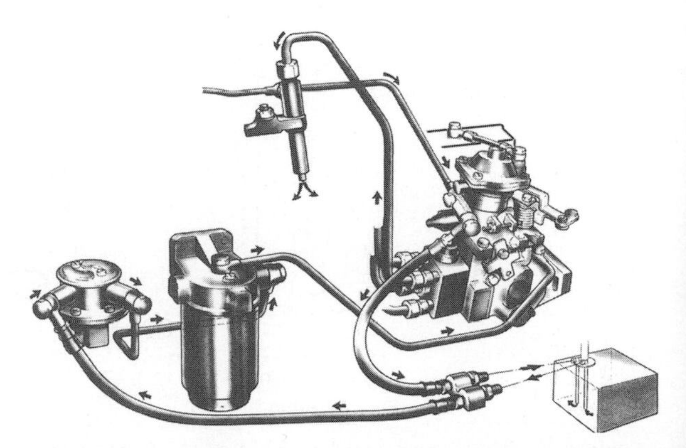

FIGURE 2-16. *Distributor-type fuel injection. (Courtesy Volvo Penta)*

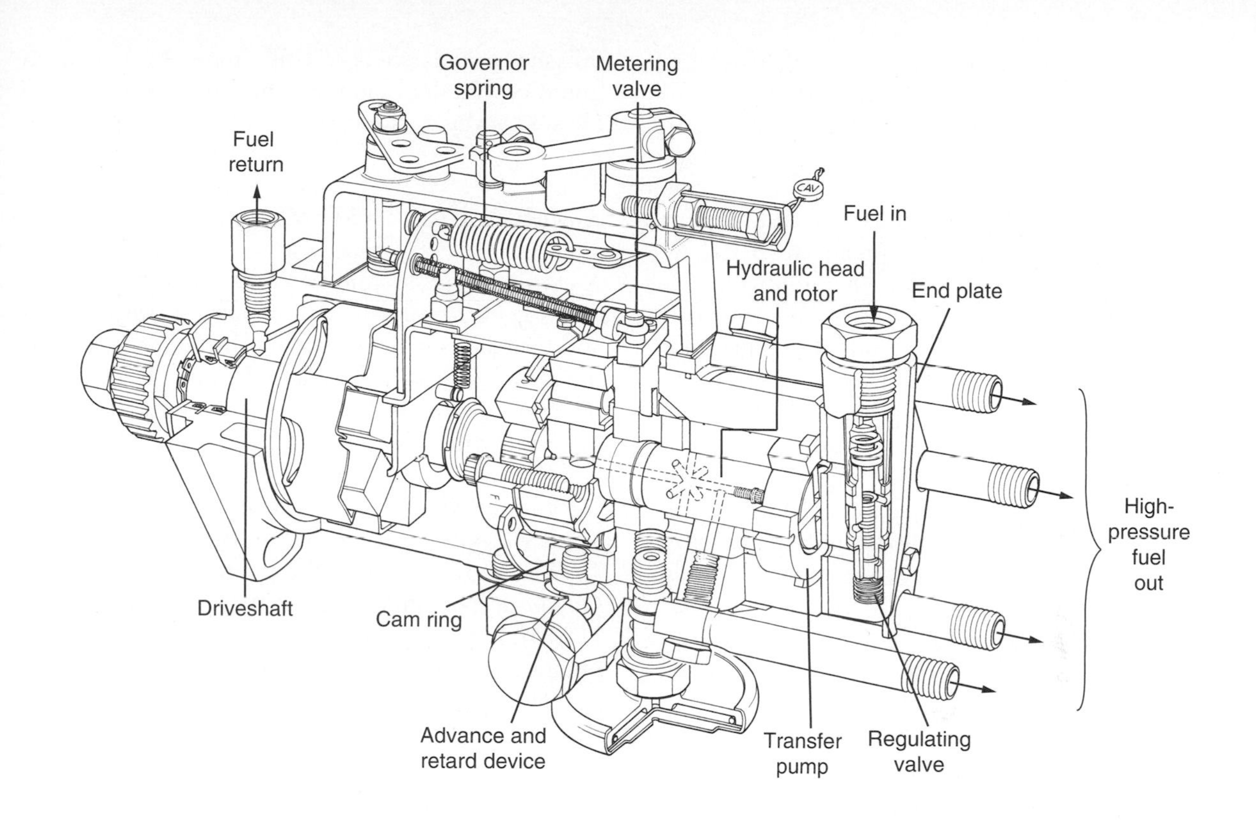

FIGURE 2-17. *Distributor-type fuel injection pump. (Courtesy Lucas/CAV)*

speeds. A metering valve on the inlet to the pump and connected to the throttle regulates the output of the pump and, consequently, the engine as well (see Figure 2-17). There also may be a smoke limiter (see above). On some engines, an electronic device connected to the engine's computer operates the throttle with the same kind of functionality as an electronically controlled jerk pump (see above once again).

UNIT INJECTORS

Unit injectors have a fuel pump that draws fuel from a tank via a primary filter, passes the fuel through a secondary filter, then discharges it continuously into a passage, or gallery, in the cylinder head at a relatively low pressure. This gallery supplies the injectors (see Figure 2-18). A pressure-regulating valve at the end of the fuel gallery holds the pressure in the system at a set point and allows surplus fuel to return to the fuel tank. Fuel flows continuously through the whole system, including the injectors, keeping it lubricated and cool.

Each injector contains its own fuel pump, similar in many ways to a jerk pump. An engine-driven camshaft

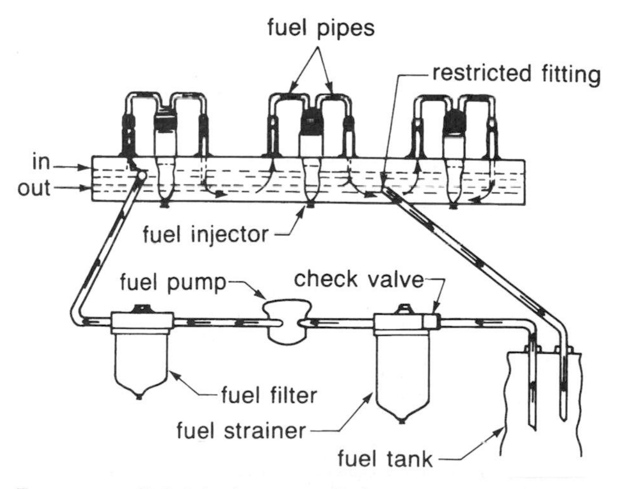

FIGURE 2-18. *Unit injection as applied on a Detroit Diesel two-cycle engine. (Courtesy Detroit Diesel)*

operates each pump, and at the appropriate moment, the pump strokes and injects fuel directly into the engine. Engine speed is controlled by varying the amount of injected fuel. On modern engines, the control mechanism is a solenoid linked to the engine's computer to optimize fuel efficiency and minimize exhaust emissions (electronic unit injector, EUI). At a minimum, the computer will be monitoring inlet air pressure and temperature, as described for electronically controlled jerk pumps, and often much more (see Section Five: Electronic Engine Controls). On some engines (notably Caterpillars), the rocker and camshaft are eliminated, and instead, high-pressure engine oil is used to generate injection pressures (hydraulic electronic unit injection, HEUI).

Unit injection is common on larger engines because it is possible to achieve extraordinarily high injection pressures (up to 30,000 psi/2,000 bar), which is important in terms of fully distributing the injected fuel inside large combustion chambers (the larger the combustion chamber, and the farther the fuel has to be projected, the higher the injection pressure needs to be). Unit injectors are rarely found on smaller engines (other than some Detroit Diesel two-cycle diesels).

Common Rail Systems

A common rail system is similar in some ways to unit injection—a pump draws fuel from the tank and circulates it continuously through a gallery in the cylinder head and back to the tank. However, in the common rail system, the fuel is circulated at full injection pressure (often 20,000 psi/1,350 bar or higher), eliminating the need for unit injectors. Instead, all that is required for injection is some kind of a valve to admit fuel to the cylinders at the appropriate time and in the appropriate quantities. On modern engines, this is typically a fast-acting electromagnetic needle valve, controlled by the engine's computer (see Figure 2-19).

The common rail system separates the two functions of generating fuel pressure and injection timing

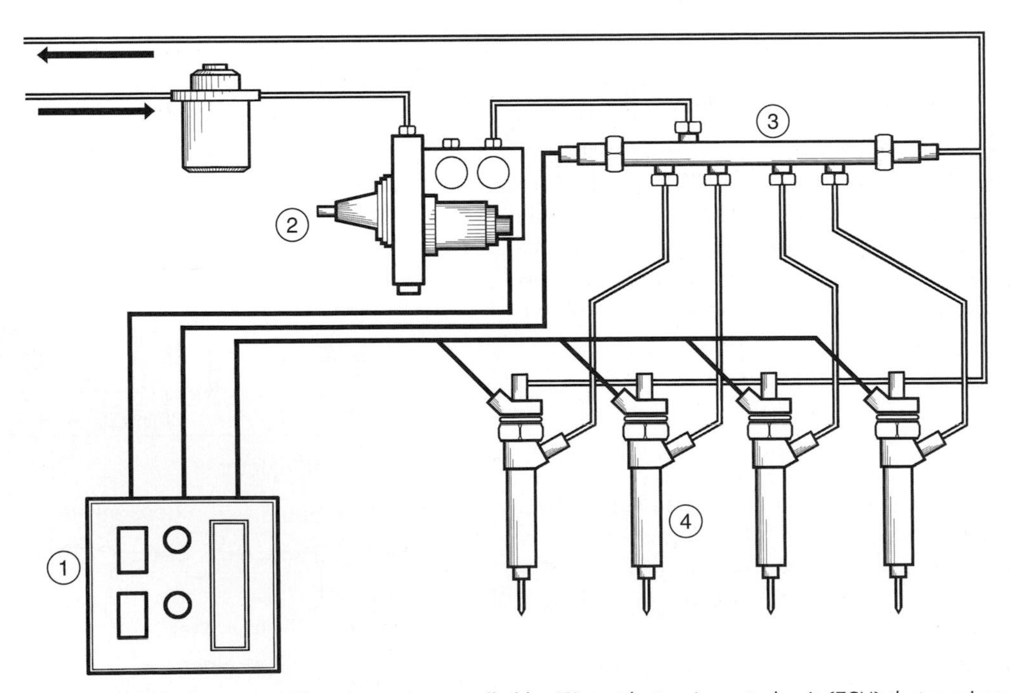

Figure 2-19. *Common rail fuel injection. The process is controlled by (1) an electronic control unit (ECU) that receives inputs from numerous sensing devices. A high-pressure pump (2) feeds the common rail (3). Although the injectors (4) resemble traditional injectors, they are, in fact, electronically operated needle valves. (Fritz Seegers)*

(which, in other systems, are combined at the injection pump or unit injector). The fuel is stored continuously at injection pressures in an accumulator rail. At the injector, an electronic valve opens and closes to allow fuel into the cylinder.

With a conventional injection pump, there is an injection lag between the moment the pump generates injection pressures and the moment of injection. This lag varies with such things as fuel temperature and viscosity, the elasticity of injection lines, and wear in the injection pump. Common rail injection is in real time, enabling extraordinarily precise control.

With a conventional system, the volume of fuel in an injection line is quite small. As soon as injection commences, there is some pressure drop in the line, which reduces the effectiveness of the injection stroke. With common rail, the volume of pressurized fuel in the accumulator relative to the volume of fuel being injected is such that it eliminates this pressure drop.

Electronically controlled common rail systems monitor numerous engine and fuel variables (see Section Five: Electronic Engine Controls), with a computer optimizing the injection process based on this information. The rate of injection is carefully regulated to create a controlled power stroke (which eliminates the characteristic clatter of a diesel engine at idle). There may be as many as five injection pulses per power stroke. The net result is improved engine efficiency over a wide range of operating conditions, lower exhaust pollution, less noise and vibration, and extended engine life. The downside is the dependence on highly sophisticated electronics.

Because of the high pressures found in a common rail system, *no amateur mechanic should ever attempt to service one of these systems*. Just slackening an injection nut (as is commonly done when troubleshooting other types of injection systems) can result in a spray of diesel that has sufficient force to penetrate the skin and cause blood poisoning. The incredibly finely atomized diesel is also a fire hazard. (To minimize fire risks in the event of damage to system piping, all piping is double walled with an alarm system that goes off if the inner wall gets ruptured.)

Although common rail engines are finding their way into some lower hp marine applications, neither Volvo Penta or Yanmar will be using this technology in marine diesel engines below 100 hp anytime soon.

INJECTORS

With conventional fuel injection, every time an injection pump (jerk or distributor type) strokes, it drives fuel through the delivery pipe at high pressure to an injector. A unit injector has the injection pump built into it, operating directly off the engine's camshaft, and thus eliminates the delivery pipes. The common rail system does away with conventional injectors altogether.

Inside an injector, a powerful spring (the thrust spring) holds a needle valve against a seat in the injector nozzle. The needle valve is designed to lift off its seat, allowing fuel to enter the engine, when the pressure in the injector reaches a certain level (see Figure 2-20). Some injectors have two thrust springs (two spring injectors), resulting in a two-stage

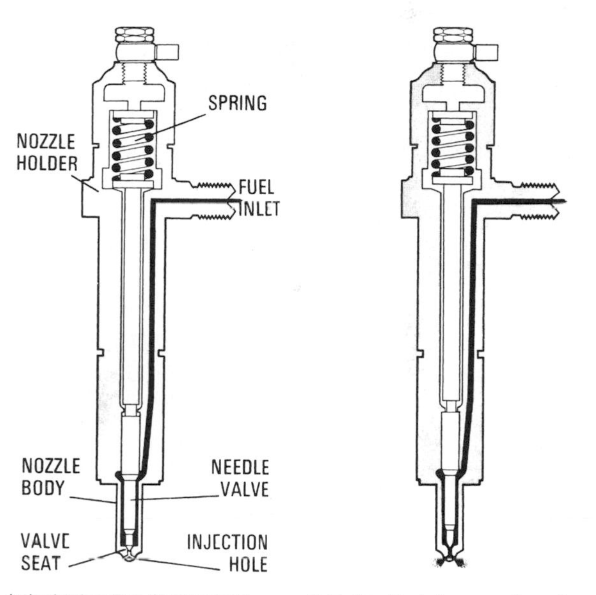

In the closed condition, the spring presses the needle valve downwards so that the tapered end of the needle fits exactly on the valve seat and seals off the injection holes

At injection, rising fuel pressure acting on the shoulders of the needle valve causes the latter to lift and so uncover the injection holes through which fuel is sprayed into the combustion chamber

FIGURE 2-20. *Traditional fuel injectors. Fuel entering the injector passes through galleries in the body and nozzle to a chamber surrounding the nozzle valve. The valve spring tightly holds the valve closed until fuel pressure from the injection stroke overcomes the spring's tension and lifts the valve, letting fuel under high pressure pass through the holes and tip of the spray nozzle. This happens instantaneously. At the end of the injection, fuel pressure rapidly falls and the spring returns the valve to its seat, ending the flow of fuel into the combustion chamber. (Courtesy Lucas/CAV)*

injection process. The first spring lifts to allow a small preliminary injection that creates turbulence and increases cylinder pressures ahead of the second (main) injection pulse. This results in a smoother, quieter, cleaner, more efficient burn. (Note that these injectors cannot be reconditioned in the field and must be exchanged for new ones.)

With jerk pumps and distributor pumps, a small amount of diesel is allowed to find its way past the stem of the needle valve to lubricate the injector. The extra fuel then returns to the fuel tank via a leak-off pipe. Fuel continuously circulates through the body of a unit injector.

With electronically controlled jerk and distributor pumps (see Section Five: Electronic Engine Controls), the injector on the #1 cylinder may include a *needle lift sensor*—a solenoid coil wound around a magnetic core—that senses the moment at which the needle valve lifts off its seat and sends a signal to the control unit. This is the moment at which injection commences. The control unit uses this signal, along with the engine speed signal, to adjust injection timing. If either signal is lost, the control unit will limit engine output. (With Volvo Pentas, if the needle lift signal is lost, the top engine speed is lowered; if the engine rpm signal is lost, the engine is stopped. Note that these injectors cannot be reconditioned in the field and must be exchanged for new ones.)

In a common rail system, the fuel is always at injection pressures. A special type of solenoid-operated valve is opened (electrically) to inject fuel into a cylinder.

When thinking about what kind of injection may be found on an engine of a given age, it's worth bearing in mind that the evolution of electronic controls over the past twenty years (1985 to 2005) has been more or less in the order that follows:

1. Electronically adjusted jerk and distributor pumps, on which the fuel rack is moved by an electrical device and based on a relatively simple computer program that monitors inlet air pressure and temperature.
2. Injectors with needle lift sensors that enable the injection timing to be adjusted according to engine speed.
3. Electronically controlled unit injectors based on increasingly sophisticated computer software receiving data inputs from an ever-increasing number of sensing devices on engines.
4. Common rail systems with, once again, increasingly sophisticated computer controls.

The addition of any electronics to the injection process, especially complex electronics, more or less precludes the owner/operator from doing any service work on the injection system.

LIFT PUMPS

Traditional fuel systems use a lift pump to move diesel from the fuel tank to the injection pump. Some engines use electric diaphragm pumps similar to the units found on many automobiles, but the majority of jerk and distributor injection systems use a mechanically driven diaphragm pump fitted to the engine block or the side of the injection pump (see Figure 2-21; for more on lift pumps, see pages 103–5). A unit injection system requires a custom-made gear pump very specifically designed for this system, as does a common rail system.

FIGURE 2-21. *A typical mechanical lift pump fitted to the side of an engine block.*

SECTION FOUR: GOVERNORS

The output of a diesel engine is controlled by regulating the amount of fuel injected into the cylinders. In marine use, you normally want the engine to run at a specific speed, regardless of the load placed on it. This cannot be done by simply pegging the throttle at a certain point because every time the load increases or decreases, the engine slows down or speeds up. Constant-speed running with a conventional diesel engine is achieved by connecting the fuel-control lever on the injection pump or unit injector to a governor. On engines with electronic engine controls, the function of a traditional governor is taken over by the engine's computer (see below).

SIMPLE GOVERNORS

The most basic governor consists of two steel weights, known as flyweights, attached to the ends of two hinged, spring-loaded arms (see Figure 2-22). The governor's driveshaft is mechanically driven by the engine, and as it spins, the flyweights spin with it, pushed outward by centrifugal force. A speeder spring counterbalances this centrifugal force.

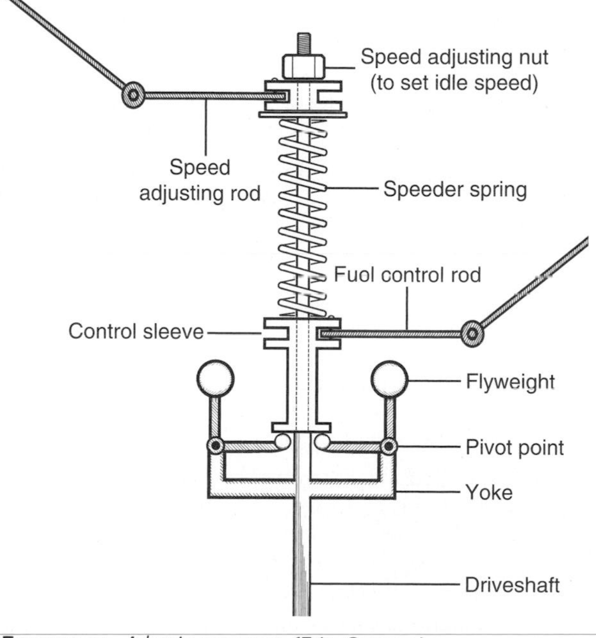

FIGURE 2-22. *A basic governor. (Fritz Seegers)*

Let's assume that the engine is running, the governor is spinning, and the flyweights are in equilibrium with the speeder spring at a certain position. If the load decreases and the engine speeds up, the governor will spin faster, and the flyweights will move out under the increase in centrifugal force. As the flyweights move out, their arms push up against the control sleeve, which compresses the speeder spring until sufficient counterbalancing pressure restores equilibrium.

The control sleeve is connected by a series of rods to the injection pump's fuel-control lever, and when the sleeve moves up the governor's driveshaft, it cuts down the injection pump's rate of delivery. The reduction in fuel slows the engine to the speed at which it was originally set.

If the load increases and the engine slows down, the centrifugal force on the flyweights decreases, and they move inward under the pressure of the speeder spring. In moving inward, the flyweight arms allow the speeder spring to push the control sleeve down the driveshaft, operating the injection pump's fuel-control lever. This causes more fuel to be injected, which returns the engine to its preset speed.

The engine can be run at any set speed by adjusting the tension on the speeder spring via the speed adjusting rod. The greater the pressure on the spring, the more the flyweights are held in, the more fuel is injected, and the faster the engine runs. Reducing the pressure on the speeder spring causes the flyweights to move out more easily, therefore sooner. When the fuel injection rate is reduced, the engine runs at slower speeds.

On larger conventional engines, the simple mechanical governor just described is likely to be replaced by a complex hydraulic one, though the principles are the same.

In the past, some small marine diesels have had the governor installed in the engine block, but normally it is in the back of the fuel injection pump. It rarely malfunctions (troubleshooting is covered on pages 152–53). Beyond the occasional need to adjust the tension of the speeder spring to set up the engine's idle speed, you should not need to know any more about a mechanical governor than the information given here.

Vacuum Governors

You may occasionally run across a vacuum governor, which operates as follows. A butterfly valve (a pivoting metal flap) is installed in the entrance to the air-inlet manifold. A vacuum line is connected between the air-inlet manifold and a housing on the back of the injection pump. Inside this housing is a diaphragm that is connected to the injection pump's fuel-control lever, or rack.

The throttle operates the butterfly valve, and when the throttle is shut down, the butterfly valve closes off the air inlet to the engine. The pumping effect of the pistons attempting to pull in air then creates a partial vacuum in the air-inlet manifold. This vacuum is transmitted to the housing on the fuel injection pump via the vacuum line, sucking in the diaphragm against a spring. The diaphragm pulls the injection pump's control lever to the closed position.

If the load increases, the engine slows, the vacuum declines, and the spring in the fuel injection pump pushes the diaphragm, and with it the fuel-control lever, to a higher setting. The engine speeds up. If the load decreases, the engine speeds up and the vacuum increases, sucking the diaphragm to a lower fuel setting and slowing the engine.

When the throttle is opened, the butterfly valve opens, the manifold vacuum declines, the diaphragm moves back under pressure from the spring, the fuel-control rack increases the fuel supply, and the engine speeds up to the new setting.

Aside from leaks in the vacuum line and around the diaphragm housing or a ruptured diaphragm (both are covered on page 154), little can go wrong with this system. The engine's idle speed is set with a screw that adjusts the minimum closed position of the butterfly valve.

SECTION FIVE: ELECTRONIC ENGINE CONTROLS

First-generation electronic engine controls replaced the functions of the governor and speeder spring with an engine speed sensor wired to an electronic control unit (ECU). The ECU outputs a signal to a solenoid-operated actuator that moves the fuel rack in the injection pump to vary the rate of fuel injection. The rate of injection is adjusted to maintain the desired engine speed.

Over time, ECUs have been designed to act upon information from various other sensors (such as a turbocharger's air pressure and temperature), resulting in improved engine performance. The latest generation of electronic engine controls (with unit injectors and on common rail systems) now operate at the individual injector level, rather than at the injection pump level.

ECUs have become extremely complex and sophisticated, incorporating data from an ever-increasing array of sensors (see the Engine Sensors sidebar on page 32). Their functions have been extended from simple engine speed control to include transmission shifting and steering, resulting in what is known as fly-by-wire operation: there is only a data cable (no mechanical link) between the throttle, gearshift lever, and steering wheel and the devices they control (see Figure 2-23).

Electronic Engine Control Acronyms

There are literally dozens of acronyms cropping up in the field of electronic engine controls (EEC). Here are a few:

MC = mechanical control
EVC = electronic vessel control
ECU = engine (or electronic) control unit
EC = electronic control
PCU = power train control unit
HCU = helm control unit
IPS = Inboard Performance System (this is unique to Volvo)

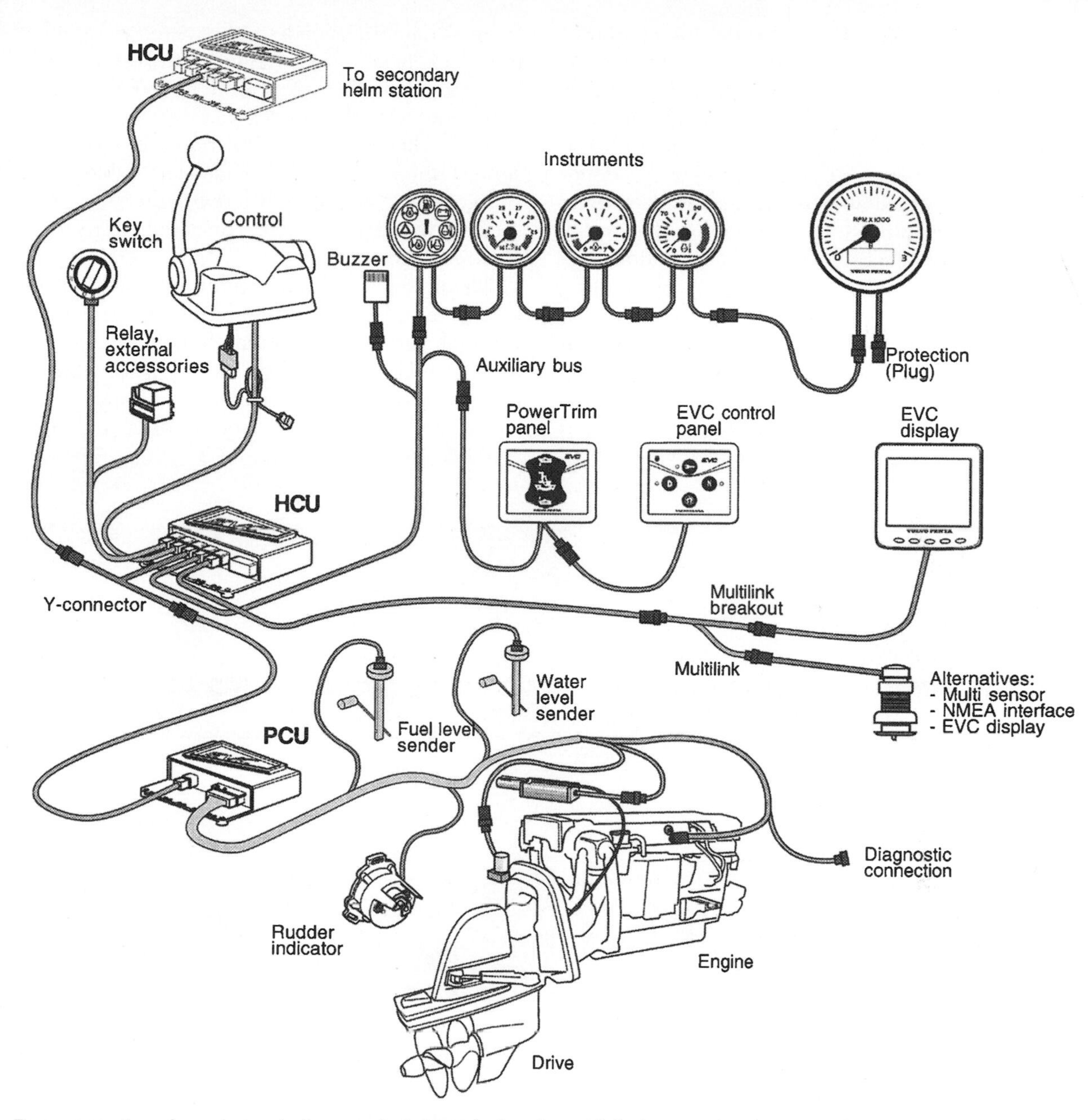

FIGURE 2-23. *A modern, electronically controlled, networked engine, with fly-by-wire throttle, gear shifting, and steering. (Courtesy Volvo Penta)*

NETWORKING

Today, almost all ECUs operate on some variant of a protocol known as CAN (controller area network) that has come out of the automotive industry, and which is found on most modern cars. (When you take your car in for servicing and the technician plugs it into a computer to check its operating history and diagnose problems, the computer is being plugged into the CAN system.) CAN-based systems on marine diesels have increasingly powerful diagnostic and troubleshooting capabilities.

Latest-generation navigational electronics also share data via the CAN system. At the time of writing

(2006), we are beginning to see the integration of navigational electronics with engine electronics. At its simplest, this may be no more than engine data being displayed on navigational screens and vice versa, but in many instances, there is beginning to be functional interaction. For example, if the depth sounder shows shallow water or the radar shows an approaching land mass, the engine may be slowed. This process of integration is certain to continue.

BENEFITS

Electronic engine control systems result in a considerable reduction in the wiring associated with an engine. In essence, all the engine sensors have point-to-point wiring to the ECU unit, which is generally mounted on or near the engine. From there all data travel on a single twisted-pair cable to the remotely mounted engine controls and displays, as opposed to the wires from individual sensors being routed to the gauges that display their information. The gauges are invariably mounted in a serial fashion—with the single data cable looped from gauge to gauge. Increasingly, the gauges themselves are giving way to on-screen displays (see Figure 2-24).

The net result is more and more powerful electronic controls with tremendous diagnostic, failure warning, and troubleshooting functions, along with a considerable amount of stored history, but the controls are beyond the capability of any amateur shade-tree mechanic to troubleshoot and repair (see Figures 2-25, 2-26, and 2-27).

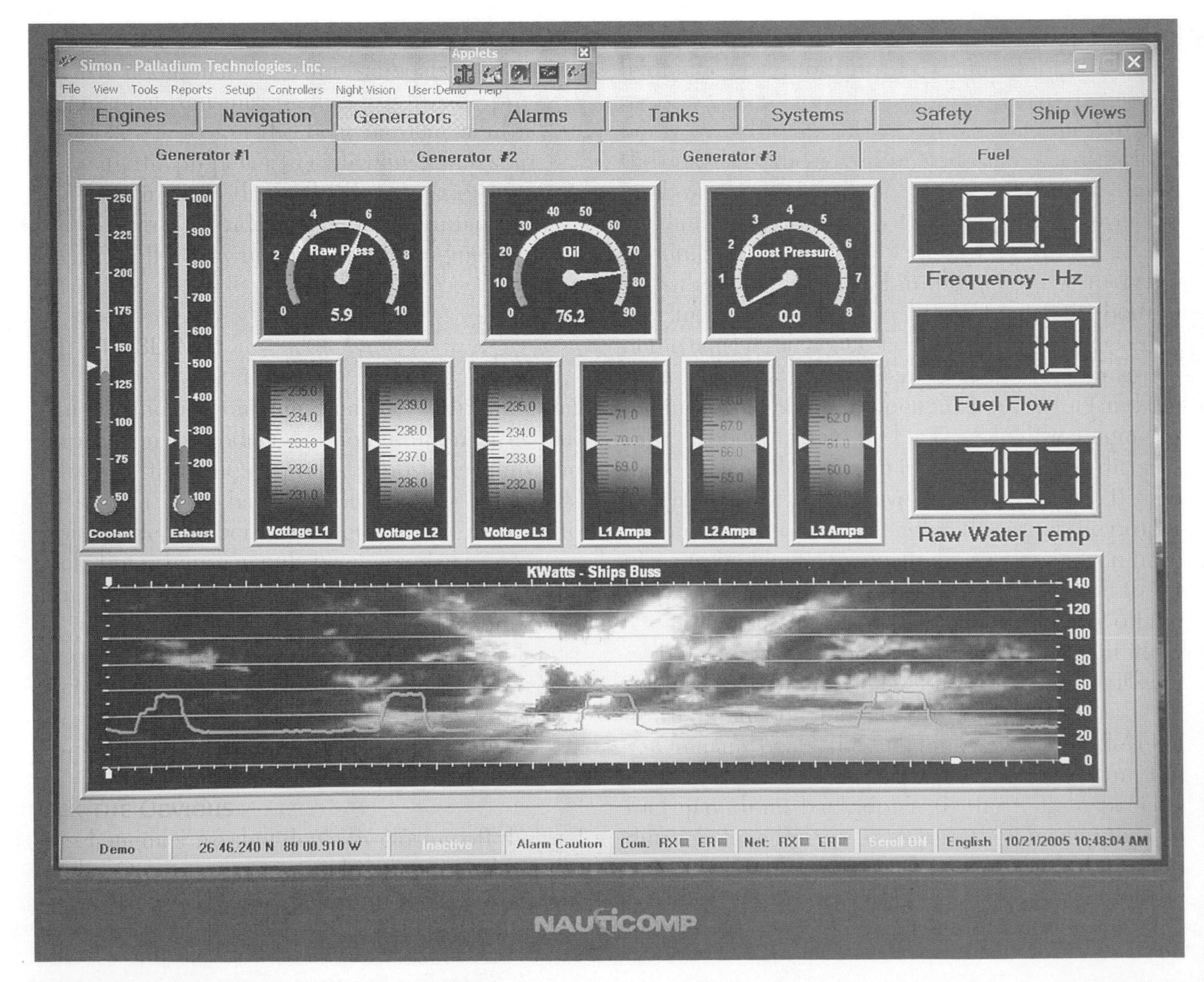

FIGURE 2-24. *Integrated onboard electronic display.*

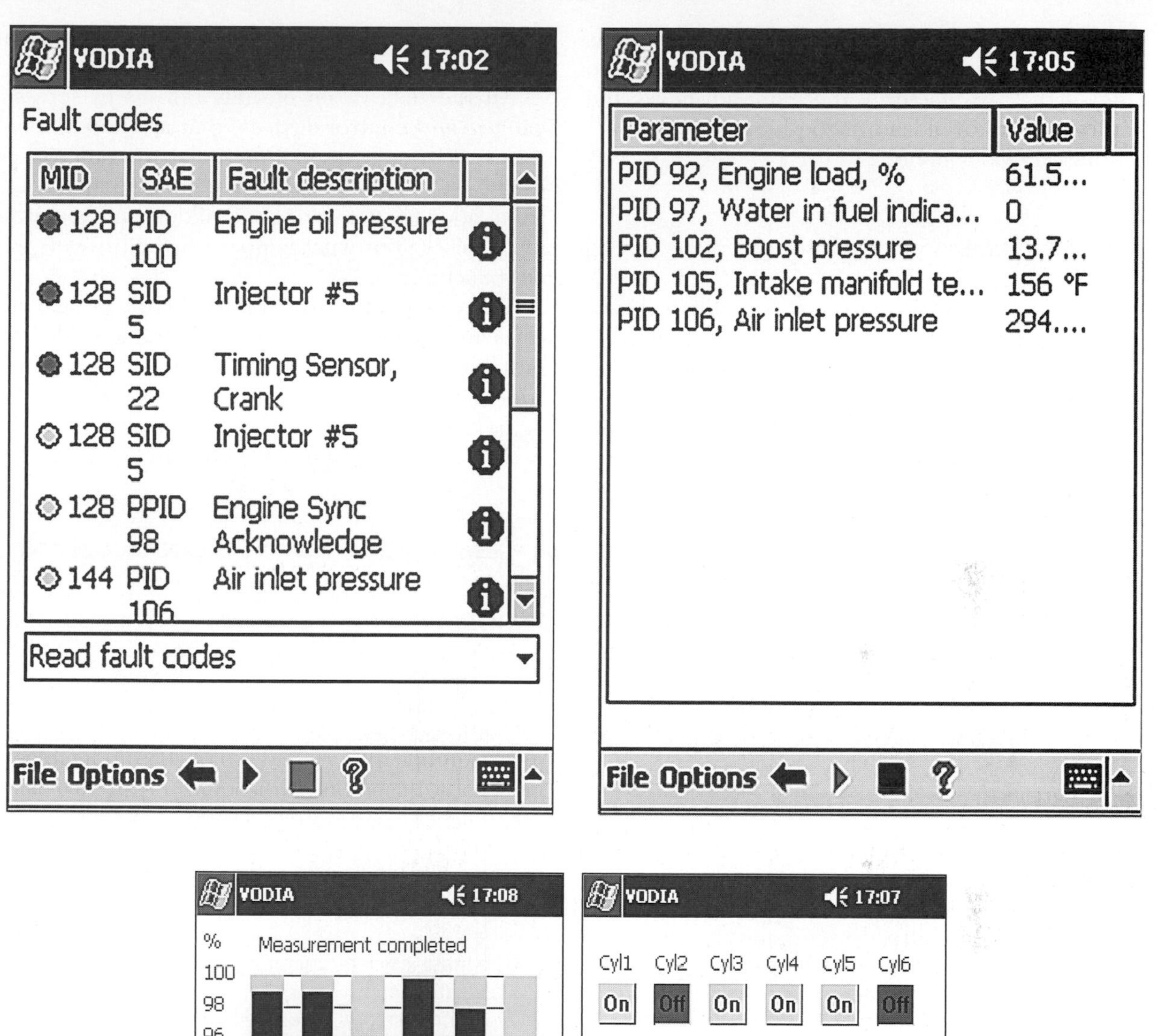

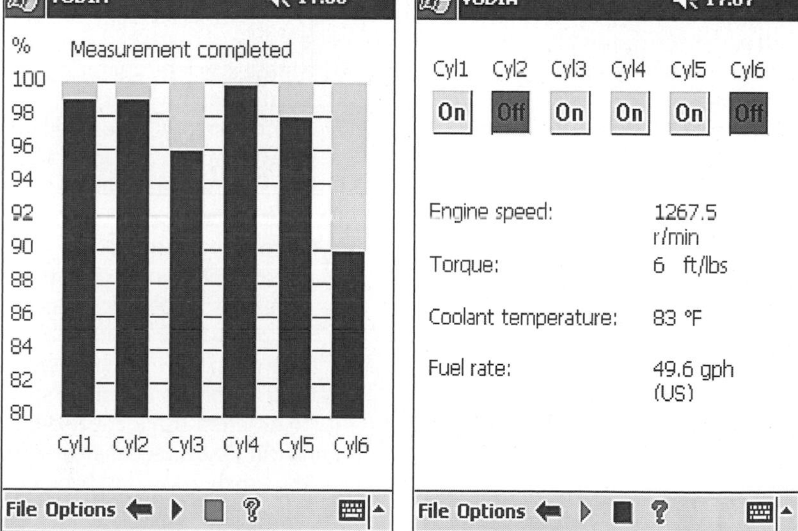

Figures 2-25, 2-26, and 2-27. *Screen shots illustrating the range of information available to the operator of a modern, electronically controlled engine. (Courtesy Volvo Penta)*

It is worth noting that electronic engine controls take a lot of resources to develop. In general, the bigger the manufacturer, the more advanced, the more sophisticated, and the better tested the electronics are likely to be.

Limping Home

The year 2006 saw the introduction of the first fully networked boats with more or less complete integration of networked navigational electronics, engine monitoring and control systems, steering and gear shifting, and the rest of a boat's electrical distribution system (using remotely operated electronic circuit breakers—what is known as a *distributed power system*).

All the data, from literally dozens of sensors, gauges, and control devices, that are necessary to make all these systems functional run on a single data bus! This is powerful technology, with some tremendous benefits for boatbuilders and boatowners alike . . . but what happens if the bus gets put out of action?

There is an obvious need to identify critical circuits and to make them as invulnerable to disruption as possible and, if necessary, provide a manual override of some kind if the bus goes down altogether. This is known as a *limp-home* capability, and it varies from manufacturer to manufacturer. It is

Engine Sensors

The following is a list of the sensors that are likely to be found on a modern, electronically controlled, diesel engine, and which will be integrated into the operation of the engine's ECU:

- Pressure in the raw-water system
- Seawater temperature
- Seawater depth
- Seawater depth alarm
- Pressure in the freshwater system
- Freshwater temperature
- Fuel pressure
- Fuel rate
- Fuel tank levels
- High-pressure injection line leakage (common rail systems)
- Engine oil pressure
- Engine oil filter differential pressure
- Engine oil temperature
- Transmission oil pressure
- Transmission oil temperature
- Turbocharger pressure
- Inlet air temperature
- Exhaust temperature
- Engine rpm
- Engine operating hours
- Engine power trim
- Battery voltage
- Rudder angle
- Throttle position
- Boat speed

ECU functions may include such things as:

- Sensor failure warnings
- Calculated fuel burn rate, range under power, etc.
- The most efficient operating speed
- Calculated load
- Operating history, including how many hours at what speed and load and the date and time of "events" such as a high exhaust temperature
- Synchronizing engines
- Programmable monitoring that can be customized so that if, for example, the engine temperature rises or the air inlet temperature rises, the engine slows down
- Setting of idling and trolling speeds according to different fish identified by a fishfinder
- Enhancing low-speed maneuvering by allowing the clutch to slip on a hydraulic transmission
- Settings for a limp-home function in the event of serious failures in the electronics of fly-by-wire steering, throttle, and transmission shifting

one of those things that anyone considering buying a fully electronic engine, or a distributed power system for the rest of the boat, should research before choosing a system.

Regardless of how well a system is *hardened* against radio frequency interference/electromagnetic interference (RFI/EMI), lightning strikes, and so on, all systems require a minimum power supply to remain operational. If the voltage drops below a certain level (generally around 5.0 to 6.0 volts), there will be a *brownout*, in which unpredictable responses may happen, followed by a complete shutdown of the network as the voltage drops lower. It's a good idea to have some kind of uninterrupted power supply (UPS) for the critical circuits on the network (e.g., the engine, steering, transmission shifting, and bilge pumps).

SECTION SIX: KEEPING THINGS COOL

Diesel engines generate a great deal of heat, only one-third of which is converted to useful work. The remaining two-thirds must somehow be released to the environment so that temperatures in the engine do not become dangerously high. Excessively high temperatures can break down lubricating oils, causing the engine to seize or the cylinder head to crack. A little less than one-half of the wasted heat goes out with the exhaust gases; a similar amount is carried away by the cooling system. The balance radiates from the engine's hot surfaces.

Three principal types of cooling systems are used in boats: raw water, heat exchangers, and keel coolers.

Raw-Water Cooling

Raw-water cooling systems draw directly from the body of water on which the boat floats. This water enters through a seacock; passes through a strainer; circulates through oil coolers, any aftercooler, and the engine; and finally discharges overboard. On most four-cycle diesels, after the water circulates through the engine, it enters the exhaust pipe and discharges with the exhaust gases. This is a wet exhaust (see Figure 2-28); a dry exhaust has no water injection.

A raw-water cooling system is simple and economical to install, but it has a number of drawbacks. Over time a certain amount of silt inevitably finds its way into the engine and coolers and begins to plug cooling passages. Regulating the engine's temperature is difficult. First, the temperature of the raw water may range from the freezing point in northern climates in the winter to 90°F (32°C) in tropical climates in the summer. Second, the system frequently has no thermostat because the inevitable bits of trash and silt can clog it and make it malfunction.

In salt water, scale (salts) crystallizes out in the hottest parts of the cooling system, notably around cylinder walls and in cylinder heads (see Figure 2-29). This leads to a reduction in cooling efficiency and localized hot spots (General Motors states that 1/16 inch of scale on 1 inch of cast iron has the insulating effect of 4½ inches of cast iron!). The rate of scale formation is related to the temperature of the water, and it accelerates when coolant temperatures are above 160°F (71°C). As a result, raw-water-cooled engines must generally be operated cooler

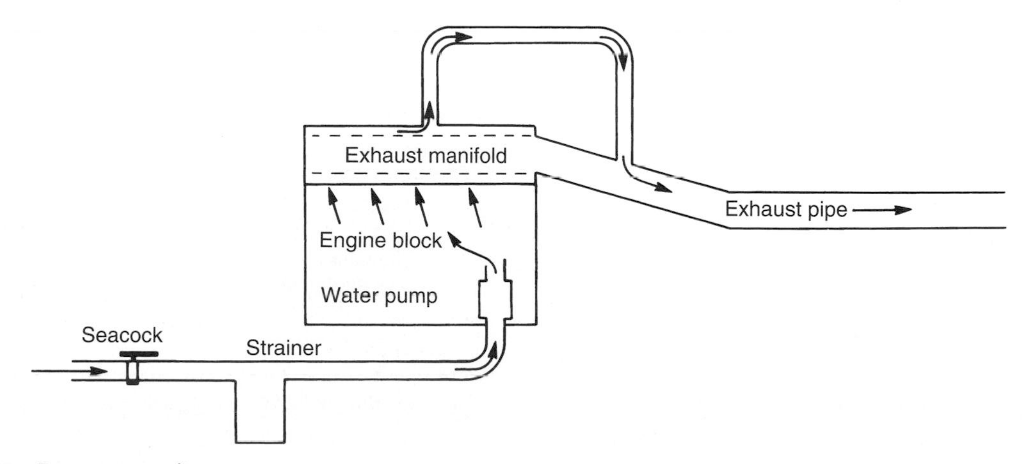

FIGURE 2-28. *Raw-water cooling.*

FIGURE 2-29. *Corrosion in a raw-water-cooled cylinder head. Note the scale partially plugging the water passage on the right-hand side.*

than other engines (around 140°F to 160°F/60°C to 71°C, compared with 185°F/85°C or higher). This lowers the overall thermal efficiency of the engine and can also cause water and harmful acids to condense out in the engine oil (see pages 51–52).

The combination of heat, salt water, and dissimilar metals is a potent one for galvanic corrosion. An engine using raw-water cooling should be made of galvanically compatible materials (e.g., a cast-iron block should have a cast-iron cylinder head, not an aluminum one). Sacrificial zinc anodes must be installed in the cooling circuit, and they must be inspected and changed regularly (see pages 61–63).

The engine cannot be protected with antifreeze in cold weather; therefore, it will have to be drained after every use. To facilitate draining, piping runs and pumps need to be installed without low spots.

Today's powerful, high-speed, lightweight diesels demand a greater degree of cooling efficiency than that provided by raw-water cooling. Almost all these power plants use a heat exchanger or keel cooler.

HEAT EXCHANGER COOLING

An engine with a heat exchanger has an enclosed cooling system. The engine cooling pump circulates the coolant from a header (expansion) tank through oil coolers, maybe an aftercooler, and the engine. The coolant then passes through a heat exchanger to lower its temperature and is circulated once again (see Figure 2-30).

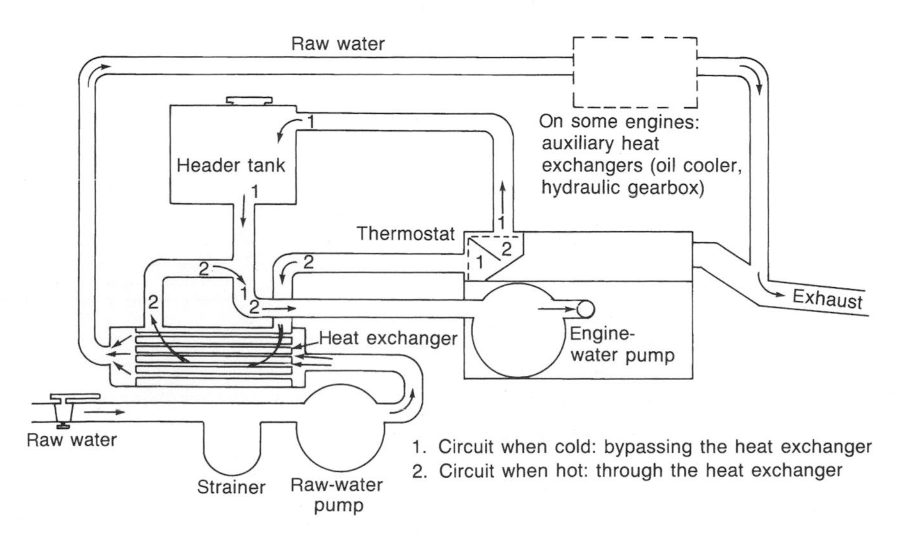

FIGURE 2-30. *A heat-exchanger schematic.*

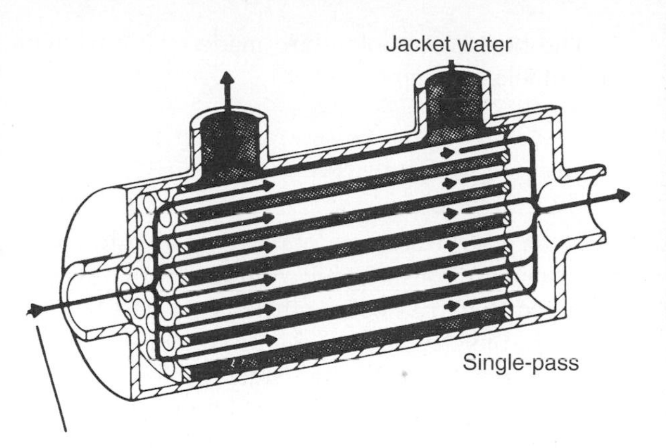

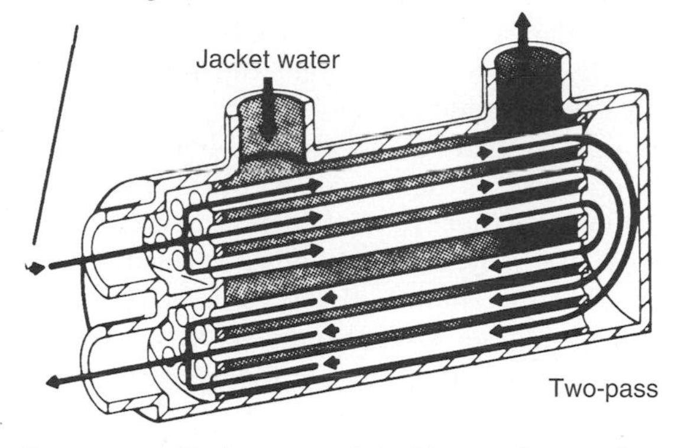

FIGURE 2-31. *Single-pass and double-pass heat exchangers. (Courtesy Caterpillar)*

A heat exchanger consists of a cylinder with a number of small cupronickel tubes running through it (see Figures 2-31 and 2-32). The hot engine coolant passes through the cylinder while cold raw water is pumped through the tubes. The raw water carries off the heat from the coolant, then discharges overboard, either directly or via a wet exhaust.

A heat exchanger is expensive and requires more piping than a raw-water system as well as an extra pump on the exchanger's raw-water circuit. On the other hand, antifreeze and corrosion inhibitors can be added to the coolant on the engine side to protect the engine against freezing and corrosion. No silt finds its way into the engine, and no salts crystallize out of the coolant around cylinder liners and in the cylinder head. Also the engine can be operated at higher temperatures, which is more thermally efficient.

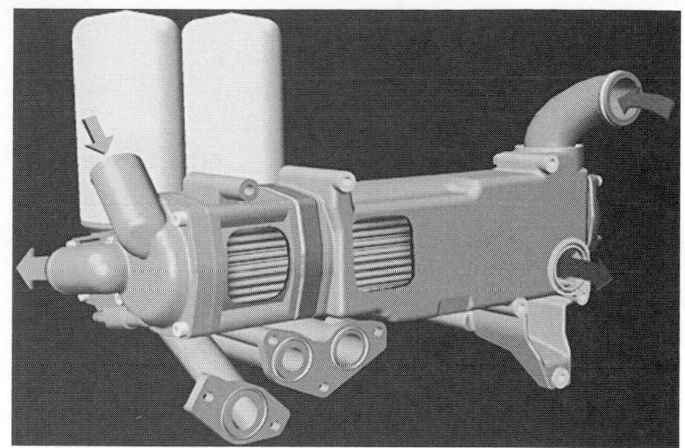

FIGURE 2-32. *A combined engine cooling water heat exchanger (right) and oil cooler (left). (Courtesy Volvo Penta)*

The expansion tank will have a pressurized cap, as on an automobile's radiator. When the pressure of the coolant is increased, so too is its boiling point. If the pressure is raised by 10 psi, its boiling point rises to approximately 240°F (116°C). Allowing the pressure to rise in a closed cooling system greatly reduces the risk of localized pockets of steam forming and causing damage at hot spots in the engine.

The raw-water side of a heat exchanger circuit will still suffer from many of the problems associated with raw-water cooling, particularly silting up of the heat-exchanger tubes and the potential for damage from corrosion and freezing. Sacrificial zincs are essential once again if there are any dissimilar metals, and the piping must have no low spots so that it can be easily drained in cold weather. (Note that many modern engines do not have sacrificial zincs; the manufacturers ensure that all the metals in the raw-water circuit are galvanically compatible and as such are unlikely to corrode.)

KEEL COOLING

Keel cooling does not require a raw-water circuit. Instead of placing a heat exchanger in the boat and bringing raw water to it, a heat exchanger placed outside the boat is immersed directly in the raw water (see Figure 2-33). This is normally done in one of two ways: (1) running a pipe around the keel of the boat and circulating the engine coolant through it, or (2) installing an arrangement of cooling pipes on the outside of the hull. Steel boats sometimes have

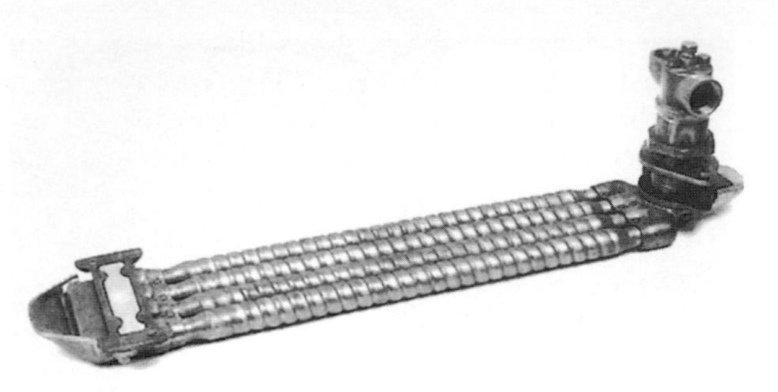

FIGURE 2-33. *A keel cooler. (Courtesy The Walter Machine Co.)*

channel-iron passages welded to the outside, or a double skin with the engine coolant circulating between the two skins. Neither of these is recommended because each adds a good deal of weight, and repairs can be very expensive if the channel iron or either skin rusts or is damaged.

A keel cooler has all the advantages of a heat exchanger plus a few more. It does not suffer any problems with silting, corrosion, or freezing on any part of the cooling circuit. If the engine has a dry exhaust, it does not need a raw-water pump or circuit. If the engine has a wet exhaust, a separate raw-water pump is necessary to supply the water injected into the exhaust. This pump often also circulates water through an aftercooler and perhaps the oil cooler for the engine or transmission.

The best keel coolers are made of bronze with cupronickel tubing.

Wet and Dry Exhausts

Exhaust gases exit cylinders at very high temperatures and in considerable volume. They must be removed from the boat as efficiently as possible. Any pressure buildup in the exhaust system *(back pressure)* interferes with the smooth flow of gases through the engine and reduces performance. This is another component of the pumping losses discussed earlier (see page 13). Although a large-diameter straight exhaust pipe will remove the gases with as little interference as possible, the noise level is totally unacceptable.

Noise is a rather complicated business, but one of its major causes is the velocity with which gases exit an engine. Another is the sudden pressure changes created as each cylinder discharges its exhaust gases. Decreasing the volume of the gases or expanding them into a larger area reduces velocity. A certain amount of back pressure in the exhaust system smoothes out pressure changes.

This is where a wet exhaust comes into its own (see Figure 2-34). The high temperature of the exhaust gases causes some of the water injected into the system to vaporize (boil). In vaporizing, the

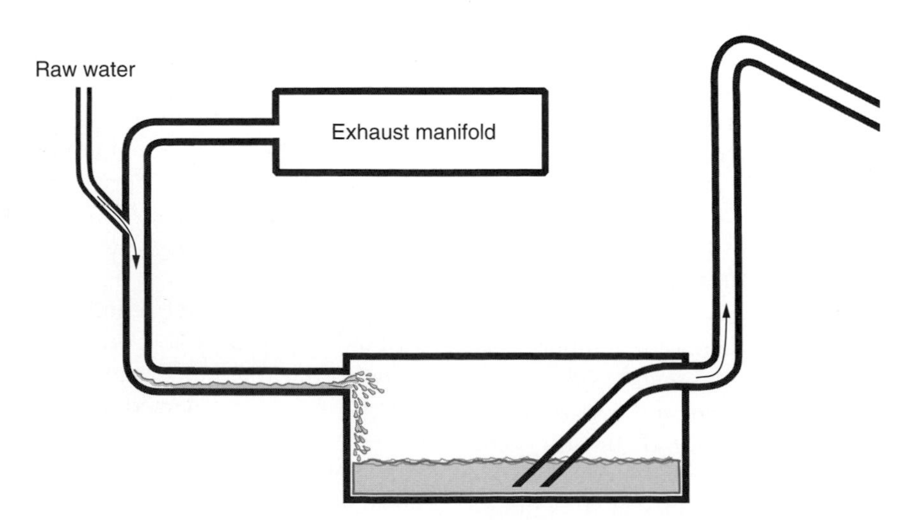

FIGURE 2-34. *This exhaust is cooled and silenced by water. (Fritz Seegers)*

water absorbs a good deal of heat from the exhaust (latent heat of vaporization), sharply lowering the temperature. The fall in temperature produces a corresponding decrease in exhaust gas volume, and this reduces the velocity.

A wet exhaust cools the gases and partially silences the system with no increase in back pressure. What is more, beyond the water injection point, a suitable rubber hose can be used for the exhaust pipe. This absorbs some of the noise that would otherwise radiate from the metal pipes necessary to withstand the heat of dry exhausts.

Silencing in a wet exhaust can be further improved with only a small increase in back pressure by using a water-lift silencer. Here's how it works. After the water is injected into the exhaust, the water and gases flow into an expansion chamber, which has an exit near its base. The unvaporized water builds up until it blocks this exit, at which point the exhaust gases blow it out of the exhaust pipe (see Figure 2-35). The expansion chamber and a small degree of back pressure combine to even out the pressure changes in the exhaust system and so reduce noise.

A further substantial reduction in noise can be achieved by running the water-cooled exhaust into a water-separation box, with the water drained off and discharged through a below-the-waterline through-hull, and the exhaust gases (which are now cool enough to not require any further cooling) discharged as a dry exhaust above the waterline.

Although water-lift silencers are very effective and now nearly universal with small marine diesel engines, several precautions regarding their installation must be observed or water may find its way into the engine. This has ruined many engines and has even sunk a few boats. These precautions are covered in Chapter 9.

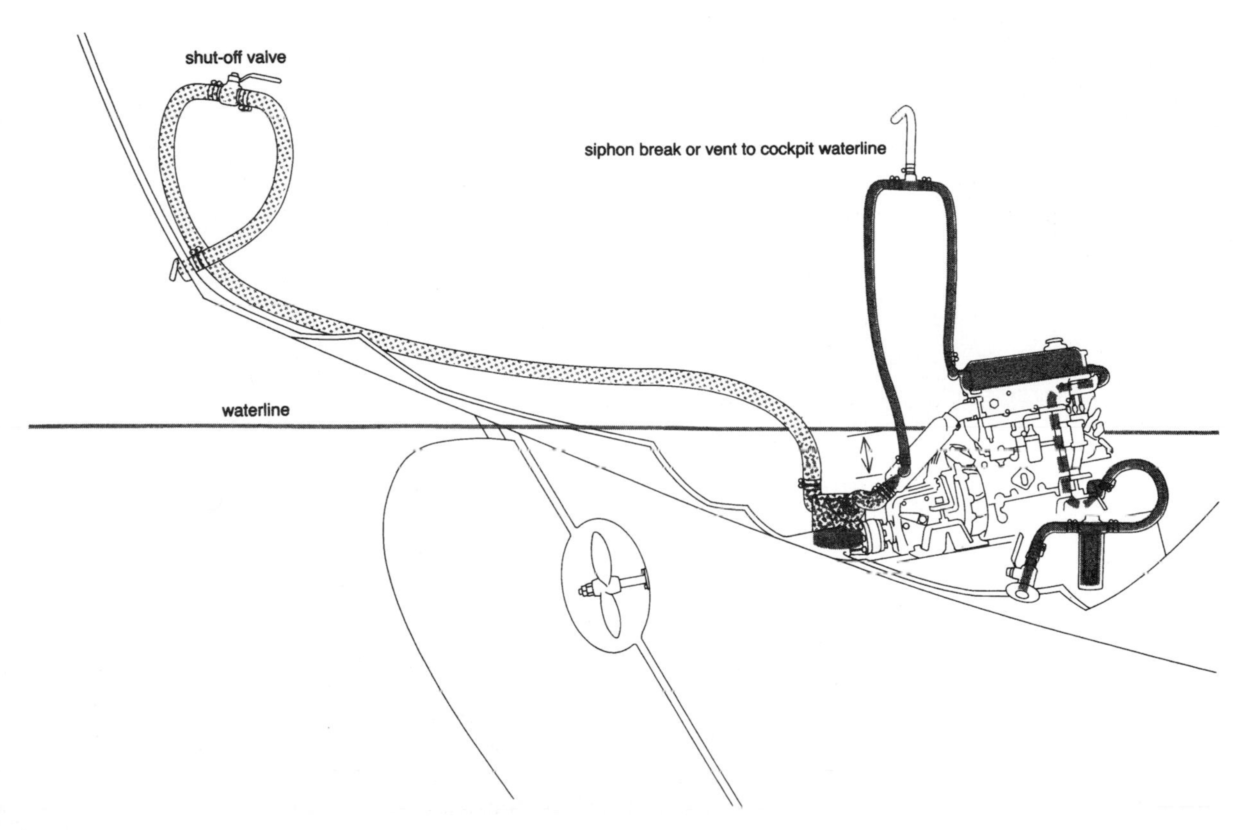

FIGURE 2-35. *Overview of the exhaust system. (Jim Sollers)*

Two-Cycle Engines

The exhaust gases of a four-cycle engine are pushed out in part by the piston on its fourth stroke. High exhaust back pressure will not, therefore, prevent the removal of the gases, but it will cause the engine to work harder, run hotter, and lose power. The exhaust gases of a two-cycle diesel, however, are cleaned out by the pressure of the scavenging air blown into the cylinder. If the exhaust back pressure is high enough, it will completely stall this flow, and the engine simply will not run.

For this reason, most two-cycle diesels have fairly direct dry exhausts that are considerably noisier than their four-cycle counterparts. This noise is exacerbated by the relatively short duration of the exhaust cycle, which requires the exhaust valves to open and close more rapidly, creating greater pressure changes in the exhaust gases. There is a limit to what can be done to muffle this racket without affecting the engine's performance. (The most effective silencing will probably be achieved with the "modified" wet silencer described in Chapter 9, but even so the engine will still be relatively noisy.)

Dry exhausts run much hotter than wet exhausts. They require thorough insulation, especially where they pass through bulkheads or where there is a danger of contact with the crew.

CHAPTER 3

ROUTINE MAINTENANCE: CLEANLINESS IS NEXT TO GODLINESS

Diesel engines are remarkably long-lived and reliable—two of their principal attractions. What is more, they require little routine maintenance, but the maintenance they do need is absolutely essential. Carelessness and inattention to detail can lead to very expensive damage in only a few seconds.

A diesel engine must have clean air, clean fuel, and clean oil, and the engine itself must be kept clean. If this book does no more than provide you with an understanding of why cleanliness is so important, and instill in you a determination to change air, fuel, and oil filters at the specified maintenance intervals, it will have been a success (see Figure 3-1 and Table 3-1).

Note that to keep track of maintenance intervals an engine hour meter or the maintenance of an accurate engine hour log is essential. In point of fact, a better indicator of service intervals is volume of fuel burned, so if it is possible to track fuel consumption (as it is on many electronically controlled engines), you can substitute this for hours of use.

SECTION ONE: CLEAN AIR

As we have seen, about 1,500 cubic feet of air at 60°F (15.6°C) is required to completely burn 1 gallon of diesel fuel. But even this figure considerably understates the actual amount of air used by a diesel engine. The pistons of a naturally aspirated engine pull in about the same amount of air at each inlet stroke, regardless of engine speed or load (on supercharged and turbocharged engines, the air intake will vary with speed and load). At low speeds and loads, very little fuel is injected, and the oxygen in the cylinder is only partly burned. As the load or speed or both increase, more fuel is injected, until at full load enough fuel is injected to burn all the oxygen (including the pressurized air from a supercharger or turbocharger). When the engine consumes all the oxygen, it is at its maximum power output. (In practice, however, the maximum fuel injection is generally kept to a level at which only 70% to 80% of the oxygen is burned. This ensures complete combustion and keeps down harmful exhaust emissions.)

Thus, at light loads, only a small proportion of the air drawn into an engine is burned. Even at full loads, there is a margin of unburned air to ensure complete combustion of the diesel fuel and to keep exhaust pollutants to a minimum. Turbocharged engines use twice as much air as naturally aspirated models, and two-cycles as much as four times more. This adds up to an awful lot of air. (Figure 3-2 gives some idea of the volume of air used by a naturally aspirated, four-cycle diesel.)

On many boats, engines are crammed into tight spaces and walled in with soundproofing insulation, with little thought given to the air supply. If the air supply is restricted, it will lower efficiency, particularly at higher engine speeds and loads, and shorten engine life.

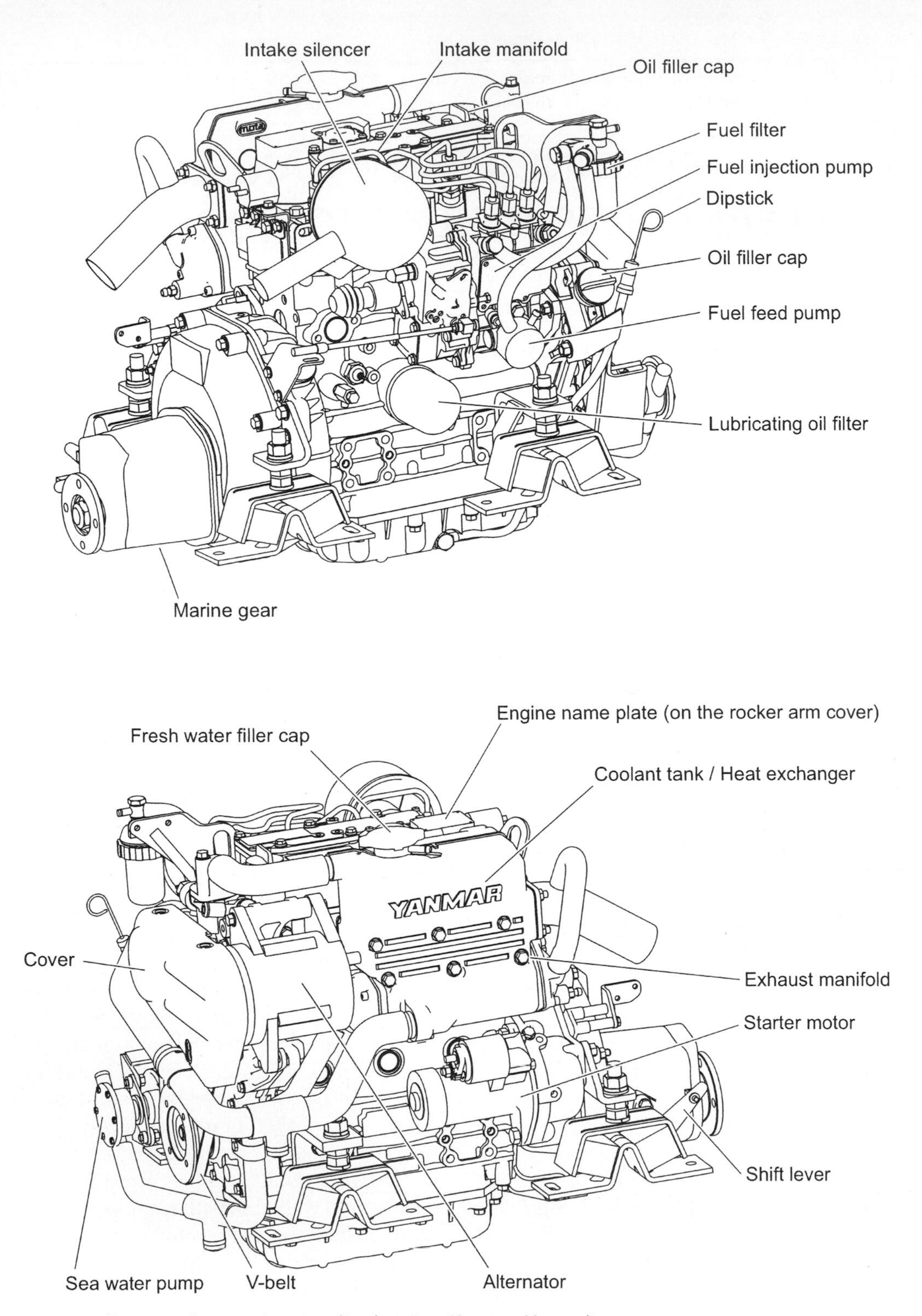

FIGURE 3-1. *External features of a typical marine diesel engine. (Courtesy Yanmar)*

TABLE 3-1. Basic Preventive Maintenance for Marine Diesel Engines

Immediately after start-up	Daily (when in regular use)	Weekly (when in regular use)	Semiannually (or more often)	Annually (or more often)
Check oil pressure. Check raw-water flow from the exhaust (unless the engine has dry exhaust).	Check engine oil level. Check freshwater coolant level in the header tank. (Do not open when hot!)	Check transmission oil level. Check pulley belt tensions. Clean raw-water strainer if necessary. When in dusty environments, check air filter and replace if necessary. **Change the engine oil and filters** every 100 to 250 operating hours as specified in the mannual (including any turbocharger oil filter).	Take a sample of fuel from the base of the fuel tank. Check for water and/or sediment. Check cooling system zinc anodes and replace as needed. When in clean environments, check air filter and replace as necessary. **Change the fuel filters** every 300 operating hours or more frequently as needed.	Check all coolant hoses for softening, cracking, and bulging. Check all hose clamps for tightness. Check the raw-water injection elbow on the exhaust for signs of corrosion. Replace as needed.

The volume of air required by a naturally aspirated 4-cycle engine running at 83% volumetric efficiency.

CID*/liters	Engine speed (RPM)						
	500	1000	1500	2000	2500	3000	Cubic feet per minute
50/0.8	6	12	18	24	30	36	
75/1.25	9	18	27	36	45	54	
100/1.6	12	24	36	48	60	72	
125/2.0	15	30	45	60	75	90	
150/2.5	18	36	54	72	90	108	

*CID = Cubic Inches of Displacement

These figures are calculated with the following formula: $\frac{\text{CID} \times (\frac{1}{2}\text{ engine speed}) \times 0.83}{12 \times 12 \times 12}$

FIGURE 3-2. *Air consumption table.*

One way to see if the air supply is adequate is to close up the engine room and run the boat's engine (and any other engines that are in the same space, such as an AC generator) at full speed and full load for a couple of minutes. Then, with the engine(s) still running, open an engine door or hatch and see if there is any pressure compensation (i.e., does the door get sucked in, or is it difficult to open out?). If so, you have a problem that needs correcting. Precise pressure measurements can be made with a manometer (see the Measuring Exhaust Back Pressure sidebar in Chapter 9).

Routine Maintenance

The air entering an engine must be clean. The efficient running of a diesel absolutely depends on its maintaining compression. Even small amounts of fine dust passing through a ruptured air filter or a leaking air-inlet manifold will lead to rapid piston-ring wear and cylinder scoring, both of which pave the way to expensive repairs. What is more, once dirt gets into an engine, it is impossible to completely clean it out. Small particles become embedded in the relatively soft surfaces of pistons and bearings, and no amount of oil changing and flushing will break them loose. This dirt then accelerates wear.

Every hour, day after day, sometimes year after year, a small diesel sucks in enough air to fill a large room, yet as little as 2 tablespoons of dust contained in that air can do enough damage to necessitate a major overhaul.

Even if a filter is not ruptured, it becomes plugged as it filters the air, thus progressively restricting airflow to the engine. This limits the amount of oxygen reaching the cylinders, and combustion, especially at higher loads, suffers. The engine begins to lose power, and on nonelectronically controlled engines, the exhaust shows black smoke from improperly burned fuel. Valves, exhaust passages, and turbochargers carbon up (as will the engine itself), further reducing efficiency and leading to other problems (see below for the effects of carbon in the oil). The engine is likely to overheat and, in extreme cases, will seize.

Air filters must be kept clean! The interval for changing air filters, however, depends on the operating conditions. In general, the marine environment is relatively free of airborne pollutants, making air filter changes an infrequent occurrence (although with pets on board, filters can get plugged with pet hair quite quickly). Since this can lead to complacency and a forgotten air filter, changing the filter at a set interval, even if it appears to be clean, is far better than forgetting it.

Air Filters

Although some small engines have no air filters, most marine diesels have a replaceable paper-element filter similar to those found in automobiles (see Figure 3-3). To change these, simply unclip, or unscrew, the cover.

Less common are oil-bath filters (see Figure 3-4), which force the air to make a rapid change of direction over a reservoir of oil. Particles of dirt are thrown out by centrifugal force and trapped in the oil. The air then passes through a fine screen, which depends on an oil mist drawn up from the reservoir to keep it lubricated and effective. In time, although the oil may still look clean, the reservoir fills with dirt, the oil becomes more viscous, less oil mist is drawn up, and the filter's efficiency slowly declines. You must periodically empty the oil from the reservoir and thoroughly clean the pan with diesel or kerosene (paraffin). At this time, also flush the screen with diesel or kerosene and blow dry.

Figure 3-3. *An air filter with a replaceable paper element. (Courtesy Caterpillar)*

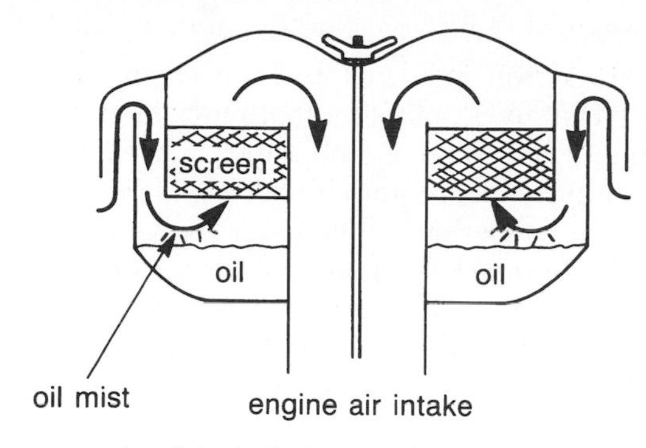

FIGURE 3-4. *An oil-bath air cleaner.*

When refilling the reservoir with oil, be careful not to overfill it—excess oil can be sucked into the engine causing damaging detonation and runaway (for more on these see pages 119 and 122).

SECTION TWO: CLEAN FUEL

A fuel injection pump is an incredibly precise piece of equipment that can be disabled by even microscopic particles of dirt or traces of water. It is also the single most expensive component on an engine, and about the only one that is strictly off-limits to the amateur mechanic. Attempts to solve problems invariably make matters worse. It is therefore vitally important to be absolutely fanatical about keeping the fuel clean. Yet many boatowners treat their fuel systems with indifference. According to Lucas/CAV, one of the world's largest manufacturers of fuel injection equipment, 90% of diesel engine problems result from contaminated fuel.

LUBRICATION

The diesel fuel flowing through a fuel injection system acts as a lubricant, with the degree of lubricity varying according to fuel grade and quality. There is no way that quality can be checked outside of a laboratory. However, it is worth noting that in the United States and other parts of the world, two grades are commonly available—No. 1 and No. 2. No. 1 has a lower viscosity and is commonly used in the wintertime in cold climates. It has less lubricity than No. 2, which reduces the life of moving parts such as injection pumps and injectors. In general, if you have a choice, use No. 2. If obliged to use No. 1, some experienced cruisers recommend adding up to a quart of outboard motor oil to every 100 gallons of diesel to improve its lubricity. Added in these low quantities, it will certainly do no harm (more is not recommended because it may lead to excess carbon deposits in the engine).

CONTAMINATION

DIRT

Any fuel can be contaminated by dirt, water, and bacteria. Even minute particles of dirt can lead to the seizure of injection-pump plungers or to scoring of the cylinders and plungers. Scoring in jerk pumps allows fuel to leak by the plungers, resulting in uneven fuel distribution. Even variations of a few millionths of a gallon in the amount of fuel injected into each cylinder adversely affects performance, causing rough running, uneven loading from one cylinder of the engine to another, and a loss of performance. If the dirt finds its way to the injectors themselves, it can cause a variety of equally damaging problems, such as plugged or worn injector nozzles.

If an engine runs unevenly, the cylinders that carry the extra load are likely to overheat, which may cause burned valves and pistons, a cracked cylinder head, or even complete seizure. Worn or damaged injectors tend to dribble, resulting in a smoky (black) exhaust. In some instances, misdirected or improperly atomized streams of fuel will wash the film of lubricating oil off a section of a cylinder wall, resulting in the seizure of that piston.

WATER

Water in the fuel opens another can of worms. It causes loss of lubrication in injection pumps and injectors, which can lead to seizure. When left for any length of time, it causes corrosion. And when injected into the cylinders, it causes misfiring and generally lowers performance. Water droplets in an injector can turn to steam in the high temperatures of a cylinder under compression. This happens with explosive force, which can blow the tip clean off an injector. Raw fuel is then dumped into the cylinder, washing out the film of lubricating oil while the injector tip rattles around, beating up the piston and valves. During extended periods of shutdown, which are quite common with most boat engines, water in the fuel system will also cause rust to form on many of the critical parts.

Note that two of the more common sources of water in the fuel are a poor seal on a deck fill fitting, and fuel tank vents located where they can get submerged when a boat is well heeled or in large following seas (see Figure 3-5). All too often deck fills are placed close to the low point on side decks where standing water can accumulate. Tank vents are frequently set in the hull side. Every year I get several e-mails from people who have just made their first ocean crossing and got water in the fuel. Invariably, what happens is they get into larger seas than they have ever experienced before, with each wave surging past the boat and driving a small slug of water up the vent.

If you're having a new boat built, it is worth ensuring that deck fills and vents will not be underwater at any time under any condition. I like to put the fill fitting in the cockpit and to mount the vents in cockpit coamings.

Bacteria

Bacteria grow in even apparently clean diesel, creating a slimy, smelly film that plugs filters, pumps, and injectors. The microbes mostly live in the fuel-water interface, requiring both liquids to survive. If you keep water out of the fuel tank, you will pretty much eliminate the potential for serious bacterial contamination. Given water, bacteria find excellent growth conditions in the dark, quiet, nonturbulent environment found in most fuel tanks.

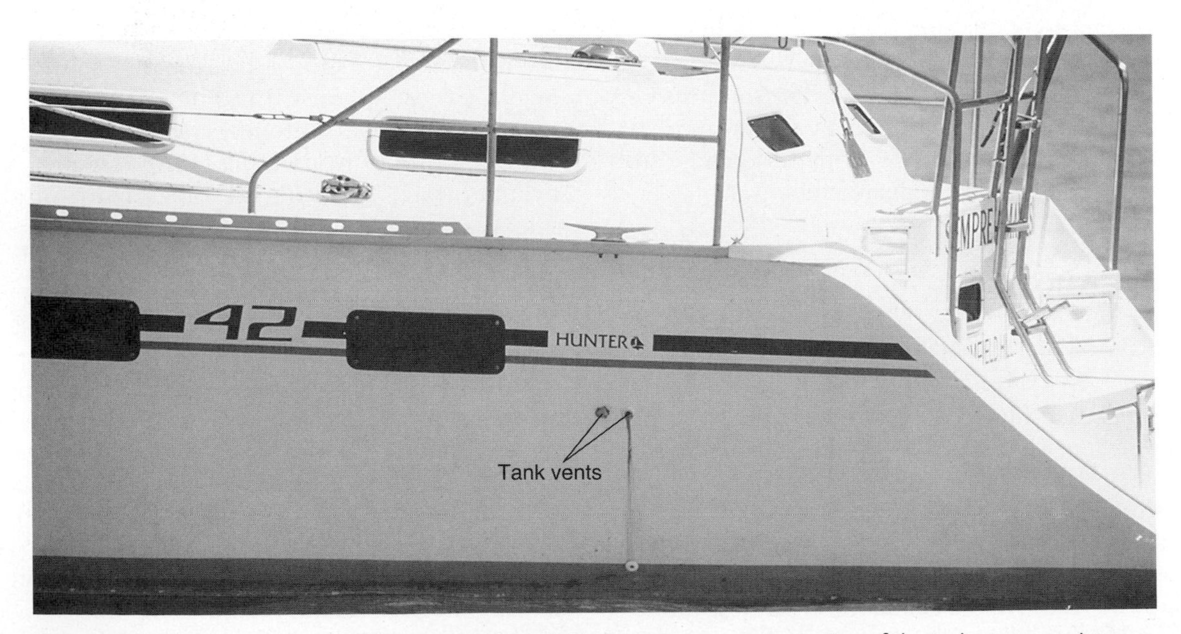

FIGURE 3-5. *Tank vents in the side of a hull are an open invitation for saltwater contamination of the tank at some point.*

Two types of biocide—water soluble and diesel soluble—are available to kill these bacteria; the latter is preferred. Follow the instructions on the can when you add these chemicals to a tank.

Fuel Handling

Various other diesel-fuel treatments on the market are not generally recommended by fuel injection specialists. Some, for example, contain alcohol (to absorb water), but this attacks O-rings and other nonmetallic parts in some fuel system equipment. Rather than treat your fuel after you have a problem, you should try to forestall any problems at the source by doing the following:

- Ensure that all cans used for carrying fuel are spotlessly clean.
- If taking on fuel from a barrel (which happens from time to time in less-developed countries—see Figure 3-6), first insert a length of clear plastic tubing to the bottom of the barrel, plug off the outer end with a finger, and then withdraw the tube. It will bring up a sample of fuel from all levels of the barrel, enabling you to see serious contamination.
- Filter all fuel using a funnel with a fine mesh, or one of the multistage filter funnels (commonly known as Baja filters) now available through various marine catalogs and at some marine chandlers. Even better, at a fraction of the price, are the Teflon-coated aircraft filters now sold by West Marine and others. If you see any signs of contamination, stop refueling at once.
- Take regular samples from the bottom of the fuel tank to check for contamination and remove any water (see Figures 3-7 and 3-8). If there is no accessible drain valve, find some means of pumping out a sample of fuel. At the first sign of contamination or water, drain the tank or pump out the fuel until no trace of contamination or water remains. Discard completely any especially dirty batch of fuel—it's not worth risking the engine for the sake of a tankful of fuel. (On our own boats, and all

Figure 3-6. *It is inevitable when cruising that once in a while fuel will be taken on from dubious sources. On this occasion, we were bumming some diesel from a Mexican lighthouse keeper.*

Figure 3-7. *Pumping a fuel sample from the base of a tank.*

others with which I am involved in the design process, I always include a fuel-sampling pump that is permanently plumbed to the lowest point in the main fuel tank. This allows me to withdraw a fuel sample at any time and pump it into a see-through container, such as an old plastic milk jug or jam jar: all I need is 1 cup of fuel, and serious contamination is immediately apparent. This inexpensive measure will do more to protect your fuel system than anything else you can do. We have twice found contamination that we were able to remove before it did any damage.)

- When leaving the boat unused for long periods (e.g., when it is laid up over the winter), fill the fuel tank to the top. This eliminates the air space and cuts down on condensation in the tank. Also add a biocide. When returning the boat to service

FIGURE 3-8. *The sample. This tank needs cleaning!*

A Simple Siphon

Once in a while, many sailors end up having to pour or siphon fuel from a jerrican into a fuel tank. Unless you have a dedicated siphon pump, this can be a messy business that sometimes results in a mouthful of diesel! Here's a way to avoid this (see Figure 3-9):

1. Place the siphon hose in the jerrican and run it all the way to the bottom. Put the other end in the fuel tank fill fitting.
2. Add another short length of hose or tubing to the top of the jerrican, then seal the opening of the jerrican by wrapping a cloth tightly around the siphon tube and the second tube.
3. Blow into the free end of the second tube. This will pressurize the jerrican sufficiently to drive fuel up and out of the siphon tube. Once the fuel starts to flow, remove the cloth and second tube. The fuel will continue to flow freely.

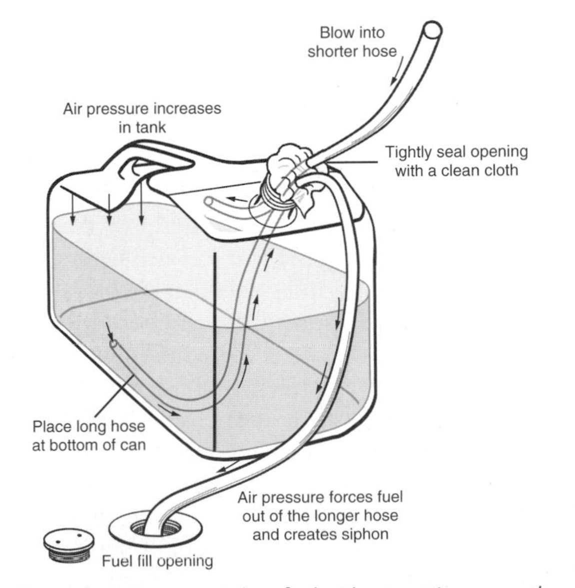

FIGURE 3-9. *A way to siphon fuel without getting a mouthful! (Fritz Seegers)*

and before you crank the engine, pump a fuel sample from the base of the tank to remove any condensates.
- Periodically flush the fuel tank(s) to remove accumulated sediment.

Fuel Filters

For all its precision, diesel fuel injection equipment is ruggedly constructed. Diesel fuel is a lubricant, and since all working parts of the system are permanently immersed in it, friction and wear are virtually absent. If a fuel injection system has a constant supply of clean fuel, it should give thousands of hours of trouble-free use. (Injector nozzles are the only exception because they are subjected to intense operating conditions in the combustion chambers. Consequently, you need to pull and clean injectors at more frequent intervals—say every 900 hours.) Apart from ensuring that you take on clean fuel, all the fuel system really needs is routine attention to the filters.

Most people regard fuel filters as the first line of defense against contaminated fuel. As should be clear by now, I regard them as the last line of defense, whose function is to deal with any minor contamination that escapes the protective measures designed to keep the fuel tank clean.

Without exception, every marine diesel engine should have both a primary and a secondary fuel filter. All engines come from the manufacturer with an engine-mounted secondary filter located somewhere just before the fuel injection pump. If this is the only filter, a primary filter MUST be installed. It needs to be mounted between the fuel tank and the lift pump, not after the lift pump, because any water in the fuel supply that passes through a lift pump gets broken up into small droplets that are then hard to filter out.

Primary and secondary filters do not have the same function. A primary filter is the engine's main line of defense against water and larger particles in the fuel supply, but it does not guard against microscopic particles of dirt and water. These are filtered out by the secondary filter.

Primary Filters

A primary filter needs to be a sedimenter type that is specifically designed to separate water from fuel. Sedimenters are extremely simple, generally consisting of little more than a bowl and deflector plate. The incoming fuel hits the deflector plate, then flows around and under it to the filter outlet. Water droplets and large particles of dirt settle out and are jetted out by centrifugal force (see Figure 3-10). The better-quality filters then pass the fuel through a relatively coarse (10 to 30 microns—a micron is one millionth of a meter, or approximately 0.00004 inch) filter element (see Figures 3-11 and 3-12).

A primary filter should have a see-through bowl with a drain plug or valve so that water can be rapidly detected and removed. The American Boat and Yacht Council (ABYC), which writes standards for recreational boat systems in the United States, requires valves "to be of the type that cannot be opened inadvertently, or shall be installed in a manner to guard against inadvertent opening"—this generally means putting a plug in the valve outlet—and specifies that "tapered plug valves with an external spring shall not be used." Beyond this, the filter may have an electronic sensing device that sounds an alarm if water reaches a certain level, a float device that shuts off the flow of fuel to the engine if the water reaches a certain level, or both.

Powerboats should have two or more primary filters mounted on a valved manifold that allows either filter to be closed off and changed without shutting down the engine (see Figure 3-13). That way, if you have a problem with dirty fuel but are in a situation that makes shutting down the engine dangerous, you can change the filters while the engine is running. Such an arrangement would also be a good idea on many motorsailers. A vacuum gauge mounted between the primary filters and the lift pump is also an excellent troubleshooting investment. Anytime the filters start to plug, the rising vacuum will alert you.

Secondary Filters

Secondary filters are designed to remove very small particles of dirt and water droplets. They cannot handle major contamination because their fine mesh will soon plug. Secondary filters are normally the spin-on type and contain a specially impregnated paper element that catches dirt. Water droplets are also too large to pass through the paper and therefore adhere to it. As more water is caught, the droplets increase in size (*coalesce* or *agglomerate*) until they are large enough to settle to the bottom of the filter, from where they can be periodically

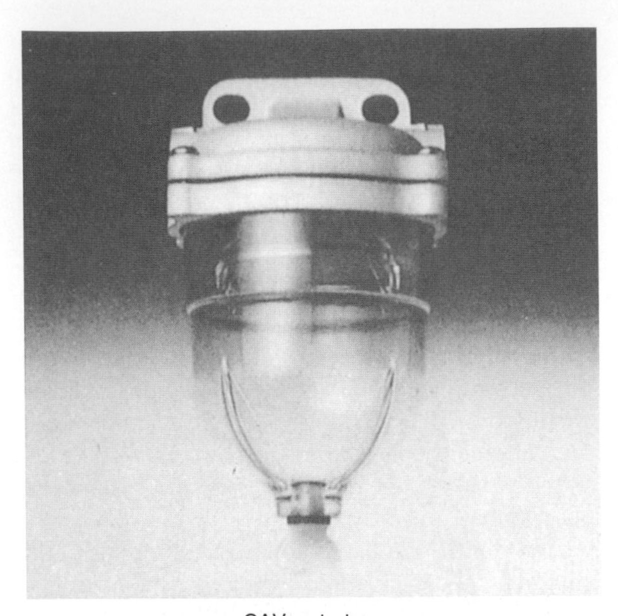

CAV watertrap

CAV waterstop

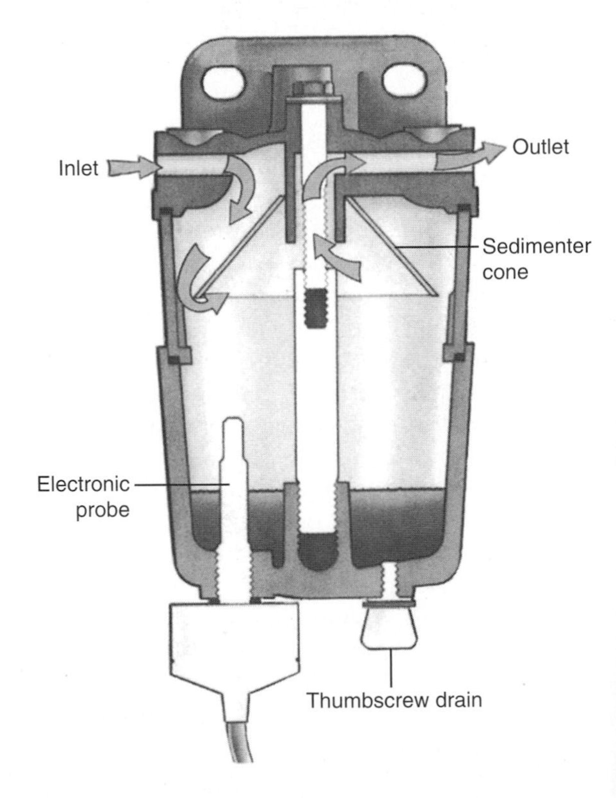

FIGURE 3-10. *Primary fuel filters. Fuel enters the CAV Waterscan (see illustration), passes over and around the sedimenter cone, through the narrow gap between the cone and the body, then to the center of the unit and out through the head and outlet connections. This radial flow allows gravity to separate from the fuel water and heavy abrasive particles, which settle at the bottom of the bowl. This filter doesn't have any moving parts. The electronic probe in the base of the filter contains two electrodes; the filter itself is the third electrode. As the level of water increases, it disturbs the balance in the system and triggers a warning that the time has come to drain the bowl. The warning can be a light, buzzer, or other device. Removing a thumbscrew opens the drain hole. The unit has an automatic circuit that triggers the warning device for a period of 2 to 4 seconds when the system is first energized. (Courtesy Lucas/CAV)*

drained. The filter mesh should no larger than 7 to 12 microns (see Figure 3-14). In certain special applications it may be as small as 2 microns.

Lift pumps also normally have a fine screen on the inlet side to filter out large particles of dirt. If you find evidence of serious contamination in a primary filter, check this screen. On engine-driven diaphragm pumps (the majority), the screen is accessible by removing a screw in the center of the pump cover (see Figure 3-15).

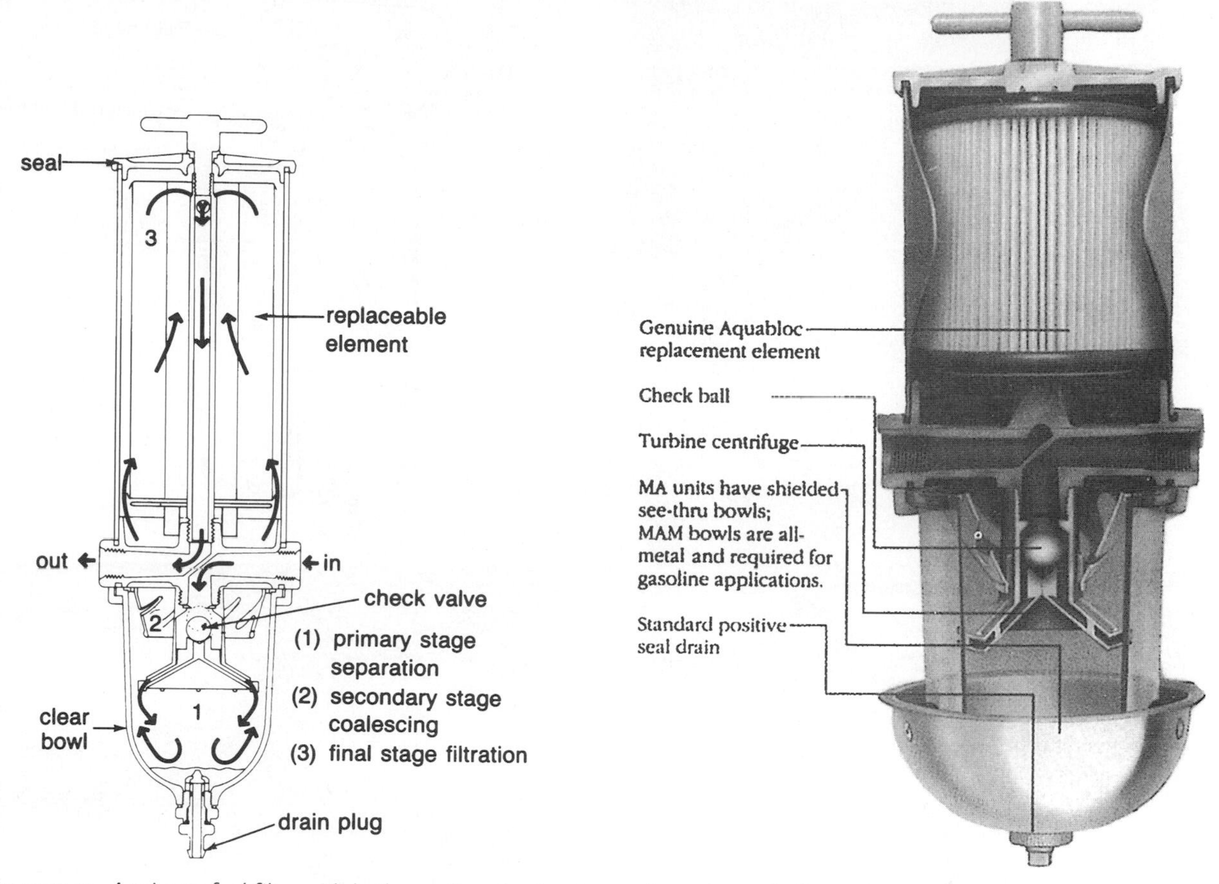

FIGURE 3-11. *A primary fuel filter with both a sedimenter function and a replaceable filter element (left). A cutaway of a good-quality primary filter (right). (Courtesy Racor)*

KEEPING THE TANK CLEAN

Note that problems with plugged filters are most common in rough weather because any sediment in a tank gets stirred up. Frequently, multiple filter changes are required. This is the worst possible time to be changing a filter once, let alone multiple times! You can completely avoid these problems by using the measures recommended above to keep the tank and fuel clean in the first place. (In over twenty-five years of cruising, I have never had even a moderately contaminated fuel filter.) If the tanks on a boat have been neglected for several years, they should be opened up and flushed out to ensure they are clean.

Once a tank is clean, regular fuel filter changes must be at the top of any maintenance schedule (for filter change procedures, see pages 54, 56–58). Just to drive home the point one more time, here's the true-life story of a friend of ours. He took on a batch of dirty fuel but failed to notice it. The contamination overwhelmed his primary filter and plugged the secondary filter. The secondary filter collapsed under

Sizes of Familiar Objects

Substance	Micron	Inch
Grain of table salt	100	0.004
Human hair	70	0.0027
Lower limit of visibility	40	0.00158
White blood cells	25	0.001
Talcum powder	10	0.0004
Red blood cells	8	0.0003
Bacteria (average)	2	0.00008

FIGURE 3-12. *How big is a micron?*

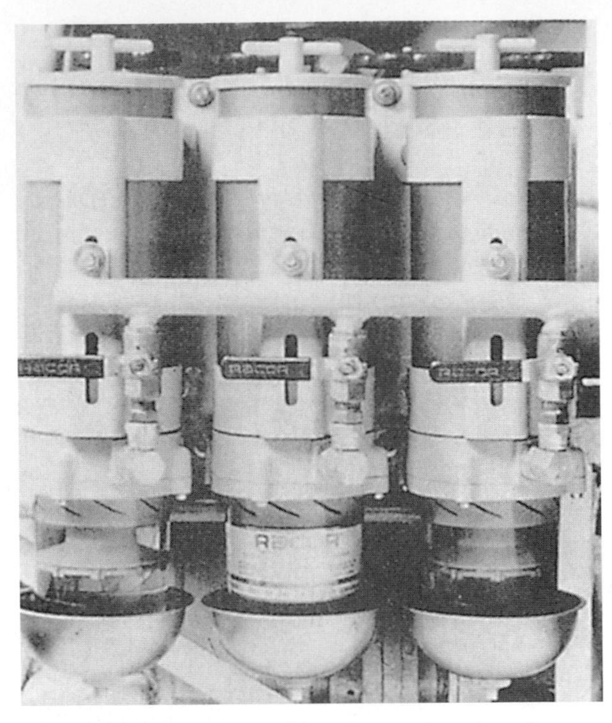

Figure 3-13. *These primary filters have valves that allow any one of them to be changed while the engine continues to run. (Courtesy Racor)*

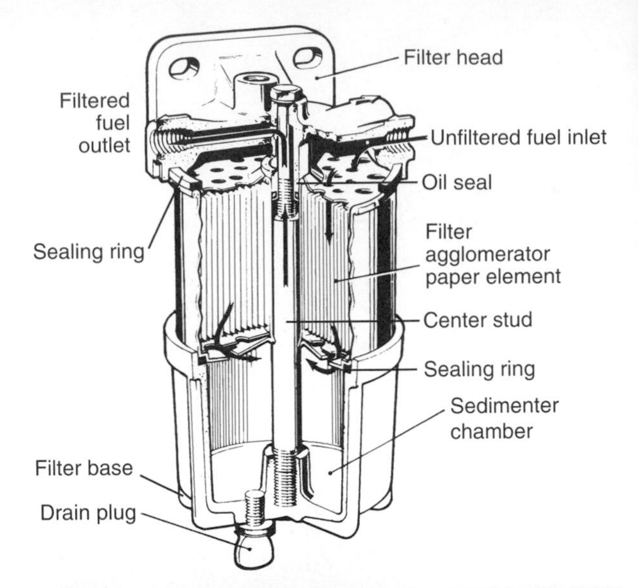

Figure 3-14. *A secondary filter with "agglomeration" capability. (Courtesy Lucas/CAV)*

the suction pressure from the injection pump and flooded the injection pump and injectors with a mass of dirty particles. The engine continued to run long enough to destroy the injection pump. Meanwhile, the messed-up injection patterns caused by the dirt resulted in the overheating and finally the seizure of two cylinders. Repairing the engine would have been more expensive than the cost of a new engine, so he bought a new engine.

"Magic Box" Devices

Every once in a while, a new device comes on the market that claims to have highly beneficial properties in terms of preventing fuel contamination or cleaning up contaminated fuel. Some of them (such as the De-Bug device and Algae-X, which are based on passing the fuel through a magnetic field) come with impressive testimonials. I am often asked whether or not they do what they claim. The answer is "I don't know!" What I do know is that if you keep the tank clean, filter all fuel taken on board, sample the fuel tank after refueling, keep the tank topped off when not in use, and maybe periodically use a biocide (however, if you keep water out of the tank, even this will probably be unnecessary), you should not have fuel problems. If you want to add additional devices, it will do no harm.

Biodiesel

Biodiesel, manufactured from soybean oil or other vegetable matter, is slowly becoming more widely

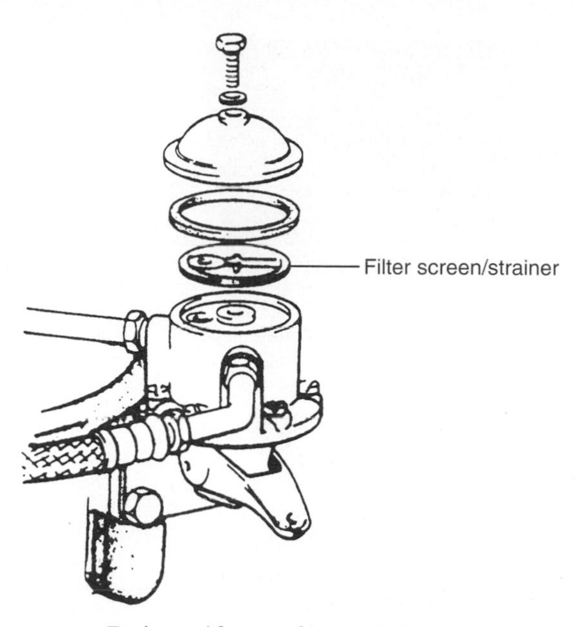

FIGURE 3-15. *To clean a lift-pump filter, undo the central retaining bolt, lift off the cover, remove the filter screen, flush it in clean diesel, and replace it. Make sure the cover gasket is undamaged and that you make a good seal when replacing the cover—any leakage here will let air into the fuel system. (Courtesy Perkins Engines Ltd.)*

available. In spite of its higher cost, some people are drawn to it on environmental grounds. I am all in favor of this, but it should be noted that it has a lower Btu content than traditional diesel fuel and, as yet, unknown long-term effects on the engine. Before using it, I would advise checking the engine warranty.

SECTION THREE: CLEAN OIL

Lubricating oil in a diesel engine works much harder than oil in a gasoline engine, owing to the higher temperatures and greater loads encountered. This is especially the case with today's lightweight, high-speed, turbocharged diesels, which contain a considerably smaller volume of oil than found in traditional engines.

What's more, diesel fuels contain traces of sulfur, which form sulfuric acid when they mix with the water that is a normal by-product of the combustion process. (Note that the sulfur content of fuels in much of the developed world is much lower than it used to be. In the United States, it is now required to be at insignificant levels—see the EPA Sulfur Rules sidebar—but high-sulfur fuels are still found elsewhere.) Many cruising boats, particularly auxiliary sailboats, compound problems with inferior fuels by running their engines without properly warming up (e.g., when pulling out of a slip) and/or for long hours at light loads to charge the batteries and run the refrigeration at anchor. The engines run cool, which causes moisture to condense in the engine. These condensates combine with the sulfur to make sulfuric acid, which attacks sensitive engine surfaces. Low-load and cool running also generate far more carbon (soot) than normal, which turns diesel engine oil black after just a few hours of engine running. This soot gums up piston rings, and coats valves and valve stems, leading to a loss of compression and numerous other problems (see Figure 3-16). Small quantities of

EPA Sulfur Rules

At one time diesel fuel in the United States contained as much as 3% to 4% sulfur. Over the years, the EPA has steadily toughened sulfur standards. The EPA's concern has not been with the effect of the sulfur on the engines themselves, but with its harmful effect on exhaust emissions. To achieve the Tier 2 emissions standards for 2004 to 2009 (see the Technology Forcing Legislation sidebar on page xiv), the sulfur content of diesel fuel had to be dramatically reduced. As a result, in 2001 and 2004 two more rounds of EPA rule making set new sulfur limits for implementation by 2006. These dropped the sulfur limit to 22 parts per million (for low-sulfur diesel, LSD), and then 15 parts per million, which is just 0.0015%! This is known as ultra-low-sulfur diesel (ULSD).

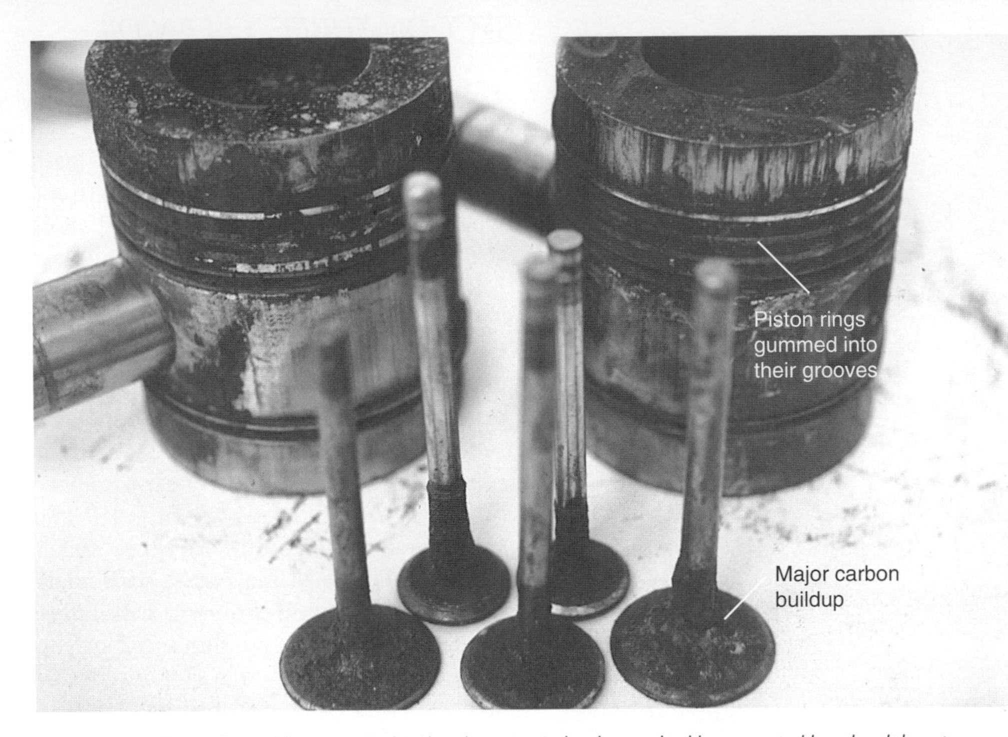

FIGURE 3-16. *This engine has only 700 hours on it, but has been completely wrecked by repeated low-load, low-temperature operation.*

soot have a disproportionate effect on oil viscosity and its lubricating qualities.

THE API "DONUT"

Diesel engine oils are specially formulated to hold soot in suspension and deal with acids and other harmful by-products of the combustion process. Using the correct oil in a diesel engine is vitally important. Many perfectly good oils designed for gasoline engines are not suitable for use in a diesel engine.

The American Petroleum Institute (API) uses the letter C (for Commercial, although I prefer to think of it as Compression ignition) to designate oils rated for use in diesel engines, and the letter S (for Service, although I prefer to think of it as Spark ignition) to designate oils rated for use in gasoline engines. The C or S is then followed by another letter to indicate the complexity of the additive package in the oil, with the better packages being given a letter later in the alphabet. Thus any oil rated CF, CF-2, CF-4, CG-4, CH-4, or CI-4 is suitable for use in diesel engines, with the CI-4 oil being the best at the time of writing (2006; the best gasoline engine oil is SM). Some CI-4 oils with higher soot capacities are labeled CI-4 Plus. Cruisers going to less-developed countries should carry a good stock of the best grade of oil that money can buy.

The API designation is found on all oil cans in the United States, and many worldwide, in the form of a donut (doughnut) symbol, which also includes, in its center, a viscosity rating developed by the Society of Automotive Engineers (SAE; see Figure 3-17). Multigrade oils have two viscosity ratings (e.g., SAE 15W-40), whereas single-grade oils have only one (e.g., SAE 30). The engine manual will specify what grade to use. If in doubt, use SAE 30 for a marine diesel.

OIL CHANGES

As the oil does its work, the additives and detergents are steadily used up. The oil wears out and must be replaced at frequent intervals, far more frequently than in gasoline engines. In particular, if you use high-sulfur-content fuels, such as those likely to be

FIGURE 3-17. *The API "donut." (Courtesy API)*

Dirt	43%
Lack of oil	15%
Misassembly	13%
Misalignment	10%
Overloading	9%
Corrosion	5%
Other	5%
	100%

FIGURE 3-18. *Major causes of bearing failures.*

found in many less-developed countries and much of the Caribbean, or the soot content increases because of extended periods of low-load and cool running, shorten your oil-change intervals to as little as every 50 hours. (A 1% increase in sulfur content will cut recommended oil-change intervals in half.) Every time you change the oil, install a new filter to rid the engine of its contaminants.

If you don't carry out regular oil changes, sooner or later the acids formed will start to attack sensitive engine surfaces, and the carbon will overwhelm the detergents in the oil, forming a thick black sludge in the crankcase and in the oil cooler (if fitted). The sludge will begin to plug narrow oil passages and areas through which the oil moves slowly, eventually causing a loss of supply to some part of the engine. A major mechanical breakdown is underway, and all for the sake of a gallon or so of oil, a filter, and less than an hour's work. One major bearing manufacturer estimates that 58% of all bearing failures are the result of dirty oil or a lack of oil (see Figure 3-18). Oil sludge in oil coolers is almost impossible to remove and generally requires a new cooler.

Changing engine oil is often complicated by the location of the drain plug in a place that you can't reach. Even if you can reach the plug, there isn't enough room to slide a container under the engine to catch the old oil. On older engines, you can slip a piece of small-bore tubing into the dipstick hole, attach a pump to it, and remove the oil this way. On newer

SYNTHETIC OILS

Synthetic oils are less volatile than mineral oils, which gives them a greater high-temperature stability, reducing sludge formation and offering better protection in high-heat situations. They also have a higher film strength, so they tend to adhere better to bearing surfaces, improving their antiwear characteristics; they flow better at cold temperatures; they resist oxidation; and they may improve fuel efficiency. However, they get a mixed reception in the marine world, largely because there are little real-life test data, and therefore, some uncertainty as to how they will perform.

Synthetic oils are expensive but they are reputed to last much longer than mineral oil, allowing oil-change intervals to be extended and thus recouping some of the additional cost. However, given the contaminants generated by marine diesels, I would not extend oil-change intervals without specific recommendations from an engine manufacturer. In the absence of such a recommendation, the added expense of synthetics is hard to justify at this time.

engines, the dipstick tube invariably goes almost to the bottom of the crankcase, so all you need to do is slip a tight-fitting piece of hose over the top of the tube and use this to pump out the oil. Better yet, fit the engine with a small sump pump plumbed directly to the crankcase. This may be electrically driven, but generally a hand pump is more than adequate.

Normally a copper pipe or hose fitting is screwed into the drain in the oil sump, and the sump pump is plumbed into this pipe or fitting. Remember that if the pipe, hose, or connections fail, you will suffer a catastrophic loss of engine oil, so make sure the installation is to the highest standards. If you have an electric pump, also make sure there is absolutely no way it can be turned on accidentally! Finally, any pump must be compatible with engine oil.

Change only hot oil. Hot oil has lower viscosity, making it easier to pump and ensuring that all of it drains out of the engine. If the pump is reversible, use it to pump in the new oil.

Changing Filters

Changing fuel and oil filters is straightforward enough, but note that many diesel fuel systems

Centrifuges and Bypass Filters

A typical full-flow engine oil filter has a mesh size of around 30 microns, which is relatively large (a finer mesh would plug faster and need changing more often). Particles of dirt smaller than 30 microns pass through the filter and circulate continuously with the oil. Various studies have shown that of the microscopic particles that pass through the filter, the most destructive in terms of engine wear are those in the 10- to 20-micron range.

Two devices have been developed to catch these particles: centrifuges and bypass filters. Both of these devices are tapped into the pressurized oil gallery that leads to the engine bearings, bleeding off a certain percentage of this oil (normally around 10%) and draining it back to the crankcase (see Figures 3-19 and 3-20; the amount of oil bled off must be kept low so as not to cause a damaging drop in engine oil pressure).

In the case of a centrifuge, oil is fed through a bowl mounted on bearings and then out of two small nozzles at the base of the bowl. The oil is driven through these nozzles under pressure from the engine oil pump, causing the assembly to spin at a high speed. The size of the nozzles, coupled with the engine oil pressure and the oil's viscosity, determines the rate of flow through the filter. The centrifugal force generated by the spinning bowl causes entrained particles of dirt to be thrown out of the oil onto the centrifuge's outer housing. Here the dirt accumulates as a dense, rubbery mat. Periodically the outer housing is removed and the dirt cake is dug out.

Centrifuges will remove particles down to 1 or 2 microns in size. Their principal drawback is that to date none has been manufactured to operate on the relatively low oil-flow rates of an auxiliary engine. Centrifuges are big-engine devices (100 to 200 hp on up—see TF Hudgins at www.tfhudgins.com).

A bypass filter, on the other hand, can be bought for any size engine. These contain a fine-mesh filter element that can filter particles down to 1 micron in size (depending on the manufacturer and the element). A restriction built into the filter at some point keeps the flow rate down to a level that will not cause a drop in the engine oil pressure. One brand of bypass filter (manufactured by Puradyn, www.puradyn.com) also includes a heating element that vaporizes any water or fuel in the oil.

The net result of either a centrifuge or a bypass filter will be cleaner oil with reduced engine wear and extended engine life. The cost, relative to the return, is low. Many larger engines come fitted with one or the other as standard equipment, but few smaller engines have either; they are worth considering as add-on equipment.

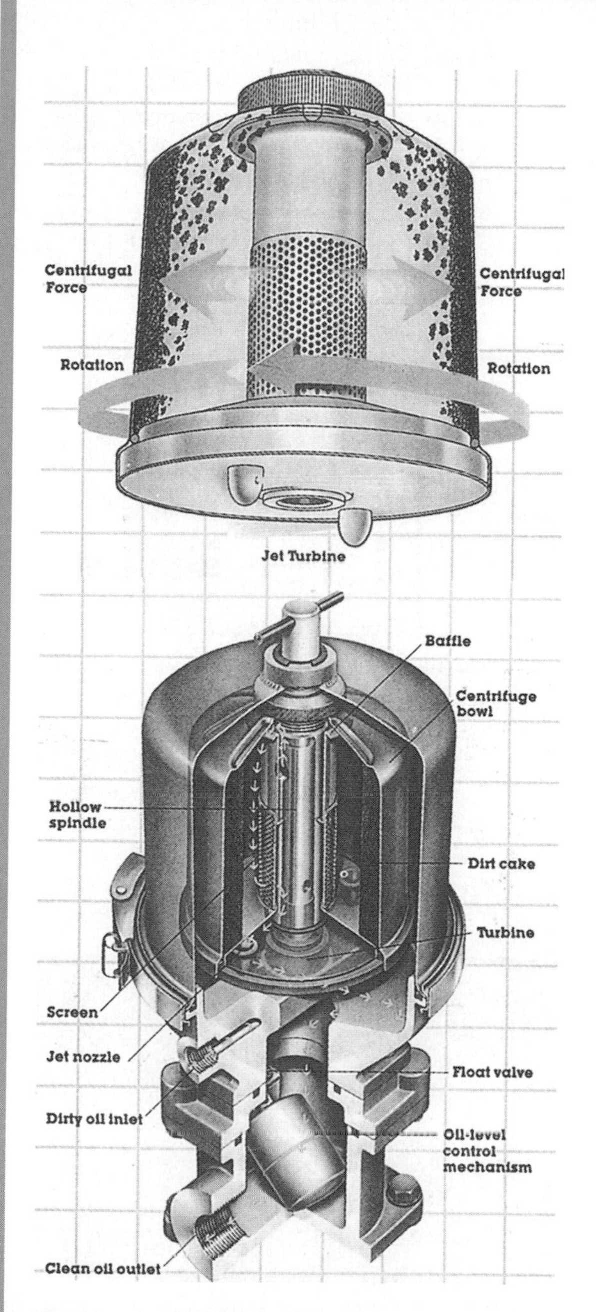

FIGURE 3-19. *The innards of the Spinner II oil centrifuge. Oil entering through the central axis is driven by the engine's oil pump out of the jets at the bottom, causing the drum to spin. Dirt is thrown outward to accumulate on the outer walls of the unit. (Courtesy TF Hudgins)*

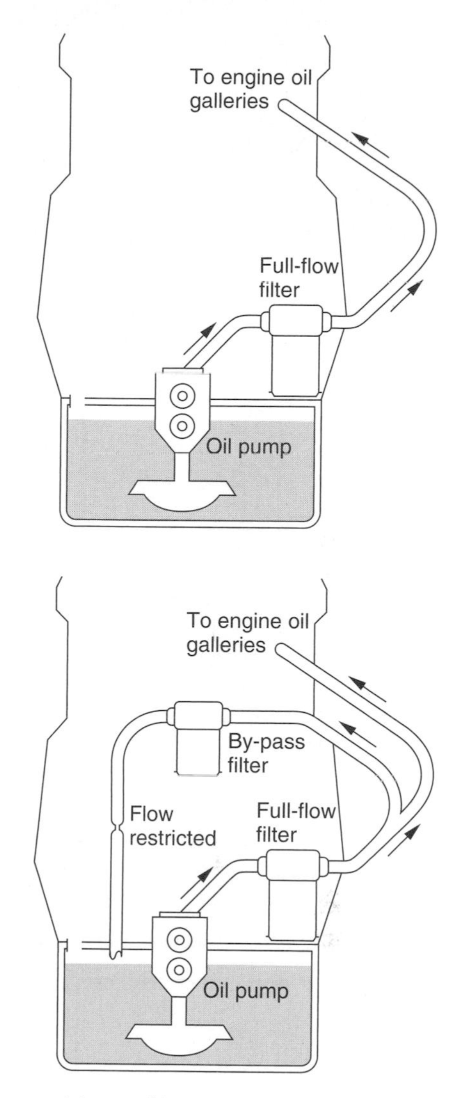

FIGURE 3-20. *A bypass filter removes smaller particles from the oil than a full-flow filter does. The bypass loop has a restriction that limits the oil flow through the circuit to around 10% of the total oil flow through the engine. After 10 to 20 minutes of operation, all the oil in the engine will have passed through the bypass filter. (Courtesy Ocean Navigator)*

need bleeding after a filter change (pages 96–101). Note also that some turbochargers have their own oil filter, which must be replaced whenever the engine oil filter is changed, and some engines have a filter inside the crankcase breather that should also be cleaned. To change a filter:

1. Scrupulously clean off any dirt from around the old filter or filter housing (see Figure 3-21).
2. Provide some means to catch any spilled fuel or oil. (I find disposable diapers—especially the ones with elasticized sides since they can be formed into a bowl shape—to be ideal! Alternatively, place a sturdy, resealable plastic bag around the filter.)
3. Most primary fuel filters have a central bolt or wing nut that is loosened to drop the filter bowl. Screw-on filters (both fuel

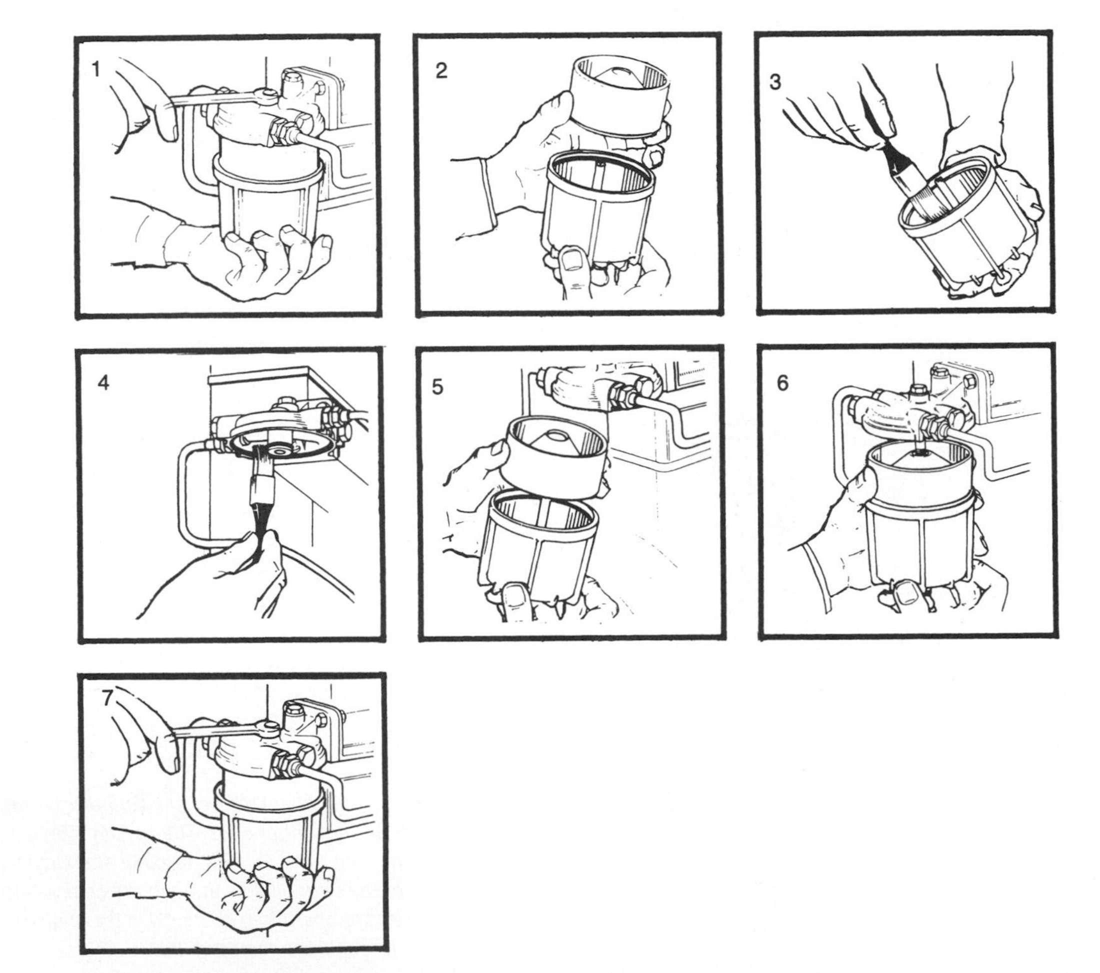

FIGURE 3-21. *Cleaning a water-trap filter. Before you begin, clean off all external dirt. If the sedimenter uses a gravity-feed supply, turn off the fuel before you dismantle the unit. Slacken off the thumbscrew in the base and drain the water and sludge. 1. Unscrew the center bolt while you hold the base to prevent it from rotating. 2. Detach the base and separate it from the sedimenter element. Inspect the center sealing ring for damage and replace it if it's not perfect. 3. Clean the base and rinse it with clean diesel fuel. Do the same to the sedimenter element. 4. Clean out the sedimenter head and inspect the upper sealing ring for damage. Replace it with a new one if it's not perfect. You can buy the sealing ring from the supplier who sold the filter element to you. 5. Make sure that the center sealing ring is correctly positioned and place the sedimenter element cone pointing upward onto the base. 6. Be sure that the upper sealing ring is correctly placed in the head, then install the assembled element and base. 7. Engage the center bolt with the central tube and make sure the top rim of the sedimenter element seats correctly before you tighten the bolt to a torque of 6 to 8 foot-pounds (0.830 to 1.106 kg m). Do not overtighten the center bolt in an attempt to stop leaks. Only hand tighten the drain thumbscrew.*

and oil) are undone with the appropriate filter wrench. This tool should be a part of your boat's tool kit (note that more than one size of filter wrench may be needed for fuel and oil filters). In the absence of a filter wrench, wrap a V-belt around the filter, grip it tightly, and unscrew the filter. Failing this, hammer a large screwdriver through the filter; it may be messy, but at least it will enable you to remove the filter.

4. If a fuel filter has a replaceable element (see Figure 3-22), take a close look at the old one. If it isn't more or less spotless—as it should be in a well-maintained fuel system—find out where the contamination is coming from and stop it before it stops the engine!
5. Fuel filters are often filled with clean diesel before installation. This practice reduces the amount of priming and bleeding that has to be done, but it also carries with it the possibility of introducing contaminants directly into the injection system. For this reason, *never fill the secondary filter before installing it*. Normally the priming can be done by operating the lift pump manually (page 98) or using a built-in electric lift pump. Sometimes on larger installations, it pays to install an additional electrically operated lift pump on a separate bypass manifold. This pump is placed before any filters and is used to push fuel through the filters, priming the system. Just as effective, with nothing to go wrong, is the installation of an outboard motor–style fuel priming bulb (the rubber kind you squeeze to prime an outboard). Install it in the fuel line between the fuel tank and the primary filter. (If you choose this option, be sure it is designed for use with diesel. West

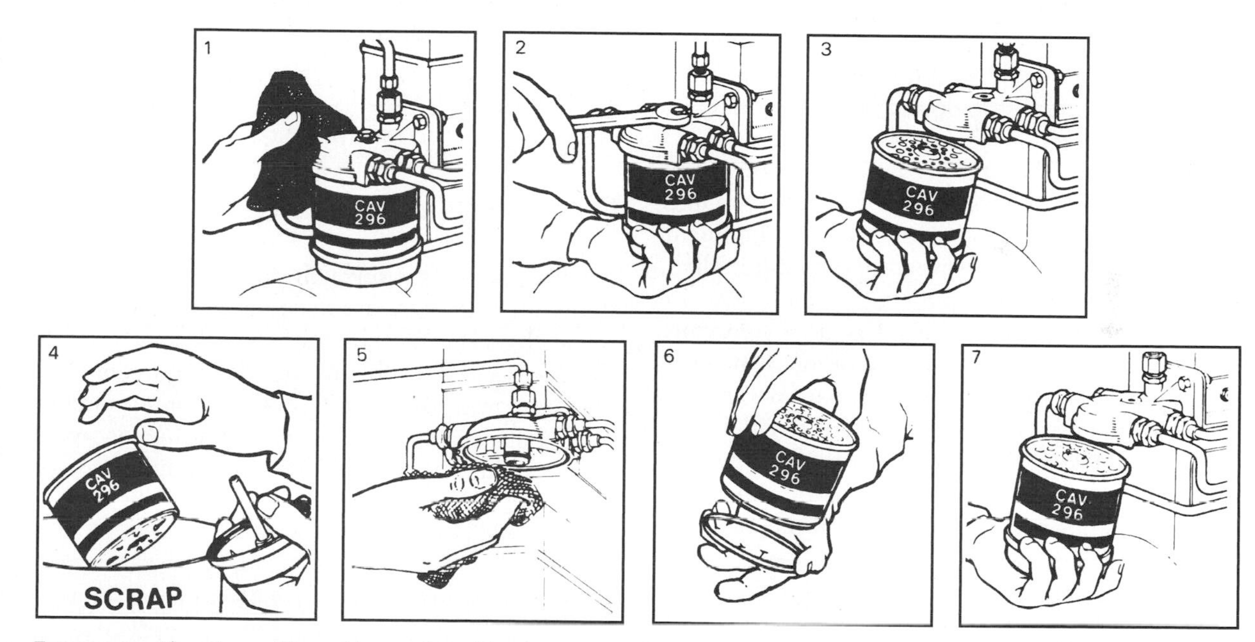

FIGURE 3-22. *Changing a filter with a replaceable element. 1. Clean all external dirt from the unit before you service the filter. Unscrew the thumbscrew in the base and drain the accumulated water and sludge. 2. Unscrew the center bolt while you hold the base to prevent it from rotating. 3. Release the filter element complete with the base by pulling the element downward while turning it slightly so that it comes free from the internal O ring. 4. Detach and discard the element. Detach and inspect the lower sealing ring for damage. Replace a defective O-ring. 5. Clean out the sedimenter base. Complete the cleaning by rinsing with clean diesel fuel. Clean the unit head and inspect both the upper sealing ring and the O-ring for damage. Replace any defective sealing rings. 6. New sealing rings are available from filter suppliers. 7. Check that the upper sealing ring and O-ring are positioned correctly in the head and fit a new filter element to the head. Rotate the element slightly when fitting it so that it slides easily over the O-ring. Ensure that the lower sealing ring is positioned correctly in the base, then install the base onto the assembled head and element. Guide the center stud through the center tube of the element and engage it with the center bolt. Make sure that the rims of the element and base are seating correctly before you tighten the center bolt. Do not overtighten. (Courtesy Lucas/CAV)*

Marine—www.westmarine.com—carries suitable priming bulbs.)
6. If the new filter has its own sealing ring, ensure that the old one doesn't remain stuck to the filter housing. If the new filter has no sealing ring, you will have to reuse the old one. To avoid this, buy a stock of rings and fit one at each filter change. Note that some sealing rings have a square cross section—they *must* go in without twisting.
7. If using a screw-on filter, lightly lubricate the sealing ring before installation. Hand tighten it, then give it an additional three-quarter turn with the filter wrench. If a fuel filter is done up with a wing nut, check closely for leaks around the nut when finished. This is a likely source of air in a fuel system and one of the first places to look if an operating problem develops immediately after a filter change.

After changing an oil filter, if the engine has a manually operated stop device, activate it and then crank the engine for 15 seconds or so (but not much more, because the starter motor may overheat). This will pump oil into the new filter and prelubricate the engine. (If you skip this step, the engine will run without oil pressure until the filter is filled.) Then go ahead and crank the engine. Once the engine is running and the oil is up to pressure, inspect the oil filter to make sure there are no leaks.

Please don't dump old oil overboard or down the nearest drain! Save it in some old oil or milk containers and take it to a proper disposal facility.

Oil and filter changes take little time and are relatively inexpensive but are too often neglected. Nothing will do more to prolong the life of an engine than this simple routine maintenance.

Oil Analysis

The other tool that can greatly help extend engine life is regular oil analysis. If you take an oil sample at each oil change, or at least once a year, and send it to a laboratory for analysis, you'll be able to detect and head off all kinds of trouble in the making at an early stage and before serious damage is done (see Figure 3-23). In the United States, the cost of analysis is between $10 and $20; the time and trouble involved in taking a sample is minimal.

To achieve the maximum benefit, you must take oil samples on a regular basis to establish the typical wear pattern for your engine, which will enable you to detect abnormalities when they occur. Take oil samples when the oil is hot, from approximately the mid-sump level so that the sample is representative of the oil in the engine. Clearly the sampling pump and bottle must be spotlessly clean to avoid misleading contamination. The laboratory will need to know what type and grade of oil is in the engine, how many hours the oil has been in service, and how much oil has been added since the last oil change. The more information you provide, the more useful the results of the analysis are likely to be.

Many marinas, marine surveyors, boatyards, and mechanics can provide the address of a local laboratory. The lab will be able to provide sample bottles, labels, and sampling pumps. Better labs will also provide some literature to help you understand the results of the analysis (www.oilanalysis.com is an excellent website for more than you ever wanted to know on this subject).

SECTION FOUR: GENERAL CLEANLINESS

Clean Water

Cooling systems, especially raw-water cooling systems, are one of the more common sources of trouble with marine diesel engines, usually as a result of a blockage and/or problems with the raw-water pump. The raw-water filter and raw-water pump impeller are regular maintenance items. Unfortunately, on too many engines the raw-water pump impeller is very hard to get at. If access is difficult, the following are worthwhile doing:

1. Replace the impeller with a Globe long-life impeller (www.globerubberworks.com), which will have a longer service life and can tolerate more abuse than a standard impeller. (However, note that not all the impellers cross-referenced in the Globe catalog are exact replacements.)
2. Replace the pump cover with a Speedseal cover (www.speedseal.com), which will make pump cover removal and replacement much easier (see Figures 3-24A and 3-24B).

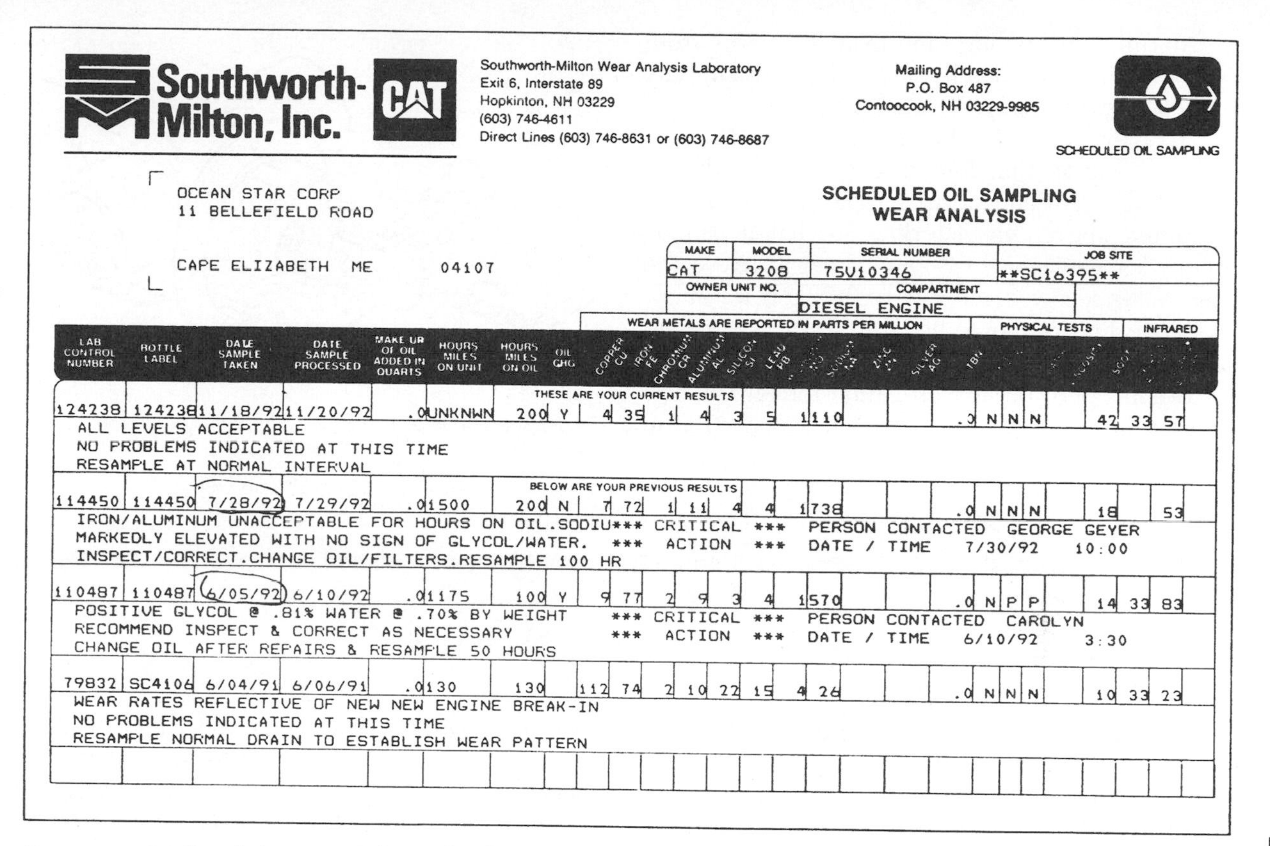

Southworth-Milton, Inc. CAT

Southworth-Milton Wear Analysis Laboratory
Exit 6, Interstate 89
Hopkinton, NH 03229
(603) 746-4611
Direct Lines (603) 746-8631 or (603) 746-8687

Mailing Address:
P.O. Box 487
Contoocook, NH 03229-9985

SCHEDULED OIL SAMPLING

OCEAN STAR CORP
11 BELLEFIELD ROAD

CAPE ELIZABETH ME 04107

SCHEDULED OIL SAMPLING WEAR ANALYSIS

MAKE	MODEL	SERIAL NUMBER	JOB SITE
CAT	3208	75V10346	**SC16395**
OWNER UNIT NO.	COMPARTMENT		
	DIESEL ENGINE		

WEAR METALS ARE REPORTED IN PARTS PER MILLION — PHYSICAL TESTS — INFRARED

LAB CONTROL NUMBER	BOTTLE LABEL	DATE SAMPLE TAKEN	DATE SAMPLE PROCESSED	MAKE UP OF OIL ADDED IN QUARTS	HOURS MILES ON UNIT	HOURS MILES ON OIL	OIL CHG	COPPER CU	IRON FE	CHROMIUM CR	ALUMINUM AL	SILICON SI	LEAD PB		SODIUM NA		SILVER AG	TBN								
THESE ARE YOUR CURRENT RESULTS																										
124238	124238	11/18/92	11/20/92	.0	UNKNWN	200	Y	4	35	1	4	3	5	1	110			.0	N	N	N		42	33	57	
ALL LEVELS ACCEPTABLE NO PROBLEMS INDICATED AT THIS TIME RESAMPLE AT NORMAL INTERVAL																										
BELOW ARE YOUR PREVIOUS RESULTS																										
114450	114450	7/28/92	7/29/92	.0	1500	200	N	7	72	1	11	4	4	1	738			.0	N	N	N		18		53	
IRON/ALUMINUM UNACCEPTABLE FOR HOURS ON OIL.SODIU*** CRITICAL *** PERSON CONTACTED GEORGE GEYER MARKEDLY ELEVATED WITH NO SIGN OF GLYCOL/WATER. *** ACTION *** DATE / TIME 7/30/92 10:00 INSPECT/CORRECT.CHANGE OIL/FILTERS.RESAMPLE 100 HR																										
110487	110487	6/05/92	6/10/92	.0	1175	100	Y	9	77	2	9	3	4	1	570			.0	N	P	P		14	33	83	
POSITIVE GLYCOL @ .81% WATER @ .70% BY WEIGHT *** CRITICAL *** PERSON CONTACTED CAROLYN RECOMMEND INSPECT & CORRECT AS NECESSARY *** ACTION *** DATE / TIME 6/10/92 3:30 CHANGE OIL AFTER REPAIRS & RESAMPLE 50 HOURS																										
79832	SC4106	6/04/91	6/06/91	.0	130	130		112	74	2	10	22	15	4	26			.0	N	N	N		10	33	23	
WEAR RATES REFLECTIVE OF NEW NEW ENGINE BREAK-IN NO PROBLEMS INDICATED AT THIS TIME RESAMPLE NORMAL DRAIN TO ESTABLISH WEAR PATTERN																										

FIGURE 3-23. *An oil analysis reporting the results of a series of tests performed on the oil of* Ocean Navigator *magazine's* Ocean Star. *The earliest analysis is at the bottom; the latest is at the top. The second analysis revealed problems with the freshwater cooling system (glycol in the oil); the third revealed problems with saltwater intrusion into the engine from a faulty exhaust installation (indicated by the high sodium levels). (Courtesy Ocean Navigator)*

FIGURE 3-24A. *A Speedseal kit with two different cover plates for different water pumps.*

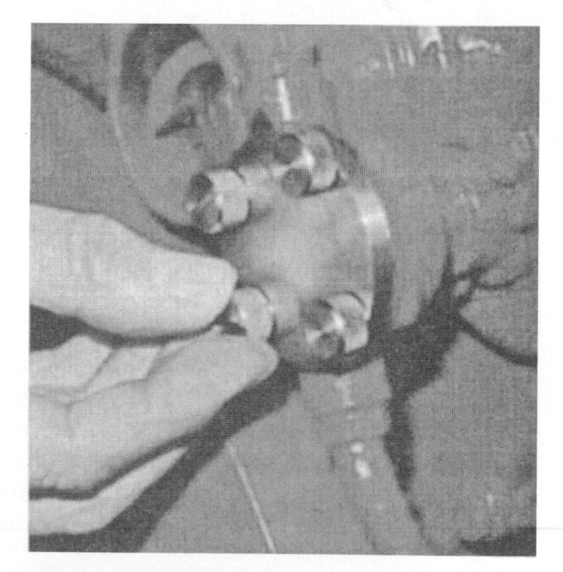

FIGURE 3-24B. *Speedseal cover being removed.*

An impeller puller (approximately $50 from West Marine and chandleries) will greatly simplify impeller removal (see Figure 3-25).

A raw-water filter should have a see-through bowl or top so that fouling can be immediately detected (see Figure 3-27). Ideally, it should be mounted above the waterline so that it can be opened and cleaned without having to close the raw-water seacock (see Figure 3-28). If it's below the waterline, it should be close to the seacock to make servicing easy.

In some parts of the world, filter fouling can be a significant problem (such as the Chesapeake Bay in

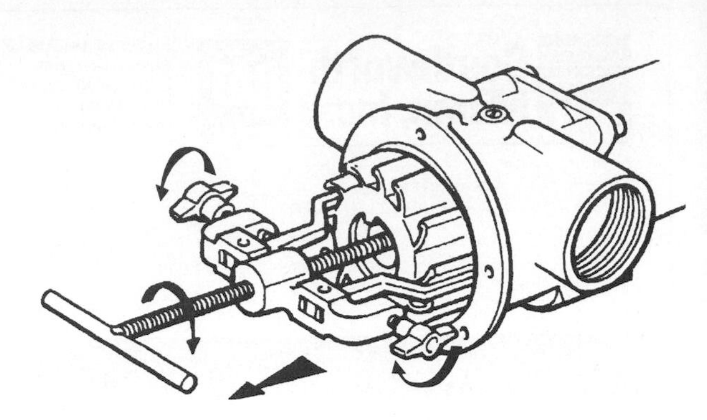

FIGURE 3-25. *An impeller puller. (Courtesy ITT/Jabsco)*

AN ALL-PURPOSE IMPELLER PULLER

Some water pumps are mounted so that it is impossible to get an impeller puller on the impeller without either removing the pump or some other part of the engine (often the starter motor). If you have such an impeller, the key to removing it is a large pair of right-angled needle-nose pliers.

The pliers need to be around 10 inches in overall length, with a jaw that will open out to as much as 2 inches. This should be adequate to bridge the hub on any impeller likely to be found on engines up to 100 hp. More powerful engines may have pumps with impeller hubs that go above 2 inches, requiring larger pliers, which will be hard to find.

When opened out, the jaws of the pliers bridge the hub of the impeller and slip between the vanes on either side of the hub. To jar the impeller loose from the pump housing, squeeze the handles shut tightly and tap them with a hammer. This may take several repeat performances (the pliers may slip off the hub), but it will do the job.

If the impeller hub is so wide that the jaws have to be opened wide to grip the impeller, the handles on the pliers will be too far apart to hold with one hand. This can make operation awkward. In this case, it is worthwhile bending the handles inward. To do this, open them out, place a spacer between them up toward the jaws (a bolt works well), heat the handles with an oxyacetylene torch, and bend them together. If you don't want to tackle this, your local garage should be able to do it for you; it takes 5 minutes (see Figure 3-26). Note that the more you bend the handles in, the farther apart the jaws will be when fully closed, so make sure you don't bend the handles to the point that you can no longer firmly grip the hub of your impeller.

FIGURE 3-26. *Heating the handles of a needle-nose pliers with an oxyacetylene torch.*

FIGURE 3-27. *A raw-water filter.*

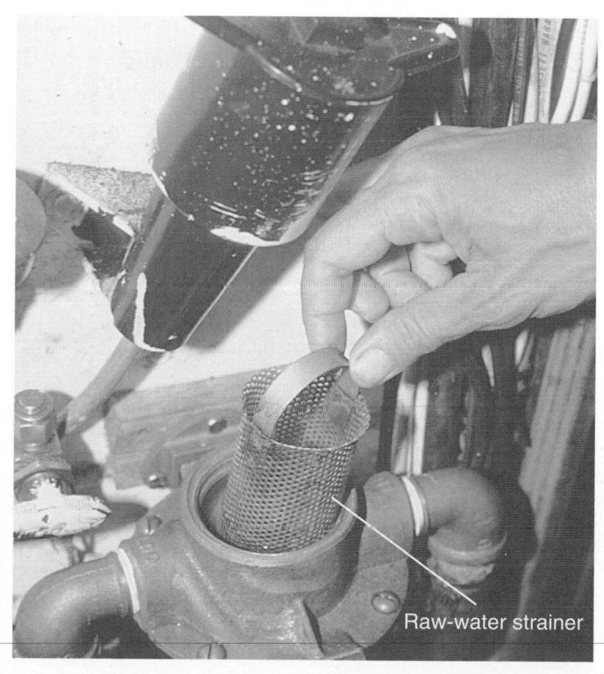

FIGURE 3-28. *Removing the raw-water strainer for cleaning.*

the summertime when the sea nettles are abundant). Cruisers in such areas may find it worthwhile to fit two raw-water inlets and filters, valved so that either can be taken out of service and cleaned while the engine runs on the other.

On the freshwater side, renew the antifreeze at least every two years (use a 50-50 mix; mix it before pouring it to ensure you get the proportions right). Antifreeze does not lose its antifreezing properties, but various corrosion-prevention and antifoaming inhibitors get used up and need replacing. *Regardless of manufacturers' claims to the contrary, to be on the safe side, replace long-life antifreeze every two years.* There are all kinds of rumors (and lawsuits) flying around about cooling systems with long-life antifreeze that have become fouled with some kind of precipitate.

SACRIFICIAL ZINCS

Another key maintenance item is any sacrificial zinc anodes or zincs (see Figures 3-29 and 3-30). An increasing number of modern engines do not have zincs because the manufacturers have succeeded in getting a tight-enough control over the metals in the engine to prevent the kind of galvanic activity that causes corrosion. But this has not always been the case. Over the years, a number of poorly made heat exchangers and oil coolers using brass, or other zinc alloys, have found their way onto marine engines

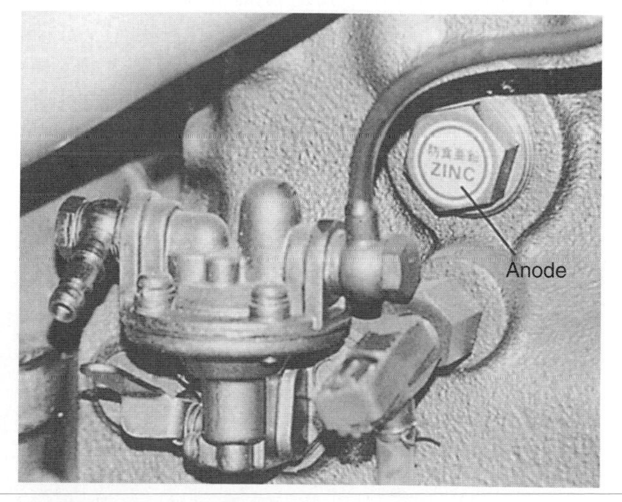

FIGURE 3-29. *A zinc anode in the cooling system of a Yamaha diesel engine.*

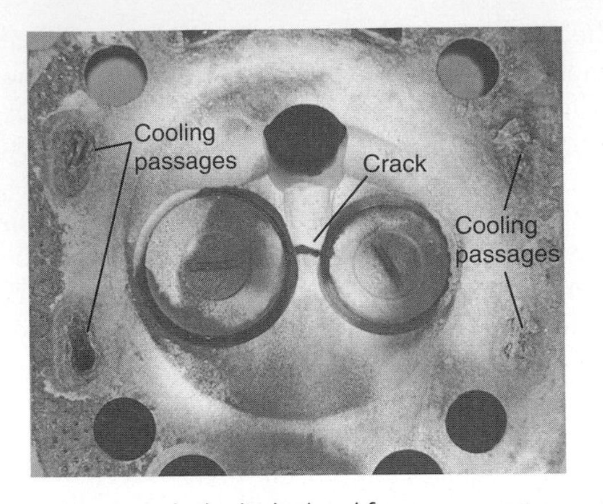

FIGURE 3-30. *A cracked cylinder head from a raw-water-cooled engine. The cooling passages were badly corroded from a failure to change the sacrificial zinc anode. The engine overheated and cracked the cylinder head*

(including some from highly reputable manufacturers). If such a heat exchanger isn't adequately protected with sacrificial zincs, its own dissimilar metals will interact in the presence of heat and salt water, destroying the zinc in a process known as *dezincification*. (*Note: The lack of zinc in the construction of a heat exchanger is no guarantee that the other metals are galvanically compatible.*) Dezincification leads to pinholes in cooling tubes and around welds and solder joints.

Unfortunately, some of these heat exchangers have been sold without sacrificial zinc anodes. If your engine does not have zinc anodes, it is important to check the past history of that particular model. If you find there have been problems, you can often retrofit zincs by drilling and tapping the raw-water side of the heat exchanger to accommodate the 1/4-inch pipe fittings that zinc pencil anodes screw in to.

For those heat exchangers that have zincs, the rate of zinc consumption is highly variable. In the tropics, you may need to replace them in as little as thirty days, so check them every month or so until you know their consumption rate. Replace them when only partially eaten away (no more than 50%; the effectiveness of a zinc is directly proportional to its remaining surface area). Don't wait until they are completely gone (see Figures 3-31 and 3-32) because by that time galvanic corrosion will already be attacking your engine (see Figure 3-33).

FIGURE 3-31. *Many heat exchangers have zinc pencil anodes. These should be replaced when they are no more than half consumed.*

If you run out of zincs while cruising, take the threaded cap from an old zinc, heat it on your galley stove until the solder in the base of the cap begins to puddle, then jam any available piece of zinc, suitably cut to size, into the solder.

Failed heat exchangers and oil coolers can result in expensive engine damage and be hard to replace, especially when you're cruising in remote areas. External corrosion on a heat exchanger will be

FIGURE 3-32. *This zinc should have been replaced long ago. The heat exchanger is likely corroding.*

FIGURE 3-33. *Persistent neglect of the zinc on this heat exchanger caused the heat exchanger to corrode through several times. It would have been easier to replace the zinc than repeatedly patch the heat exchanger!*

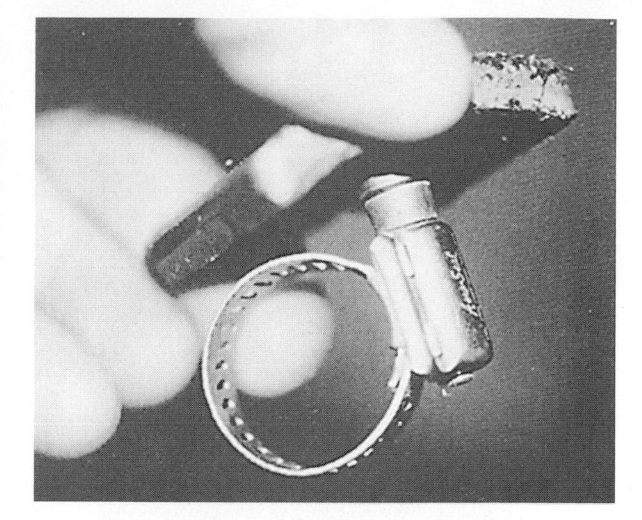

FIGURE 3-34. *An "all-stainless" hose clamp with a screw of an inferior (magnetic), corrosion-prone grade of stainless.*

obvious enough (it will leak!), but internal corrosion is harder to detect. If you suspect the tubes are leaking, drain the raw-water side and remove the end caps. If water (or oil with an oil cooler) is dribbling out of any of the tubes, you know they have failed. You can use the heat exchanger temporarily, albeit with some loss of capacity, by plugging both ends of the offending tube(s) with a piece of soft doweling.

HOSES AND HOSE CONNECTIONS

Suction-side hoses to water pumps need to be the noncollapsing type, preferably with an internal steel spring. In time, hoses that are subjected to heat soften and bulge, or crack, and need replacing.

Hose clamps must be made of all 300-series stainless steel—*including the worm screw*. There are a lot of stainless steel hose clamps on the market that use 400-series screws, which will rust in the marine environment (see Figure 3-34).

Double-clamp all hose connections located below the waterline. Periodically, you should loosen hose clamps and inspect the band where it has been inside the screw housing—this area of a band is prone to failure from crevice corrosion (see Figure 3-35). Also check hoses for soft spots, bulging, cracking, or damage in the vicinity of the hose clamps.

SIPHON BREAKS

All correctly installed water-cooled exhausts require a mechanism to prevent seawater from siphoning into the engine when the engine is at rest. This is usually achieved by using one or more siphon breaks (see Chapter 9). Siphon breaks, however, tend to plug with salt crystals, rendering them ineffective while also causing them to spray salt water over the engine and its electrical systems. To maintain a

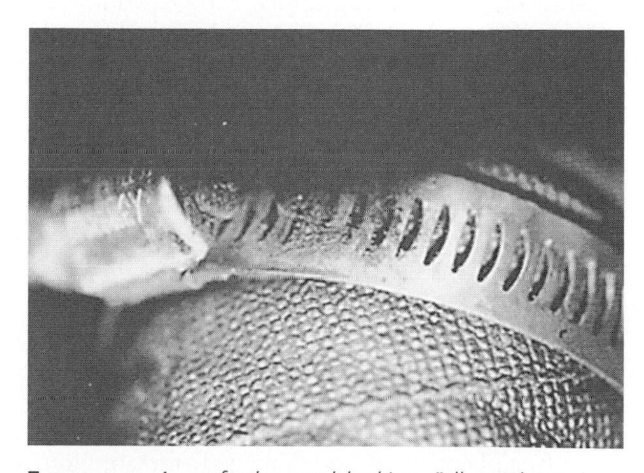

FIGURE 3-35. *A perfectly good-looking "all-stainless" hose clamp, which, when the screw was backed off a turn or two, revealed considerable corrosion inside the screw housing, caused by using an inferior grade of stainless steel for the screw.*

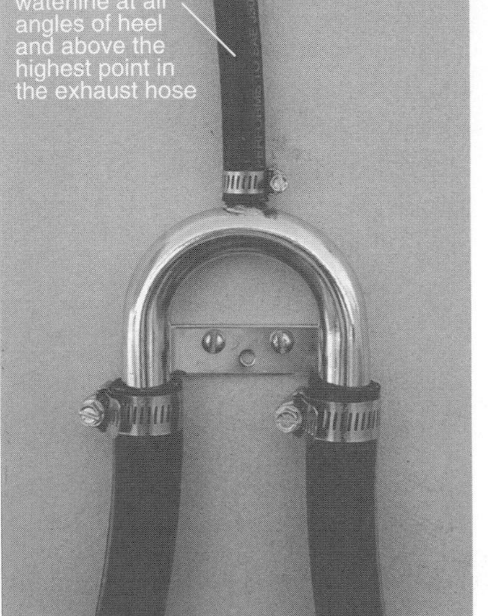

Figure 3-36. *A siphon break on an engine cooling circuit that has been adapted by removing the valve and adding a length of hose, which is vented into the cockpit.*

siphon break, periodically remove the valve and rinse it in warm fresh water.

You could also use a T-fitting in place of the siphon break. My preference is to remove the valve element altogether, add a hose to the top of the vented loop, and discharge this well above the waterline (into the cockpit works well; see Figure 3-36). If you plan to do this, you must discharge above the highest point in the exhaust hose to prevent water weeping out of the discharge fitting when the engine is running. If a vent installed in this fashion subsequently weeps, it is a sure sign of excessive back pressure in the exhaust (probably from carbon fouling), which needs attention.

Clean Electrical Systems

The starting and other electrical circuits must be kept free of corrosion and damage from vibration or contact with hot or moving parts; keep the wires neatly bundled and properly fastened. Pay particular attention to sources of potential water leaks

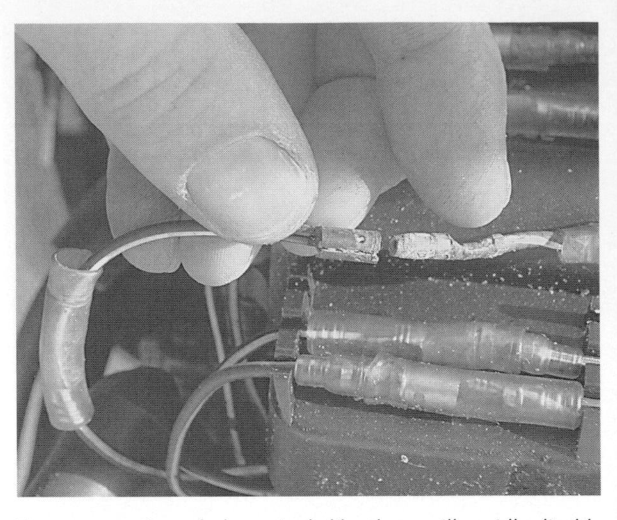

Figure 3-37. *Corroded terminals like these will rapidly disable electric circuits.*

(e.g., the raw-water injection nipple into the exhaust) and ensure that connecting plugs are not located beneath these, where they may get dripped on (see Figure 3-37). On boats with wet bilges (mostly older, wooden boats), starter motors can be a source of trouble due to water being thrown up into the pinion gear and causing rust. If this is a possibility, periodically remove the starter and lubricate the pinion gear.

Common Problem Areas

Dead batteries are the most common electrical failure in marine use. As often as not, this comes from a lack of understanding of the unusual electrical operating conditions found on most boats, which require the use of deep-cycle batteries and special charging equipment. Chapter 4 describes how to troubleshoot an engine's starting circuit; Chapter 9 goes into more detail on how to set up the electrical system in the first place so that it doesn't cause problems.

If the charging circuit is not operating, first check the alternator drive belt—it may be broken or slipping. Using moderate finger pressure, you should not be able to depress the belt by more than 3/8 to 1/2 inch in the middle of the longest belt run (see Figure 3-38). A slipping belt often produces a rhythmic squealing, especially immediately after start-up; the output (if you have an ammeter) will cycle up and down. Tighten the belt immediately.

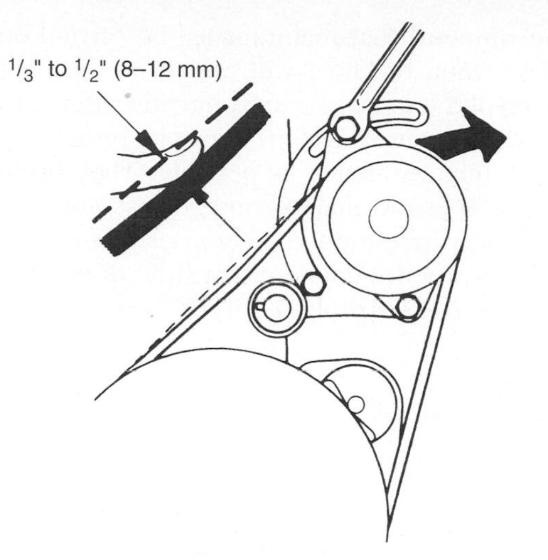

FIGURE 3-38. *Checking V-belt tension. (Courtesy Volvo Penta)*

If it is left to slip, the belt will heat up, becoming brittle and prone to rapid failure (see Figure 3-39). The alternator pulley also will heat up, which can sometimes cause the alternator shaft and rotor to become hot enough to demagnetize the rotor; this will almost completely disable the alternator's output for good.

Note that if the alternator circuit to the battery is broken at any time while the alternator is running, it will almost instantly cause the diodes in the alternator to blow out. Most boats are fitted with a battery selector/isolation switch. If this switch is opened (the batteries isolated) while the engine is running, the alternator may be open-circuited, which may destroy the diodes. Even switching from one battery to another may blow the diodes, unless the switch is the make-before-it-breaks variety—both batteries are first brought online and then one is disconnected so that the connection to the boat's circuits is never interrupted.

FIGURE 3-39. *An alternator belt destroyed by being allowed to slip.*

Connecting battery leads in reverse at any time, or accidentally grounding the alternator's output wire when it is running, will also probably destroy the diodes.

High-output alternators are common in boats (see Chapter 9). The higher the output, the more massive the cables need to be to carry the load. Heavy cables mounted to the back of an alternator tend to vibrate loose. This generates arcing, which can rapidly melt the output stud on the alternator, at which point the cable drops off. As long as the battery isolation switch is not tripped, the cable is likely to have a direct connection to the battery. As a result, it will create a dead short if it touches any grounded surface (e.g., the engine block), where it may then start a fire. Be certain to lock the cables to the alternator, preferably with locknuts, and to check this connection for tightness at periodic intervals (see Figure 3-40).

For comprehensive troubleshooting and repair procedures on batteries and alternators, see my *Boatowner's Mechanical and Electrical Manual*.

A Clean Engine

A clean engine is as much a psychological factor in reliable performance as a mechanical factor. The owner who keeps the exterior clean is more likely to care about the interior, and maintenance is less onerous. Nothing will put you off an overdue oil change more than the thought of having to crawl around a soot-blackened, dirty, greasy hunk of paint-chipped cast iron, with diesel fuel, old oil, and smelly bilge water slopping around in the engine drip pan.

Scheduled Overhauls

Engine manufacturers lay down specific schedules for overhaul procedures of such things as injectors

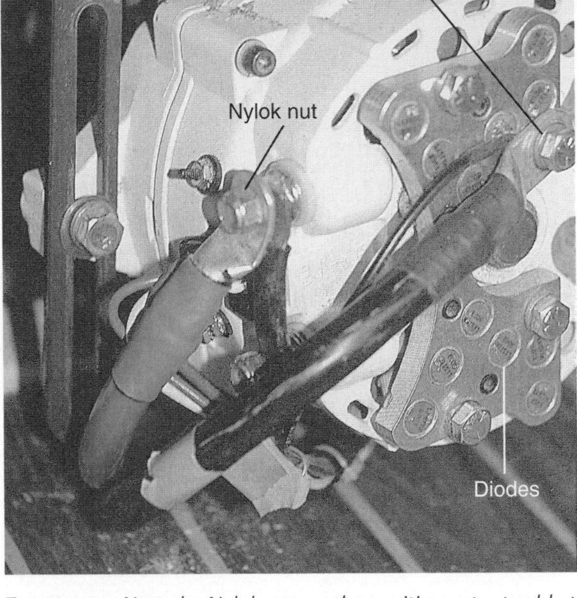

FIGURE 3-40. *Note the Nylok nut on the positive output cable to this high-output alternator, and the locking washer under the bolt holding the negative cable.*

and valve clearances. However, keep in mind that the marine environment and engine use are quite varied. One engine may need work much sooner than another, especially if subjected to a poor operating regimen, while others may have some of their maintenance intervals safely extended (such as injector overhauls). For this reason, I tend to subscribe to the philosophy "If it ain't broke, don't fix it!" At the first sign of trouble, however—whether it be difficult starting, changes in oil pressure or water temperature, a smoky exhaust, vibration, or a new noise—the problem must be resolved right away. Delay may cause expensive repair bills (see the Troubleshooting sections).

Given that the timing of overhauls of such things as injectors and valve clearances is not critical, and that such maintenance sometimes has unintended consequences (e.g., a gasket is not properly replaced and an oil leak subsequently develops), I particularly recommend that maintenance be carried out at a time when the boat will remain in the neighborhood of the mechanic for some time afterward. This way, if an issue develops, it can be resolved with as little inconvenience as possible. The corollary to this is to avoid maintenance that is not especially time sensitive immediately prior to an ocean passage or a voyage where it will be difficult to deal with secondary problems if they arise.

SECTION FIVE: WINTERIZING

The following is a reasonably comprehensive winterizing checklist (see Figure 3-41). One or two of these items will not need to be performed every year (such as breaking loose the exhaust to check for carbon buildup or removing the inner wires of engine control cables to lubricate them), but most should be:

1. Change the oil and oil filter at the beginning of the winter, not the end. The used oil will contain harmful acids and contaminants that should not be allowed to go to work on the engine all winter long. Change the transmission oil at the same time.
2. Change the antifreeze on freshwater-cooled engines at least every two years. The antifreeze itself does not wear out, but it has various corrosion-fighting additives that do. Always mix the water and antifreeze before putting it in the engine. A 50-50 water-antifreeze solution provides the best protection. (Note: Some antifreezes come premixed, in which case they should not be diluted.)
3. Do one of the following: Drain the raw-water system, taking particular care to empty all low spots. Clean the strainer. Remove the rubber pump impeller, grease it lightly with petroleum jelly, and replace. Leave the pump cover screws loose; otherwise the impeller has a tendency to stick in the pump housing. Leave a prominent note as a reminder of what has been done! Run the engine for a few seconds to drive any remaining water out

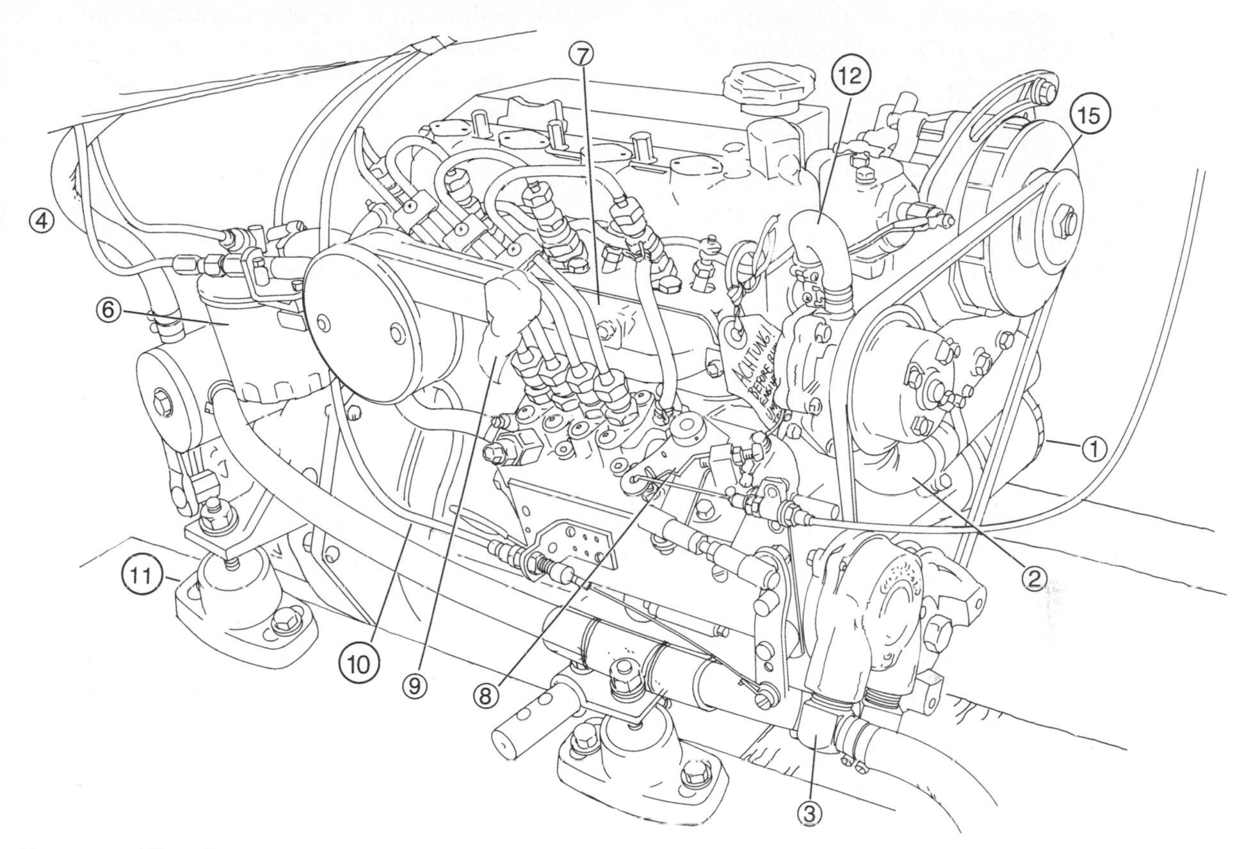

FIGURE 3-41. *Winter layup maintenance points (see the text for an explanation). (Jim Sollers)*

of the exhaust. Drain the base of a water-lift muffler.

Or: Close the raw-water seacock and make a routine inspection (as above) of the strainer and raw-water pump impeller. Disconnect the engine suction hose from the seacock, dip it in a 50-50 solution of antifreeze and water, and run the engine until the solution emerges from the exhaust. (Note: Ethylene glycol—commonly found in antifreeze—is harmful to the environment, so use propylene glycol instead, which is commonly used for winterizing drinking water systems. Propylene glycol is often sold already diluted to a 50% solution, in which case do not dilute it further.)

4. Wash the valves on any vented loops in warm water to clean out salt crystals.
5. Break the exhaust loose from the exhaust manifold or water-lift muffler (not visible in illustration) and check for carbon buildup; inspect the raw-water injection elbow for corrosion. Remove the raw-water hose from the injection nipple and check for any obstruction (this is a likely spot for scale and debris to get trapped).
6. Check the fuel filters and fuel tank for water and sediment; clean as necessary. Fill the tank to minimize condensation.
7. Squirt some oil into the inlet manifold and turn the engine over a few times (without starting) to spread the oil around the cylinder walls.
8. Grease any grease points (many modern engines have none).
9. Seal all openings into the engine (air inlet, breathers, exhaust) and the fuel tank vent.

Put a conspicuous notice somewhere as a reminder to unseal everything at the start of the next season.

10. Remove the inner wires of all engine control cables from their outer sheaths. Clean, inspect, grease, and replace. Check the sheathing as outlined on pages 213, 216–17.
11. Inspect all flexible feet and couplings for signs of softening (generally from oil and diesel leaks) and replace as necessary.
12. Inspect all hoses for signs of softening, cracking, or bulging, especially those on the hot side of the cooling and exhaust systems. Check hose clamps for tightness and corrosion, especially where the band goes inside the worm gear. (Undo the clamps a turn or two to inspect the band. When undoing clamps, push in relatively hard on the screwdriver so that the screw does not come out of its housing, leaving the band stuck to the hose!)
13. Remove the inlet and exhaust ducting (not visible in illustration) from any turbocharger and inspect the compressor wheel and turbine for excessive deposits or damage. Clean as necessary.
14. Make sure that batteries (not visible in illustration) are fully charged, and in the case of wet-cells, that they are topped off and recharged monthly.
15. Lightly spray the alternator and starter motor with WD-40 or some other moisture dispersant/lubricant; loosen belts. If there is any likelihood of salt water having contacted the starter motor pinion, remove the starter and lubricate the pinion.

RECOMMISSIONING

When you're ready to start boating again, perform the following tasks before you set sail:

1. Unseal engine openings and tighten the raw-water pump cover if loose. If a paper gasket is rubbed with grease or petroleum jelly, it will not stick to the cover or pump body next time the cover is removed (most modern pumps have O-rings).
2. Replace all zincs (if fitted; many modern engines have none).
3. Capacity-test the batteries (see my *Boatowner's Mechanical and Electrical Manual*).
4. Tighten alternator and other belts.
5. Prime the cooling system.
6. If possible, crank the engine without starting it until oil pressure is established.
7. Once running, check the oil pressure, the raw-water discharge, and the engine for oil and water leaks.
8. After the boat has been in the water for a few days, check the engine alignment (see Chapter 9).

CHAPTER 4

TROUBLESHOOTING, PART ONE: FAILURE TO START

It is useful to distinguish two differing situations when dealing with an engine that will not start. The first is a failure to crank—the engine will not turn over at all. The second is a failure to fire—the engine turns over but does not run.

SECTION ONE: FAILURE TO CRANK

When an engine will not crank at all, the problem is almost always electrical (see the Starter Motor Does Not Crank section below), but occasionally it is the result of water in the cylinders or a complete seizure of the engine or transmission. If there's an appreciable amount of water in the engine, it may show up as a high oil level on the dipstick. However, if the water is only in one or two cylinders (as is often the case), it will not show up this way.

Before checking the electrical system, see if the engine can be turned over by hand with the hand crank (if fitted) or by placing a suitably sized wrench on the crankshaft-pulley nut (turn the engine in its normal direction of rotation to prevent accidentally undoing the nut). If the engine has a manual transmission, put it in gear and turn it with a pipe wrench on the propeller shaft, but only after wrapping a rag around the shaft to avoid scarring it.

If the engine is locked up solidly, it has probably seized and will need professional attention. If it turns over a little and then locks up, or turns with extreme difficulty, water may have siphoned into the cylinders through a water-cooled exhaust (see Chapter 9). The other way water may have got in is from excessive cranking, since anytime an engine is cranked the raw-water pump will push water into the exhaust, but until the engine fires there will not be the necessary exhaust gases to lift the water out. Eventually enough water can accumulate to flood the engine. (If an engine requires extensive cranking, either periodically drain the water-lift muffler or temporarily close the raw-water inlet—but be sure to remember to open it once the engine starts.)

WATER IN THE ENGINE

Water, especially salt water, that remains for any length of time will do expensive damage to bearing and cylinder surfaces, requiring a complete engine overhaul. If the water is discovered in time, it can be eased out of the exhaust and the engine will continue to operate.

Water is not compressible, so any attempt to rapidly crank an engine with water in a cylinder that is on its compression stroke is likely to do damage (piston rings get broken and connecting rods bent). If the engine has decompression levers and a hand crank, simply turn it over several times. If it does not have decompression levers, a professional will often remove injectors and blow the water out of the injector holes, but I advise against this for amateurs. All

Troubleshooting Chart 4-1. Diesel Engine Problems: An Overview

Seizure	Excessive oil consumption	Rising oil level	Low oil pressure	High exhaust back pressure	Loss of power	Overheating	Poor idle	Hunting	Misfiring	White smoke	Blue smoke	Black smoke	Knocks	Low cranking speed	Lack of fuel	Low compression	Poor starting	
															●		●	Throttle closed/fuel shut-off solenoid faulty/tank empty
					●										●		●	Lift pump diaphragm holed
					●		●		●						●		●	Plugged fuel filters
					●		●		●	●			●		●		●	Air in fuel lines
					●		●		●	●		●	●				●	Dirty fuel
					●		●		●			●	●				●	Defective injectors/poor quality fuel
					●		●		●						●		●	Injection pump leaking by
					●	●	●		●			●	●				●	Injection timing advanced or delayed
						●						●						Too much fuel injected
	●				●		●		●	●	●					●	●	Piston blow-by
					●		●		●	●						●	●	Dry cylinder walls
					●	●	●		●	●						●	●	Valve blow-by
											●							Worn valve stems
					●		●		●	●						●	●	Decompressor levers on/valve clearances wrong/valves sticking
									●	●							●	Pre-heat device inoperative
									●			●					●	Plugged air filter
				●	●							●					●	Plugged exhaust/turbocharger/kink in exhaust hose
●			●			●												Oil level low
●	●		●											●				Wrong viscosity oil
●		●	●															Diesel dilution of oil
			●															Dirt in oil pressure relief valve/defective pressure gauge
							●	●										Governor sticking/loose linkage
							●											Governor idle spring too slack
●						●												Defective water pump/defective pump valves/air bound water lines
●						●												Closed sea cock/plugged raw-water filter or screen/plugged cooling system
●		●				●			●	●						●	●	Blown head gasket/cracked head/water in cylinders
●						●												Uneven load on cylinders
					●								●			●	●	Worn bearings
						●											●	Seized piston
					●									●			●	Auxiliary equipment engaged
														●			●	Battery low/loose connections/undersized battery cables
●					●	●						●						Engine overload/rope in propeller
		●									●							Leaking turbocharger oil seals
	●													●			●	Water in the cylinders

too often injectors are difficult to remove, and when amateurs try to remove them, damage is done. (In particular, many Volvo Penta engines have each injector in a sleeve that may come out with the injector, in which case a special tool is needed to put the sleeve back.) If the engine can be turned over in small increments, the water in those cylinders that are on the compression stroke can be forced a little at a time past the piston rings without breaking them. This is especially the case on engines with some wear on them. *The key thing is to not force the process.*

To remove water, either close the throttle so the engine will not start, or disable the engine in some

Troubleshooting Chart 4-2. Starting Circuit Problems: Engine Fails to Crank Note: Before jumping out solenoid terminals, completely vent the engine compartment, especially with gasoline engines, since sparks will be created.	
Turn on some lights *powered by the cranking battery* and try to crank the engine. Do the lights *go out*? NO ↓	YES → If the solenoid makes a rapid "clicking," the battery is probably dead; replace it. If the solenoid makes one loud click, the starter motor is probably jammed or shorted; free it up or replace it as necessary. Before taking the starter out, try turning the engine over by placing a wrench on the crankshaft pulley nut. If the engine won't turn, this is the problem; there may be water in the cylinders (pages 69–72) or the engine may be seized up.
When cranking, do the lights *dim*? NO ↓	YES → Check for voltage drop from poor connections or undersized cables. Try cranking for a few seconds, then feel all connections and cables in the circuit—battery and solenoid terminals, and battery ground attachment on the engine block. If any are warm to the touch, they need cleaning. If this fails to show a problem, use a voltmeter as shown in Figure 4-12, then try to crank the engine. If there is no evidence of voltage drop, check for a jammed or shorted starter.
Is the ignition switch faulty? NO ↓ **TEST:** With a jumper wire or screwdriver blade, bridge the battery and ignition switch terminals on the solenoid. If the starter motor now cranks, the ignition circuit is defective.	YES → Replace ignition switch or its wiring as needed.
Is the solenoid defective? NO ↓ **TEST:** Use a screwdriver blade to jump out the two heavy-duty terminals on the solenoid. (This procedure is tricky: Observe all precautions outlined in the accompanying text.) If the starter now spins, the solenoid is defective.	YES → Replace the solenoid.
Is there full battery voltage at the starter motor when cranking? NO ↓ **TEST:** Check with a multimeter between the starter positive terminal and the engine block when cranking.	YES → The starter is open-circuited and needs replacing. First check its brushes for excessive wear or sticking in their brush holders.
No voltage: The battery isolation switch is probably turned off!	

other way (this is really important!). Flick the starter motor on and off, or use a wrench on the crankshaft-pulley nut (this give more control) to turn the engine over bit by bit, turning in the normal direction of rotation and pausing between each movement. Don't rush things!

Another way to remove water involves more work but minimizes the risk of doing damage. Remove the valve cover and push a coin (such as a U.S. quarter or a UK 5 pence piece) in between the tops of the valve stems and the rocker arms for the exhaust valves (see Chapter 7 for how to identify these valves; if in doubt, do all the valves). This will crack the valves open. Now turn the motor over a little at a time, and the water will be forced out the valves. Note that on some engines the clearance between the tops of the pistons and the valve faces is quite small. If the engine locks up again when doing this, the coins may be holding the valves open too far. Take the coins out before something gets bent!

Once the engine has turned through two complete revolutions, the cylinders should be substantially free of water. Spin the engine a couple of times without starting it. Now check the crankcase for water in the oil. If any is present, change the oil and filter. Start the engine and run it for a few minutes to warm it, then shut it down and change the oil and filter again. Now give the engine a good run to drive out any remaining moisture. After 25 hours of normal operation, or at the first sign of any more water in the oil, change the oil and filter for a third time.

Put appropriate siphon breaks in the cooling and exhaust system (see Chapter 9).

Starter Motor Does Not Crank

To troubleshoot electrical problems, you first need to understand the starting circuit.

The heavy amperage draw of a starter motor requires heavy supply cables. Ignition switches are frequently located some distance from the battery and starter motor. To avoid running the heavy cables to the ignition switch, a special switch called a solenoid is installed in the cranking circuit and used to turn the starter motor on and off (see Figure 4-1). The solenoid is triggered by the (remotely mounted) ignition switch. The wiring between the solenoid and the ignition switch carries little current and as a result can be relatively light weight.

Some starting circuits include a neutral start switch, which prevents the engine from being cranked when it is in gear. Some heavy-duty circuits have two solenoids. The first is energized by the ignition switch to close the circuit to the second, which then operates the starter motor (see Figure 4-2). (This is done because the second solenoid itself draws a moderately heavy current, requiring fairly heavy cables. Adding a second light-duty solenoid enables lighter wires to be run to the ignition switch.)

A solenoid contains a plunger and an electromagnet. When the ignition switch is turned on, it energizes the magnet, which pulls down the plunger, which closes a couple of heavy-duty contacts, thus completing the circuit to the starter motor (see Figures 4-3 and 4-4).

The ground side of a starter motor circuit almost always runs through the starter motor case and engine block to a ground strap, which is connected directly to the battery's negative terminal post. However, on rare occasions an insulated ground is used, in which case the starter motor is isolated electrically from the engine block; a separate cable grounds it to the battery.

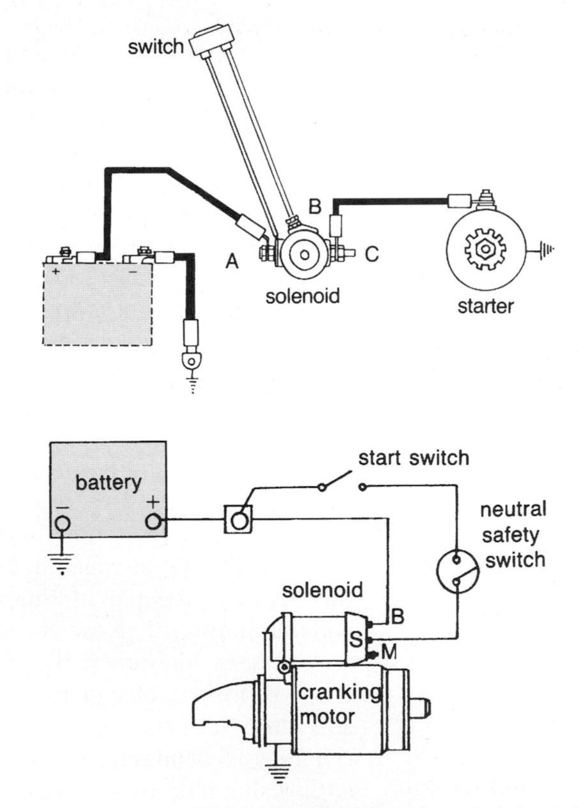

FIGURE 4-1. A starting circuit with an inertia starter (top). To bypass the switch, connect a jumper from A to B. To bypass the switch and the solenoid, connect a heavy-duty jumper (e.g., a screwdriver) from A to C. A starting circuit with a preengaged starter (bottom). To bypass the switch, connect a jumper from terminal B to terminal S on the solenoid. To bypass the solenoid and switch, connect a heavy-duty jumper (e.g., a screwdriver) from terminal B to terminal M. (Courtesy PCM)

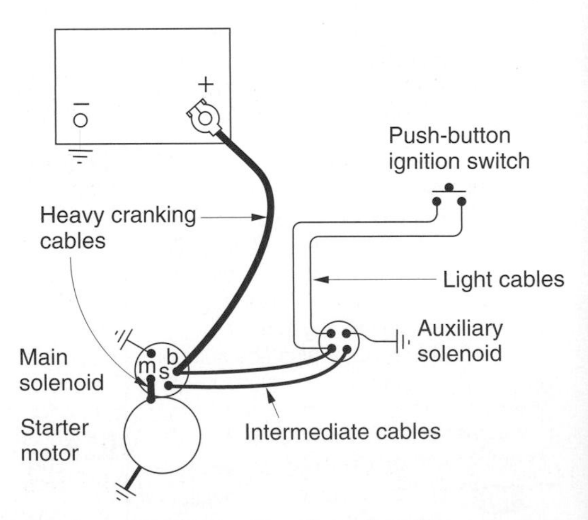

FIGURE 4-2. A starting circuit with main and auxiliary solenoids. To bypass the switch circuit, connect a jumper from B to S on the main solenoid. To bypass the solenoid and switch, connect a heavy-duty jumper from B to M.

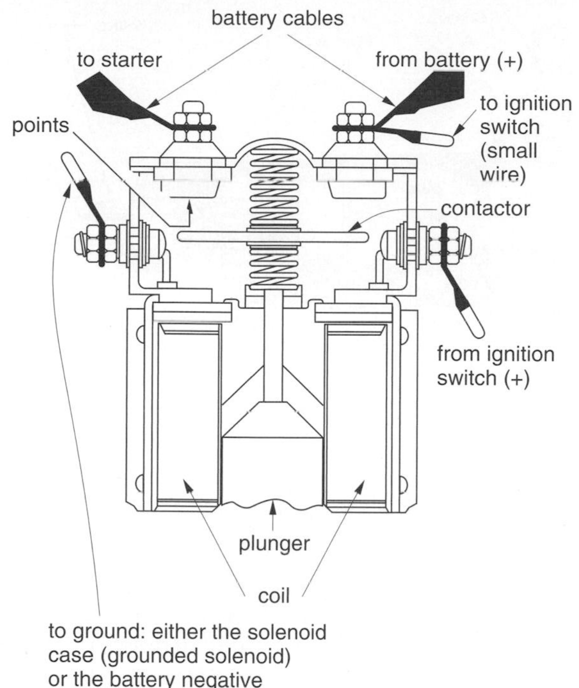

FIGURE 4-3. *Solenoid operation. Upon closing the ignition switch, a small current passing through the solenoid's electromagnet pushes the contactor up against spring tension, completing the circuit from the battery to the starter motor.*

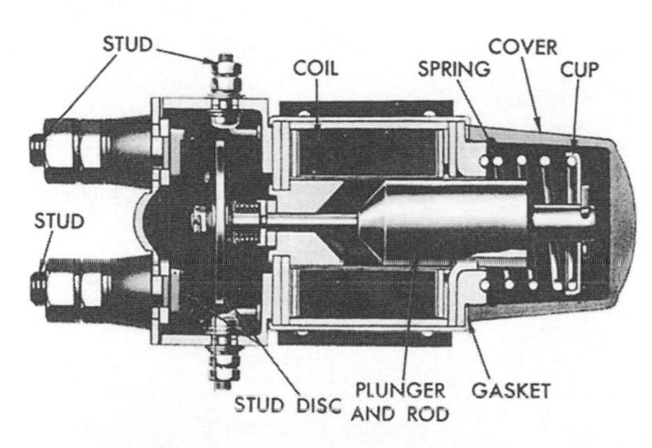

FIGURE 4-4. *A cutaway of a typical solenoid. (Courtesy Detroit Diesel)*

INERTIA AND PREENGAGED STARTERS

Most starter motors are series-wound motors although some newer ones have permanent magnets. Regardless of motor type, there are two basic kinds of starters: inertia and preengaged. Inertia starters are mostly found on older engines; as far as I know, all modern engines have preengaged starters. Some newer starter motors have a reduction gear (generally between 2:1 and 3.5:1) between the electric motor and the drive gear (the pinion gear) that engages the engine flywheel. This enables a smaller, lighter, faster-spinning motor to do the work of a larger motor.

INERTIA STARTERS

The solenoid for an inertia motor is mounted independently, at a convenient location. The pinion gear is keyed into a helical groove on the motor shaft. When the solenoid is energized, the motor spins. Inertia in the pinion gear causes it to spin out along the helical groove and into contact with the engine flywheel, which is then turned over (see Figure 4-5).

Sometimes an inertia starter's pinion will stick to its shaft and not be thrown into engagement with the flywheel. In this case, the motor will whir loudly without turning the engine over. A smart tap on its case while it is spinning may free it up. However, don't do this too often, as it's hard on the gear teeth. Also, don't do this to a permanent-magnet motor—it may crack the magnets or break them loose from the motor case (as far as I know, there are no inertia starters with permanent-magnet motors).

At other times the pinion may jam in the flywheel and not disengage. In this case you can often gain access to the end of the armature shaft via a cover, normally located in the center of the rear housing of the starter motor. With the engine shut down, turn the squared-off end of the shaft back and forth with a wrench to free the pinion.

PREENGAGED STARTERS

The solenoid of a preengaged starter is always mounted on the starter itself (a surefire indicator). When the solenoid is energized, the electromagnet pulls a lever, which pushes the starter motor pinion into engagement with the engine flywheel; the main solenoid points now close, allowing current to flow to the motor, turning it over. Preengaged starters mesh the drive gear with the flywheel before the motor spins, greatly reducing overall wear (see Figure 4-6).

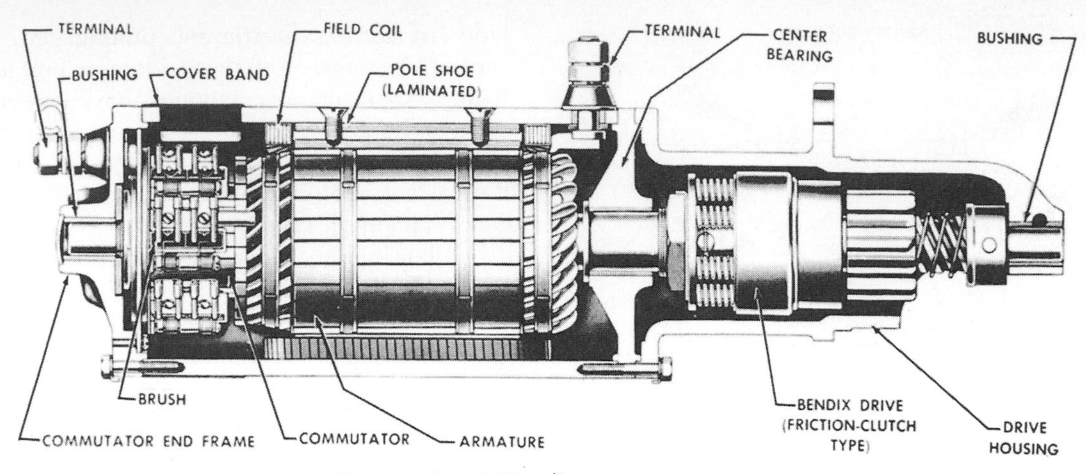

FIGURE 4-5. *An inertia-type starter motor. (Courtesy Detroit Diesel)*

Unlike an inertia starter, once the engine fires up, the pinion gear does not get thrown out of engagement with the flywheel. As the engine speeds up, it has the potential to spin the gear at a very high speed. To prevent damage to the starter, the gear is mounted on a clutch arrangement that allows it to freewheel until it is disengaged by deenergizing the solenoid.

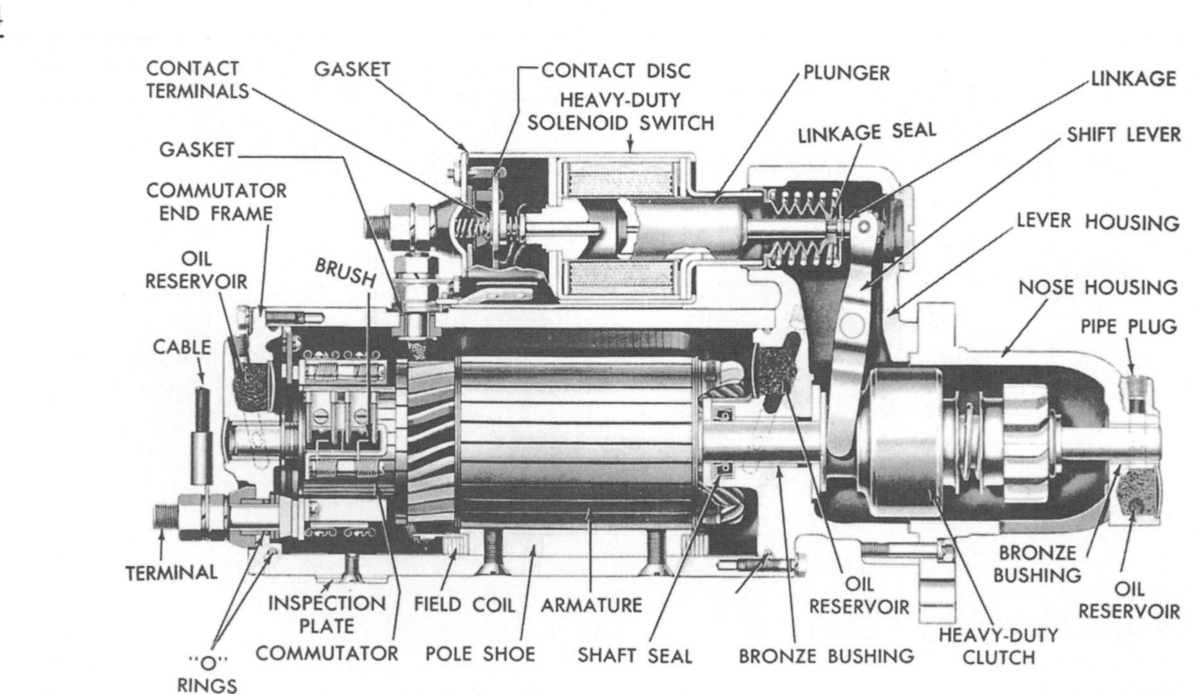

FIGURE 4-6. *A preengaged starter motor with a solenoid mounted on it. (Courtesy Detroit Diesel)*

Circuit Testing

Never crank a starter motor continuously for more than 15 seconds when performing any of the following tests. Because starter motors are designed for only brief and infrequent use, they generally have no fans or cooling devices. Continuous cranking will burn them up.

Remember that a battery that is almost dead may show nearly full voltage on an open-circuit voltage test. If a starter motor fails to work, turn on a couple of lights (switched into the cranking battery circuit, not the house battery) and try to crank the engine. The lights should dim but still stay lit. If the lights remain unchanged, no current is flowing to the starter; if the lights go out, the battery is dead or the starter is shorted. Check the battery first. (See my *Boatowner's Mechanical and Electrical Manual* for an extensive description of various methods for testing batteries.)

Note that burned coils in a solenoid or starter motor have a distinctive smell; once encountered, you will never forget it. It is not a bad idea to sniff them before starting any other detective work—this may rapidly narrow the field of investigation.

Preliminary Circuit Tests

Note: Some of these tests will create sparks. Be sure to vent the engine room properly, especially if gasoline is present.

A couple of quick steps will isolate problems in the starting circuit. The solenoid will have two heavy-duty terminals: one is attached to the battery's positive cable; the other is attached with a second cable (or short strap in the case of preengaged starters) to the starter itself. There will also be one or two small terminals. If only one of the small terminals has a wire attached to it, this is the one we need. If both small terminals have wires attached, one is a ground wire; we need the other one (it goes to the ignition switch). If in doubt, turn off the ignition and battery switches and test from both solenoid terminals to ground using the ohms function of a multimeter (R × 1 scale on an analog meter). The ignition switch wire should give a small reading; the ground wire will read 0 ohms.

Turn the battery isolation switch back on. Now bypass the ignition switch circuit (and neutral start switch if fitted) by connecting a jumper wire or screwdriver blade from the battery terminal on the solenoid to the ignition switch terminal (see Figures 4-7, 4-8, and 4-9). If the motor cranks, the ignition switch or its circuit is faulty. If the solenoid clicks but nothing else happens, the battery may really be dead (check it again), or the starter is probably bad. (But note that the solenoid on a preengaged starter will sometimes throw the pinion into engagement with the flywheel but then fail to energize the motor due to faulty main points, so always check the solenoid before condemning the motor.) If nothing at all happens, there is either no juice to the solenoid (check the battery isolation switch) or the solenoid is defective.

If the starter failed to work, use a screwdriver to jump out the two heavy-duty terminals on the

Figure 4-7. *A typical preengaged starter motor and solenoid. Jumping across 1 (B in Figure 4-1, bottom) and 2 (S in Figure 4-1, bottom) bypasses the ignition switch; 1 (B) and 3 (M in Figure 4-1, bottom) bypasses the ignition switch and the solenoid. The starter should spin but not engage the engine's flywheel.*

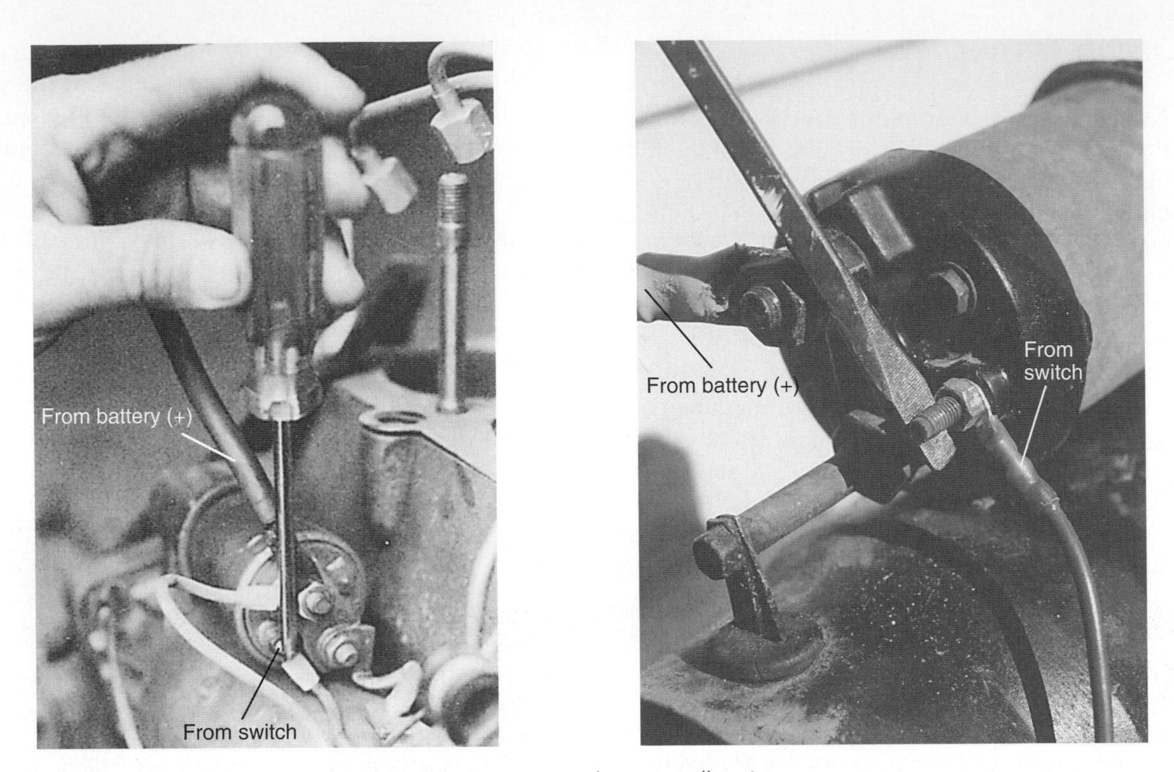

FIGURES 4-8 and 4-9. *Bypassing the ignition switch by jumping out the two smaller wires.*

solenoid (see Figures 4-10 and 4-11). *Warning:* The full battery current may be flowing through the screwdriver blade; considerable arcing is likely, and a big chunk may be melted out of the screwdriver blade! Do not touch the solenoid case or starter case with the screwdriver: This will create a dead short. Hold the screwdriver firmly to the terminals. If the starter now spins, the solenoid is defective. If the motor does not spin, the motor is probably faulty. If no arcing occurred, there is no juice to the solenoid.

A motor with an inertia starter can be cranked by jumping a defective solenoid as described. A preengaged starter, however, will merely spin without engaging the engine flywheel since the solenoid is needed to push the starter pinion into engagement with the flywheel. In this case, have someone hold the start switch on while you jump the main solenoid terminals. If the cranking problem is a failure of the solenoid points (most likely), this will crank the engine. If the problem is a failure of the solenoid coil, the pinion will still fail to engage the flywheel. In the latter case, you can remove the solenoid to access the pinion fork, allowing you to manually engage the gear while you jump out the main solenoid points. This should get the engine going in an emergency, but it may leave the pinion rattling up against the flywheel when the engine is running.

VOLTAGE DROP TESTS

The above tests will quickly and crudely determine whether there is juice to the starter motor and whether or not it is functional. More insidious is the effect of poor connections and undersized cables; these create voltage drop in the circuits, rob the starter motor of power, and result in either sluggish cranking or failure to crank at all, especially when the engine is cold. A potential result is a burned-out starter motor. (If you have to replace the starter motor, always do a voltage drop test after fitting the new one to ensure it doesn't burn out also.)

Proper marine installations require a battery isolation switch that completely removes the battery

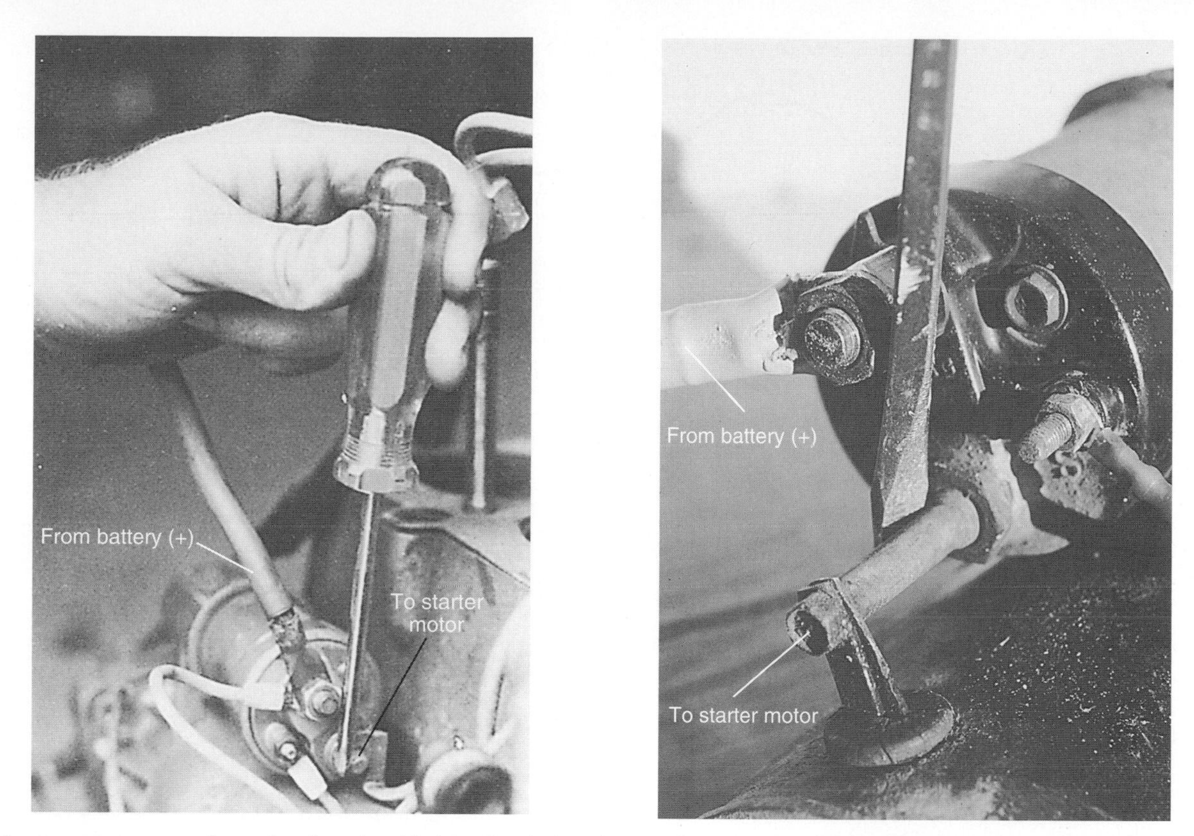

FIGURES 4-10 and 4-11. *Bypassing the solenoid altogether by jumping out the two main cable terminals.*

from the ship's circuits in the event of an emergency, such as an electrical fire. Isolation switches are often located at some distance from the battery and starter motor. The longer a cable run, the more the voltage drop for a given size of cable. All too often, the long battery cable runs found on boats are woefully undersized.

You can do a quick test for voltage drop by cranking (or trying to crank) for a few seconds and then feeling all the connections and cables in the circuit—both battery terminals, both isolation switch terminals, the solenoid terminals, the battery ground attachment point on the engine block, and the cables themselves. If any of these are warm, there is resistance and voltage drop; either the connection needs cleaning or the cable needs to be upgraded.

You can make more accurate tests with the DC volts function of a multimeter as described below (Figure 4-12 shows the connections for each test):

- Test 1: Run the positive meter probe to the battery positive post (not the cable clamp) and the negative meter probe to the solenoid positive terminal, then crank the engine (note that voltage drop tests can only be made when the circuit is under a cranking load). Switch down the voltage scales on an analog meter. Any voltage reading indicates voltage drop in this part of the circuit. Clean all connections and repeat the test.
- Test 2: Switch back up the voltage scales and perform the same test (cranking the engine) across the two main solenoid terminals. Switch down the scales. Any voltage indicates resistance in the solenoid (see below for cleaning dirty and pitted points).
- Test 3: Repeat the same test from the solenoid outlet terminal to the starter motor hot terminal. The cumulative voltage drop

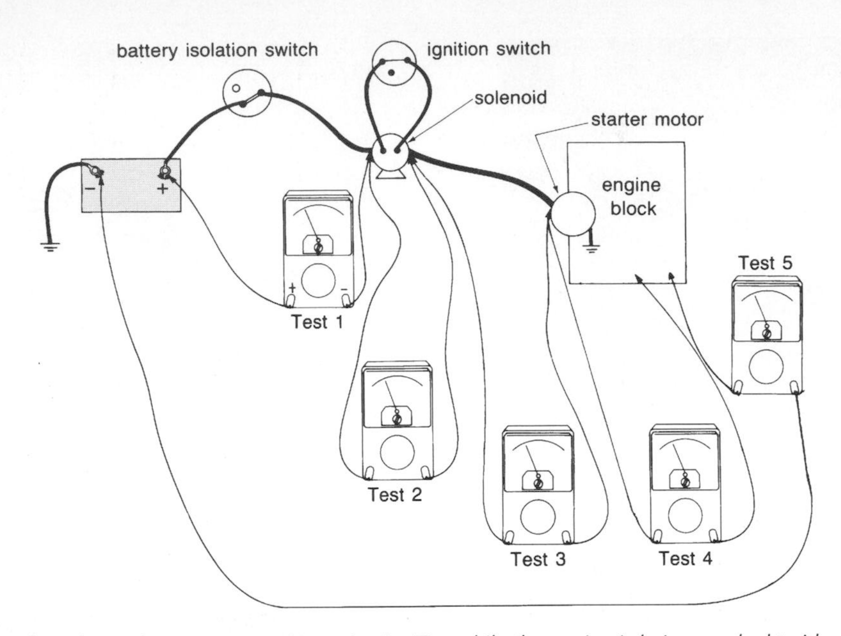

FIGURE 4-12. *Testing for voltage drop on starter motor circuits. Test while the engine is being cranked to identify areas of excessive resistance.*

revealed by these tests should not exceed 0.7 volt on a 12-volt system (1.4 volts on a 24-volt system). It is not unusual to find 2.0 volts!

- Tests 4 and 5: Test from the starter motor case to the engine block (Test 4). Then test from the engine block back to the battery negative post (not the cable clamp—Test 5). The latter test frequently reveals poor ground connections. The cumulative voltage drop revealed by these tests should not exceed 0.3 volt on a 12-volt system (0.6 volt on a 24-volt system). It is not unusual to find 1 volt or more.

The total voltage drop on both the positive and negative sides of the circuit should not exceed 1.2 volts on a 12-volt system (2.4 volts on a 24-volt system). All too often, it is much higher than this. If the voltage drop is still too high after cleaning and tightening all the connections in the circuit, and after correcting any problems revealed with the solenoid (Test 2: the voltage drop across the solenoid terminals should be below 0.3 volt), then the cranking cables and engine ground cable need to be replaced with larger cables.

AMPERAGE TESTS

Many modern multimeters incorporate a clip-on ammeter that can measure up to 200 amperes (amps) DC. Although the inrush current of even a small diesel engine starter motor will be on the order of several hundred amps, the sustained cranking current is below 200 amps. One of these meters can be used to measure the cranking current, providing a rapid indication of whether or not the starter motor is receiving the kind of amperage it needs to operate (see Figure 4-13). The test will prove far more useful

FIGURE 4-13. *Testing the amp draw of a starter motor with a clamp-on DC ammeter.*

if you have already tested the engine when it is cranking properly and made a note of the current draw to use as a benchmark. Note that the draw will vary according to the ambient temperature and the engine temperature (the colder the ambient and engine temperatures the higher the draw), so testing in different conditions is recommended.

Motor and Solenoid Disassembly, Inspection, and Repair

Before removing a starter motor from an engine, isolate its hot lead, or better still, disconnect it from the battery. The starter will be held to the engine by two or three nuts or bolts. If it sticks in the flywheel housing, a smart tap will jar it loose (but check first to see that you didn't miss any of the mounting bolts, and don't whack a permanent-magnet motor!).

Preengaged Starters

To check the solenoid points on a preengaged starter (see Figures 4-14A to 4-14K), remove the battery cable, the screw or nut retaining the hot strap to the solenoid, and all other nuts on the stud(s) for the ignition circuit wiring. Undo the two retaining screws that hold the end housing. The end housing (plastic) will pull off to expose the contactor and the points. (Note: A spring will probably fall out; it goes on the center of the contactor.) Check the points and contactor for pitting and burning and clean or replace as necessary. A badly damaged contactor can sometimes be reversed to provide a fresh contact surface. In a similar way, the main points can sometimes be reversed in the solenoid end housing, but take care when removing them because the housing is brittle.

To remove the solenoid coil (electromagnet), undo the two screws at the flywheel end of the solenoid and turn the coil housing through 90 degrees. The coil will now pull straight off the piston. The piston and fork assembly generally can be removed only by separating the starter motor case from its front housing (see below).

If you need to undo the pivot pin bolt on the solenoid fork, mark its head so you can put it back in the same position. Sometimes, turning this pin adjusts how far the pinion is thrown when it engages the flywheel.

Remove the two starter motor retaining screws from the rear housing and lift off the rear cover; it comes straight off and contains no brushes (see Figures 4-15A to 4-15D). This will expose the brushes and commutator.

Inertia Starters

Access and check the solenoid points, as above. To disassemble an inertia starter, remove the metal band from the rear of the motor case. Undo the locknut from the terminal stud in the rear housing, but do not let the stud turn; if necessary, grip it carefully with Vise-Grips (mole wrenches). Note the order of all washers; they insulate the stud, and it is essential that they go back the same way.

Undo the two (or four) retaining screws in the rear housing. If tight, grip with a pair of Vise-Grips from the side to break them loose. Lift off the rear housing with care; it will be attached to the motor case by two brush wires. Lift the springs off the relevant brushes and slide the brushes out of their holders to free the housing.

All Starters

The motor case can now be pulled off (see Figures 4-16A to 4-16M). If the end cover was held with four short screws as opposed to two long ones, there will be four more screws holding the motor case at the other end. Note that both end housings will probably have small lugs so that they can be refitted to the motor case in only one position.

Inside the case will be the field windings—densely packed copper coils enclosed in an insulating material. On the motor shaft will be the armature—another series of densely packed copper coils. Various tests can be performed on these coils with the ohms function of a multimeter, but these are beyond the scope of this book (see *Boatowner's Mechanical and Electrical Manual*).

Inspect the commutator (the segmented copper bars on the aft end of the armature shaft—see Figure 4-17) and brushes (the spring-loaded carbon blocks that bear on the commutator) for signs of arcing or excessive wear (burned or pitted areas on the commutator).

You can clean a commutator by pulling a strip of fine sandpaper (400- to 600-grit wet-or-dry) lightly back and forth until all the segments are uniformly

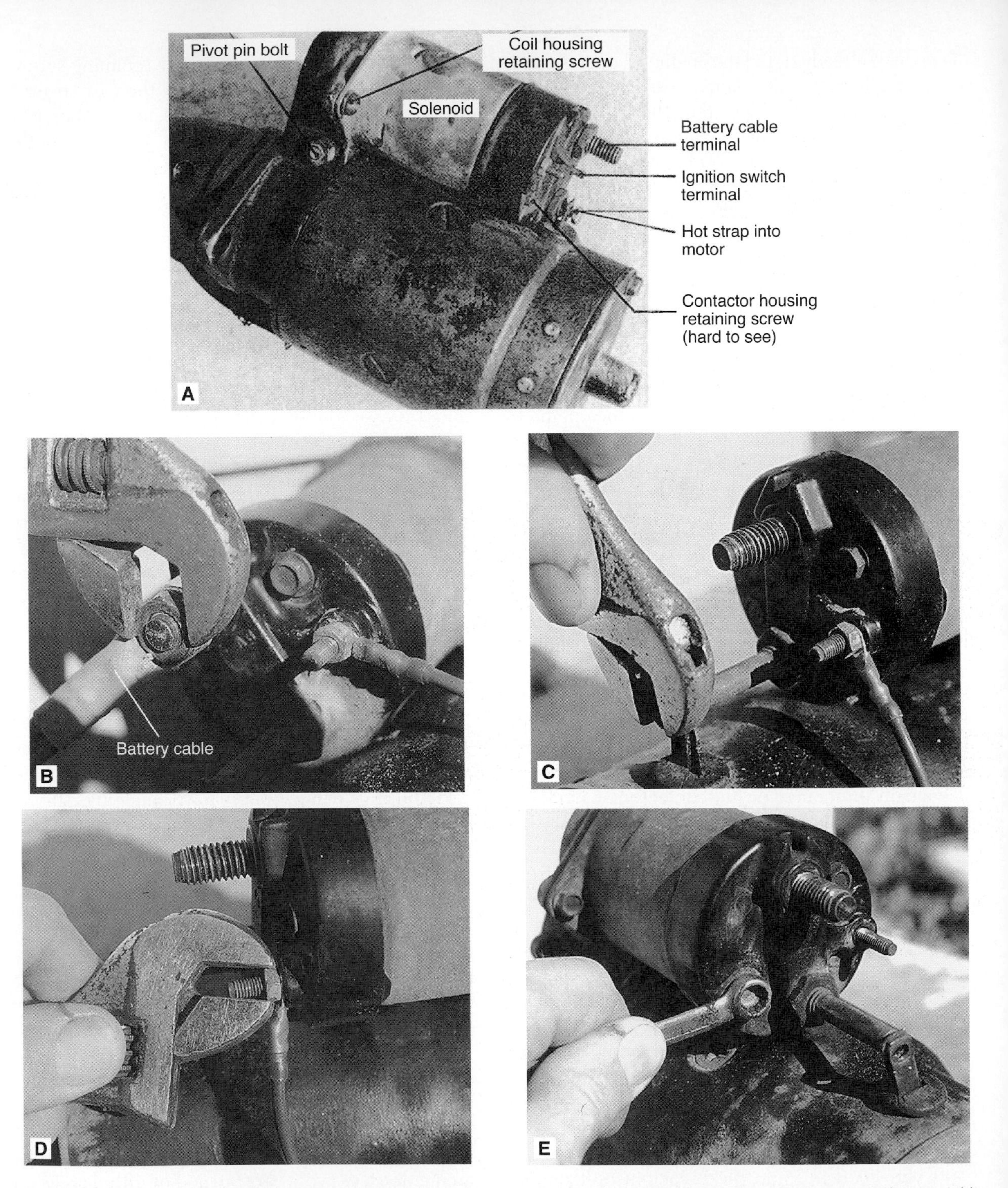

FIGURES 4-14A to 4-14K. *Disassembling a preengaged starter's solenoid. First, identify the main components (A). Remove the battery cable (B). Disconnect the solenoid from the starter motor (C). Disconnect the cable to the ignition switch (D). Take the cover bolts off the solenoid (E and F). Bend the starter motor strap out of the way so the solenoid cover can be pulled off (G). Remove the solenoid cover (H); there is a spring in here that may fall out. A view of the contactor plate, showing the burned contactor (I). Note the severe pitting on the surfaces of the contactor (the circular disc) and the points (J). Sometimes the contactor can be flipped over to reveal an unused surface, and as in this case, the points can be rotated in their housing to bring a new surface into action (K). Otherwise, they can be filed and sanded smooth.*

F

Starter motor strap
G

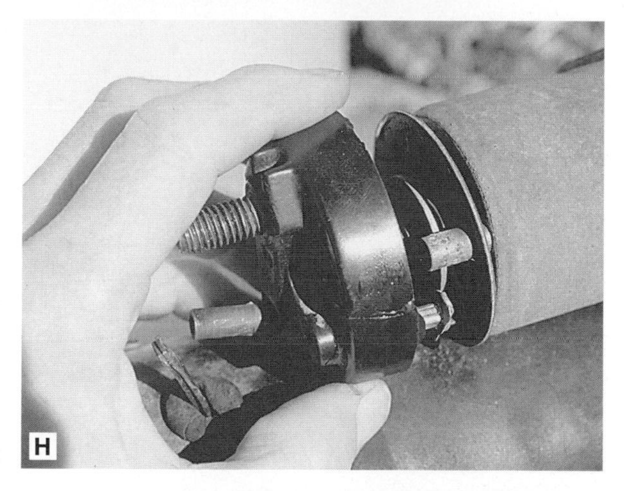
H

Contactor
I

Burned point that makes
contact with contactor
J

Burned point
Unused portion
of point
K

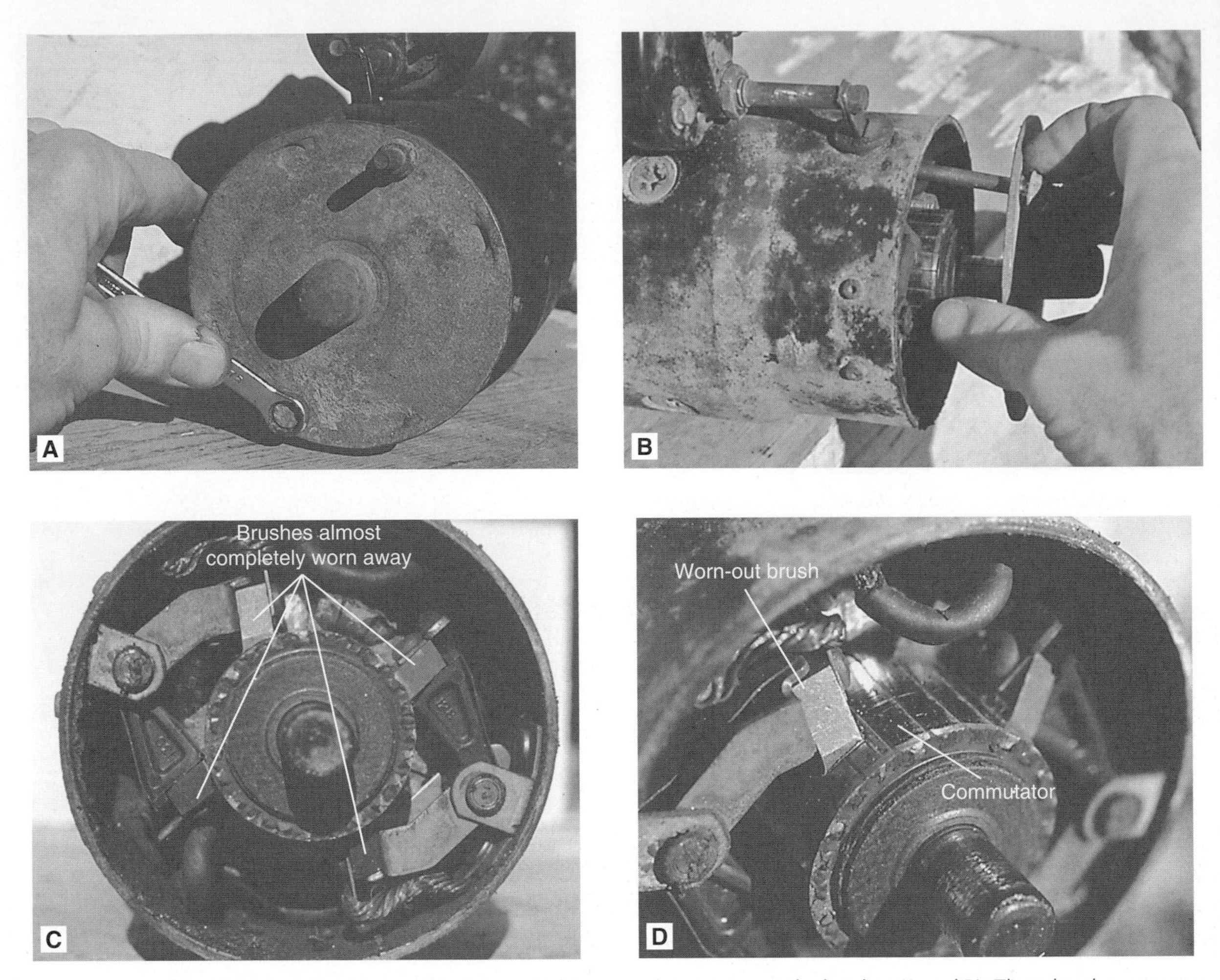

FIGURES 4-15A to 4-15D. *Starter motor disassembly. Removal of the rear plate to expose the brushes (A and B). These brushes are worn almost completely down (C and D). This starter motor is pretty well worn out!*

shiny (see Figure 4-18). Cut back the insulation between each segment of the commutator to just below the level of the copper segments by drawing a knife or sharp screwdriver across each strip of insulation. Take care not to scratch the copper or burr its edges (see Figure 4-19). Use a triangular file to bevel the edges of the copper bars. Always renew the brushes at this time.

Brushes, in any case, need replacing if their length is much less than their width. Some brush leads are soldered in place; others are retained with screws. To replace soldered brushes, cut the old leads, leaving a tail long enough to solder to. Tin this tail and the end of the new brush wires and solder them together. The new brushes must slide in and out of their holders without binding; if necessary, very gently sandpaper them down to make this possible.

The commutator ends of new brushes will need bedding in. Wrap fine sandpaper (400- or 600-grit wet-or-dry) around the commutator under the brushes, with the sanding surface facing out (see Figure 4-20). Spin the armature until the brushes are bedded to the commutator; they should be almost shiny over their whole surface. Remove the

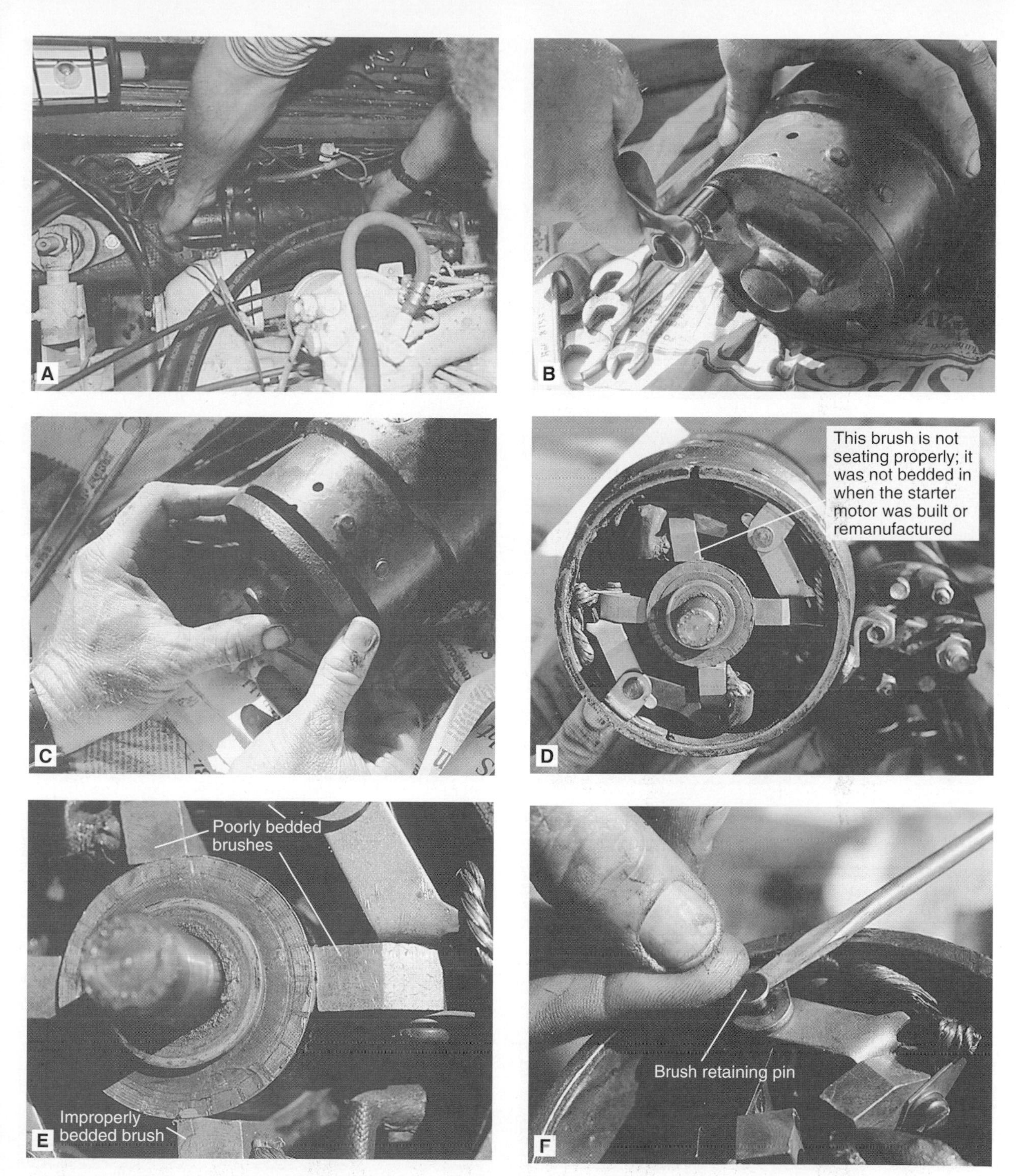

FIGURES 4-16A to 4-16M. *Removing and disassembling another recalcitrant starter motor. Pulling out the starter (A). Unbolting the rear housing (B). Pulling off the housing (C) to reveal the brushes (D). Note how poorly the brushes are seating on the commutator (E); this is a recently reconditioned starter motor that was clearly not put together properly. Removing a brush retaining pin (F), which revealed a partially broken brush wire that had arced and burned the brush (G). A properly seated brush compared to one that was never bedded in (H). Cleaning the commutator (I); there is a fair amount of pitting in evidence as a result of arcing from the poorly seated brushes. Disassembling the solenoid (J) revealed a burned contactor (K) and points (L). Cleaning the points with a piece of emery cloth (M).*

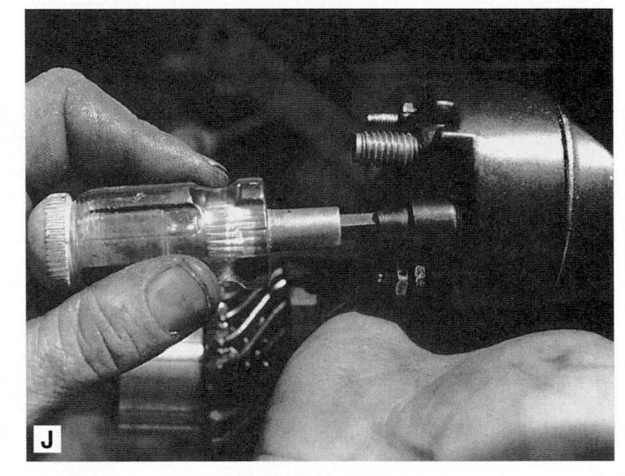

FIGURES 4-16A to 4-16M. *(continued)*

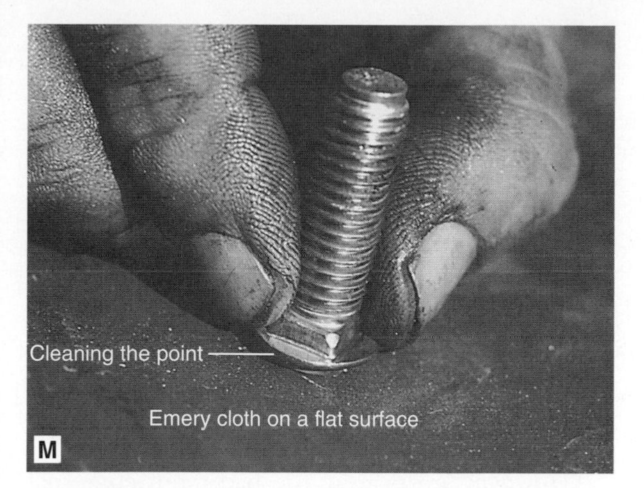

FIGURES 4-16A to 4-16M. *(continued)*

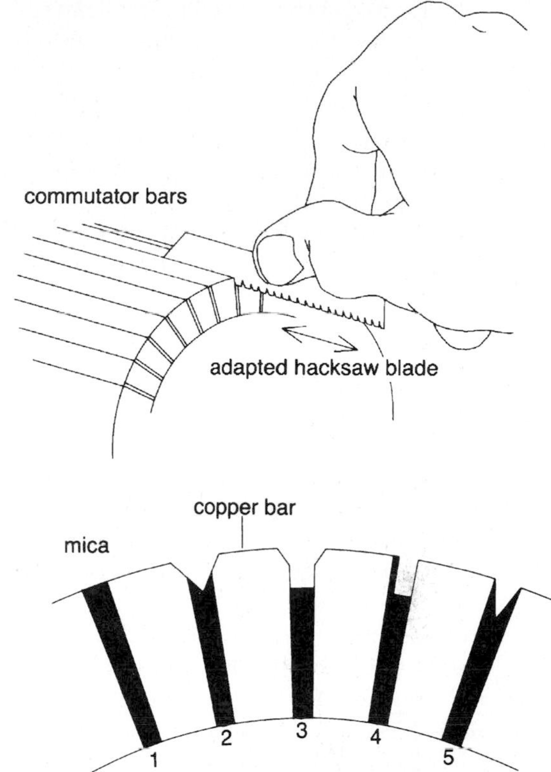

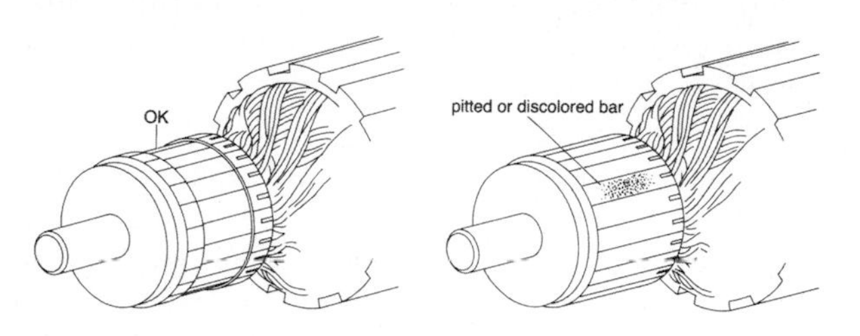

FIGURE 4-17. *Maintenance procedures for starter motors. A worn and grooved commutator is OK as long as the ring is shiny. A pitted or dark bar results from an open or short circuit in the armature winding. (Jim Sollers)*

FIGURE 4-19. *To cut back the insulation on a commutator, modify a hacksaw blade as shown and run it between the commutator bars. When you start, the mica insulation will be flush with the commutator bars (1). Cut back as in 2 (good) or 3 (better). Avoid cutting back as in 4 or 5. (Jim Sollers)*

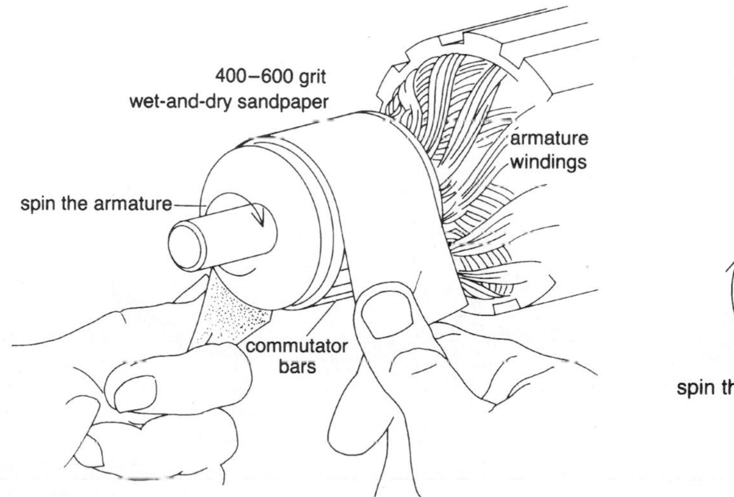

FIGURE 4-18. *Polishing a commutator. (Jim Sollers)*

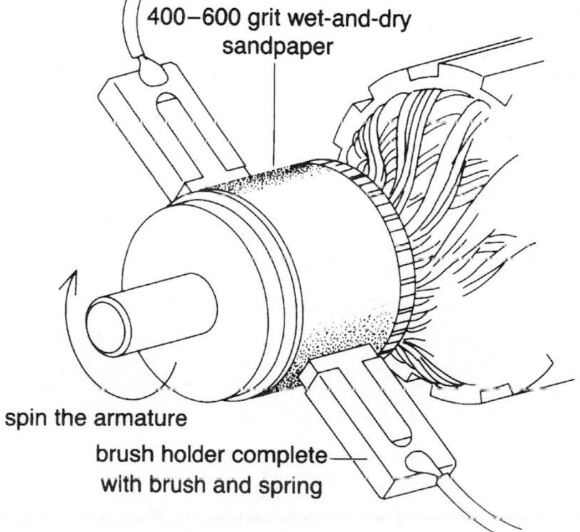

FIGURE 4-20. *Bedding in new brushes. (Jim Sollers)*

sandpaper and blow out the carbon dust, taking care to blow it away from the armature and not into its windings!

The brushes attached to the field windings are insulated; those to the end housing or motor case are uninsulated. When replacing the end housing of an inertia starter, hold the brushes back in their holders and jam them in place by lodging the brush springs against the side of the brushes. Once the plate is on, slip the springs into position with a small screwdriver, and then put the metal band back on the motor case.

The Pinion

Given the intermittent use of boat engines and the hostile marine environment, the pinion (or drive gear, often known as a bendix) can be especially troublesome. Pinions are particularly prone to rusting, especially where salt water in the bilges has contacted the flywheel and been thrown all around the flywheel housing.

Inertia starters are more prone to trouble than preengaged starters. Dirt or rust in the helical grooves on the armature shaft will prevent the pinion from moving freely in and out of engagement with the flywheel.

A small resistance spring in the unit keeps the pinion away from the flywheel when the engine is running. If this spring rusts and breaks, the gear will keep vibrating up against the flywheel with a distinctive rattle.

In automotive use, the shaft and helical grooves are not oiled or greased since the lubricant picks up dust from any clutch plates and causes the pinion to stick. In marine use, however, there are no clutch plates in the flywheel housing, and the shaft should be greased lightly to prevent rusting.

The pinion assembly is usually retained by a spring clip (a snap ring or circlip) on the end of the armature shaft, occasionally by a reverse-threaded nut and cotter pin (split pin). An inertia starter has a powerful buffer spring that must be compressed before the clip can be removed. Special tools are available for this, but a couple of small C-clamps or two pairs of adroitly handled Vise-Grips usually will do the trick.

Check the clutch unit on a preengaged starter by ensuring that it locks the pinion gear to the motor in one direction of rotation and freewheels in the other.

Flywheel Ring Gear

Sometimes gear teeth get stripped off a section of the flywheel ring gear. Whenever the engine stops with this section of the flywheel lined up with the starter motor pinion, the starter will simply spin without cranking the engine.

If there are only a few teeth missing from the ring gear, as a temporary expedient you can turn the engine over fractionally by hand, either with a hand crank (if fitted) or else with a wrench on the crankshaft pulley nut (turn in the normal direction of rotation). This will line up some good teeth with the starter motor pinion and should enable you to start the engine.

However, every time the bad teeth come past the starter pinion, the starter motor will momentarily race and its pinion will crash into the good teeth on the ring gear as the flywheel's momentum brings them around. This will steadily tear up more and more teeth, so you need to attend to the problem as soon as possible.

To change a starter pinion, see the preceding sections on starter motor disassembly. The ring gear is an interference fit on its flywheel. To remove it, you must unbolt the flywheel from its crankshaft, and to do this, you have to take the transmission out of the boat. Then place the flywheel on a stand and heat the ring gear evenly with a propane torch until it has expanded enough to allow you to drive it off with a hammer and copper bar (or something similar—see Figure 4-21). To fit a new ring gear, pack the flywheel in ice and evenly heat the new

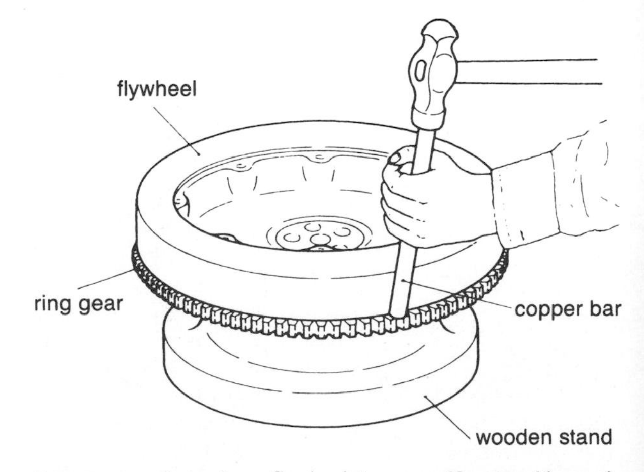

Figure 4-21. *Renewing a flywheel ring gear. (Courtesy Yanmar)*

gear, which should allow it to drop into place. Position the new gear in exactly the same spot as the old one.

SECTION TWO: FAILURE TO FIRE

When I was working on oil platforms in the Gulf of Mexico, the most common emergency call I received went something like this: "Such and such engine won't start. It was working fine the last time I used it, but now it just won't run." I usually asked three questions before calling up a boat or helicopter to go out and investigate:

1. "Have you checked to see if it has any fuel?"
2. "Have you checked to see if the fuel filter is plugged?"
3. "Have you left any life in the battery?"

"Oh, sure," was the usual answer, but probably half the time I found the engine out of fuel, the filter plugged, the battery dead, or a combination of these. Fuel can be added easily, and a clean filter can be fitted, but you can't get around a dead battery.

If an engine will not start as usual, *you need to stop cranking and start thinking!* Those extra couple of cranks in the hope that some miracle will happen very often flatten the battery to the point at which the engine cannot be started at all. Most starting problems are simple ones that can be solved with a little thought—and frequently in a whole lot less time than it will take you to recharge a dead battery!

A diesel engine is a thoroughly logical piece of equipment. If the airflow is unobstructed, the air is being compressed to ignition temperatures, and the fuel injection is correctly metered and timed, the engine more or less has to fire. Troubleshooting an engine that won't start, therefore, boils down to finding the simplest possible procedure to establish which of these three preconditions for ignition is missing.

An Unobstructed Airflow

A failure to reach ignition temperatures or problems with the fuel supply are the most likely causes for starting failures. The airflow is the easiest to check, however, so investigate it first.

Does the engine have air? This may seem to be a stupid question, but certain engines (notably Detroit Diesel two-cycles) have an emergency shutdown device—a flap that completely closes off the air inlet to the engine and guarantees that no ignition will take place. Once the flap is activated, even if the remote operating lever is returned to its normal position, the flap will remain closed until manually reset at the engine. (Note that stopping an engine by closing the air flap will soon damage the supercharger's air seals and should be done only in an emergency, such as engine runaway—more on this later.)

If the engine doesn't have an air flap, what about the air filter? It may be plugged, especially if the engine has been operated in a dusty environment or if there are pets on board. It may have a plastic bag stuck in it or even a dead bird (which I found on one occasion). If the boat has been laid up all winter, a bird's nest may be in there. Or maybe someone placed a plastic bag over the air inlet when winterizing the engine.

Does the engine have a turbocharger? Poor oil change procedures or operating behavior (particularly racing the engine on start-up and just before shutdown—more on this later) may have caused the shaft to seize in its bearings. Remove the inlet ducting and use a finger to see if the compressor wheel spins freely.

The other side of the airflow equation is the ability to vent the exhaust overboard. Starting problems, particularly on Detroit Diesel two-cycles, may sometimes be the result of excessive back pressure in the exhaust. The most obvious cause would be a closed seacock. Other possibilities are excessive carbon buildup in the exhaust piping or in the turbocharger. In cold weather, there could be frozen water in a water-lift-type muffler, which has the same effect as a closed seacock.

Achieving Ignition Temperatures

If there is no obstruction in the airflow, perhaps the air charge is not being adequately compressed to achieve ignition temperatures. Although numerous variables may be at work here, an attempt must be made to isolate them in order to identify problems.

Troubleshooting Chart 4-3. Engine Cranks but Won't Fire **Note:** See Chart **4-2** if engine won't crank.		
Is the air flow obstructed? Check any air flaps, air filter, and exhaust seacock for blockage or closure. NO ↓	YES →	Open air flap; replace air filter element; open exhaust seacock.
Is the engine cranking slowly? Note: Stop cranking and save the battery! NO ↓	YES →	Check for: low battery, voltage drop, improper oil viscosity. Try the five methods for boosting speed listed in the text under "Cranking Speed". If slow cranking is due to cold, see below. If these fail, recharge the batteries.
Is the engine too cold? NO ↓ Check cold-start devices. If glow plugs and manifold heaters are working, the cylinder head will be noticeably warmer. Plugs can be tested by using a multimeter, or unscrewing the plug and holding it against a good ground. See text under "Cold Start Devices" for details.	YES →	Replace faulty glow plugs or manifold heaters; warm the engine, inlet manifold, fuel lines, and battery using a hair dryer, light bulb or kerosene lantern. Raise temperature slowly and evenly—concentrated heat can crack the engine castings.
Is the compression inadequate to achieve ignition temperature? NO ↓ (a) Suspect inadequate cylinder lubrication or piston blow-by. . . . (b) Suspect valve blow-by or blown cylinder head gasket. . . . (c) On Detroit Diesels, is the blower defective?	YES →	**FIX:** On engines with custom-fitted oil cups on the inlet manifold, fill cups with oil and then crank engine. On others remove air filter and squirt oil into the inlet manifold as close to the cylinders as possible *while* cranking. See "Compression" in text. Check for incorrect valve clearance or head gasket blown. **FIX:** If valves are poorly seated, a top-end overhaul is needed. **FIX:** Consult a Detroit Diesel manual.
Is the fuel level too low? Check the fuel level in the tank. NO ↓	YES →	Add fuel. It may also be necessary to bleed the fuel system (see pages 96–101).
Is the fuel delivery to the engine obstructed? NO ↓ Check to see that no kill devices are in operation; all fuel valves are open; no fuel filters are plugged; the remote throttle is actually advancing the throttle lever on the engine; and any fuel solenoid valve is functioning.	YES →	If stop or kill control has been pulled out, push it in. Check power supply to and operation of fuel shutdown solenoid valve by connecting it directly to the battery with a jumper wire. If see-through fuel filters are plugged, change filters. Open the throttle wide. Push in the fuel knob on a Detroit Diesel hydraulic governor.
Is the fuel delivery to the injectors obstructed? NO ↓ **TEST:** Open throttle wide, loosen an injector nut and crank the engine. (But not on Detroit Diesels—see text).	YES →	If no fuel spurts out, check primary, secondary, and lift pump filters and bleed the system. Check fuel lift pump for diaphragm failure. If fuel still does not flow, go back and check system for fuel level, blockages, and air leaks. Only after all else has been eliminated, suspect injection pump failure.
Note: If fluid spurts out when conducting previous test, make sure it is fuel, not water.		
If you have exhausted these tests, you can suspect incorrect timing, a worn fuel-injection pump, worn or damaged injectors.	YES →	**FIX:** Replace pump or injectors. Timing problems indicate a serious mechanical failure; correction requires a specialist.

Cold-Start Devices

The colder the ambient air, the lower its temperature when compressed, and the harder it is to get it up to ignition temperatures. As if this were not enough, cold thickens engine oil, which makes the engine crank sluggishly. Slower cranking gives the air in the cylinders more time to dissipate heat to cold engine surfaces and to escape past poorly seated valves and piston rings. In addition, a fully charged battery that puts out 100% of its rated capacity at 80°F (27°C) will put out 65% at 32°F (0°C), and only 40% at 0°F (–18°C) (see Figure 4-22).

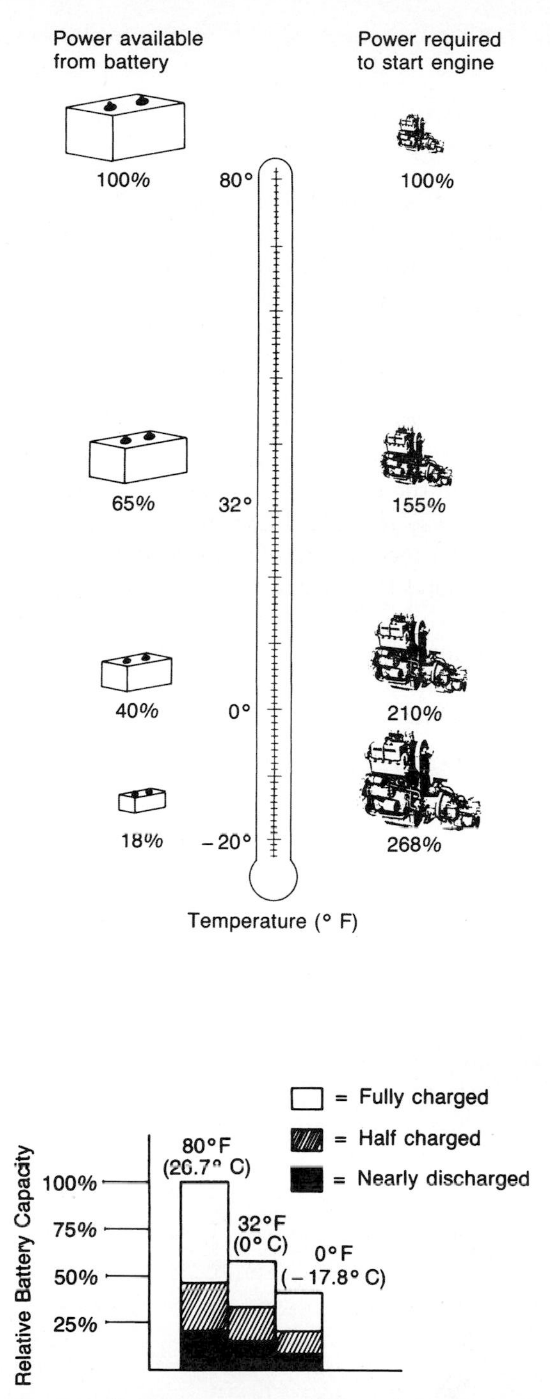

FIGURE 4-22. *As the temperature falls, a battery's starting power (cold cranking amps) declines, while the energy required to crank an engine increases. A cranking battery must have a sufficient energy reserve to deal with the worst anticipated situation (top). The effects of temperature and state of charge on battery capacity (bottom). (Courtesy Battery Council International)*

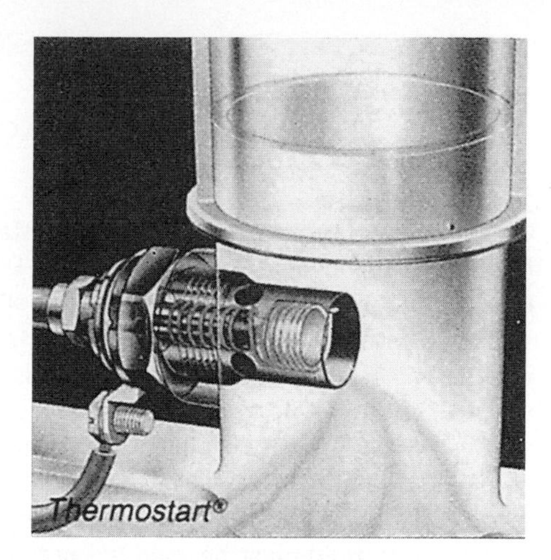

FIGURE 4-23. *A Thermostart flame primer. (Courtesy Lucas/CAV)*

Cold is a major obstacle to reliable engine starting. As a result, many engines incorporate some form of a cold-start device to boost the temperature of the air charge during initial cranking. The most common device is a *glow plug*—a small heater installed in a precombustion chamber. Glow plugs are run off the engine-cranking battery, becoming red hot when activated.

Direct combustion engines generally cannot use glow plugs because the combustion chamber doesn't have sufficient clearance above the piston crown. In this case, the incoming air charge may pass over some kind of a heater in the air-inlet manifold—perhaps a heating element or a *flame primer* (a device that ignites a diesel spray in the inlet manifold, thus warming the entering air; see Figure 4-23). Sometimes a carefully metered shot of starting fluid is used to trigger the initial combustion process (some Detroit Diesels). Note that starting fluid should not be used in most instances (see the Starting Fluid and WD-40 sidebar).

If glow plugs and manifold heaters are working, the cylinder head or manifold will be noticeably warmer near the individual heating devices. If they are not working, first check the wiring in the circuit. There is frequently a solenoid activated by the ignition switch, or a preheat switch, in the same way that a starter motor solenoid is activated by the ignition switch. You can apply the same kinds of circuit tests to this circuit as to a starting circuit (see the Starter Motor Does Not Crank section above, but substitute "glow plugs" for "starter motor"; see Figure 4-24).

Starting Fluid and WD-40

Unless specific provision for the use of starting fluid is made on an engine (e.g., some Detroit Diesels and Caterpillars), do not use it at all. It is sucked in with the air charge and being extremely volatile, will often ignite before the piston is at the top of its compression stroke. This can result in serious damage to pistons and connecting rods. For some reason, Detroit Diesels seem to tolerate starting fluid better than four-cycle diesels.

If starting fluid must be used, do not spray it directly into the air-inlet manifold. Rather, spray it onto a rag and hold this to the air intake. This will control the amount being drawn in. Adding a little diesel to the inlet manifold will also help the engine to pick up once the starting fluid fires.

Diesels will run on WD-40 (don't ask me why!) at far less risk of premature detonation than that possible with starting fluid. In fact, if the battery is low but it is necessary to do some extended cranking—such as when having problems purging a fuel system—you can open the throttle wide and spray a continuous stream of WD-40 into the air inlet while someone else cranks. The engine will fire and continue to run as long as the spray is maintained. The engine speed can be controlled by varying the rate of spray until the fuel system is bled and the engine takes over. This is also an effective way to get a drowned engine running again, after the water has been driven out of the cylinders and the oil and filter changed. Note, however, that WD-40 should never be used in conjunction with manifold heating devices since it may ignite in the manifold and blow back, causing a fire.

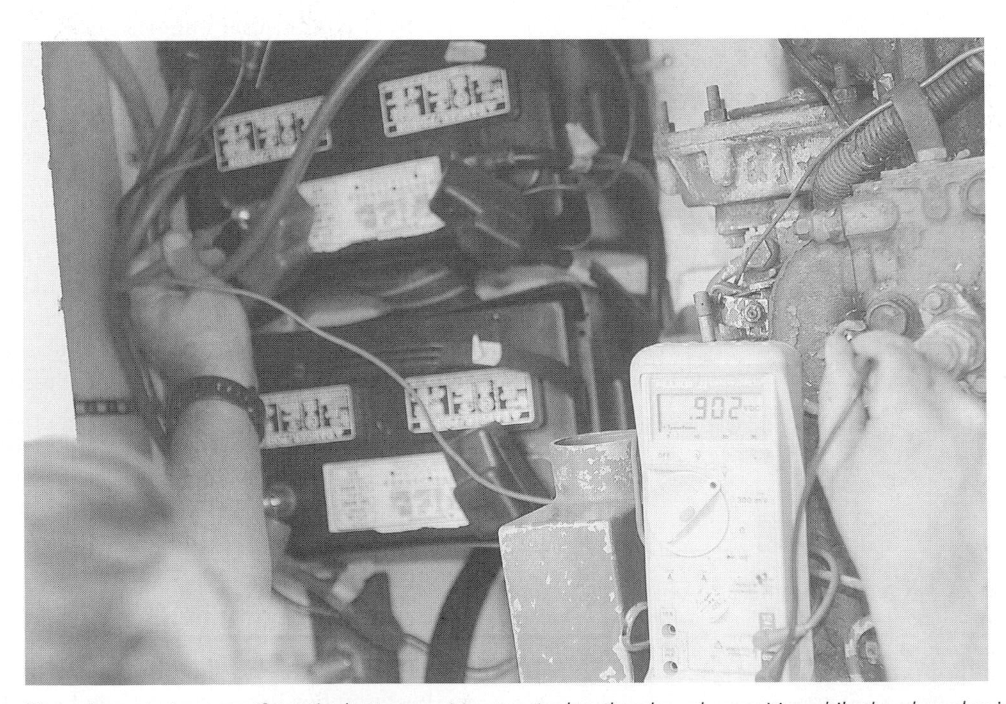

Figure 4-24. *Glow plug volt drop test (from the battery positive terminal to the glow plug positive while the glow plug is activated). Although this 0.9 volt drop is not unusual, it is on the high side and results from poor connections and undersized wiring.*

Check glow plugs with an ammeter (they should draw about 5 to 6 amps per plug on a 12-volt system) or an ohmmeter (around 1.5 ohms per plug). To test the amp draw in the absence of a suitable clamp-on DC ammeter, place a DC ammeter in the power supply line between the main hot wire and each glow plug in turn (not in the main harness itself, since this may carry up to 40 amps on a six-cylinder engine). Test resistances by disconnecting the hot wire from each glow plug and checking from the hot terminal on the plug to a ground, using the most sensitive ohms scale on the meter (see Figure 4-25). Only a good ohmmeter will be accurate enough to distinguish between a functioning glow plug and a shorted plug.

You can unscrew a glow plug, hold it against a good ground (the engine block), and turn it on. It should glow red hot (see Figure 4-26; be careful not to burn yourself). However, if the plug is seized in the cylinder head, soak it in penetrating fluid and leave it for a while before trying to get it out—it is not uncommon for heavy-handed mechanics to break them off!

If a flame primer is not working, check its electrical connections and fuel supply. If the unit has been removed, test the heating coil by jumping it from the battery. Also check the fuel discharge and atomization, but not at the same time since this may result in an uncontrolled flare-up.

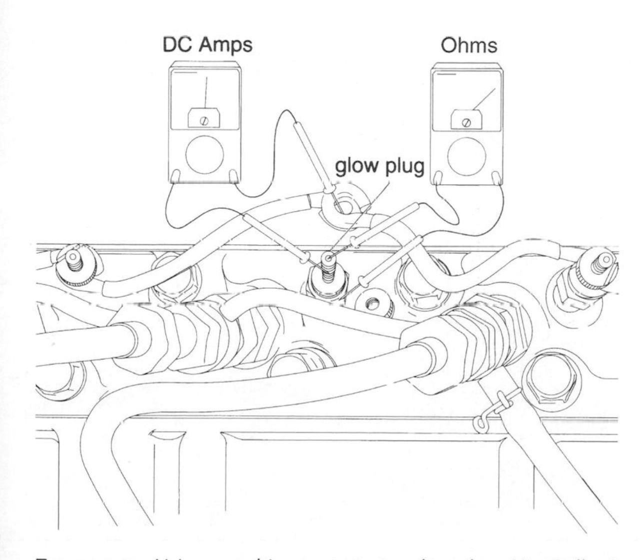

FIGURE 4-25. *Using a multimeter to test a glow plug. (Jim Sollers)*

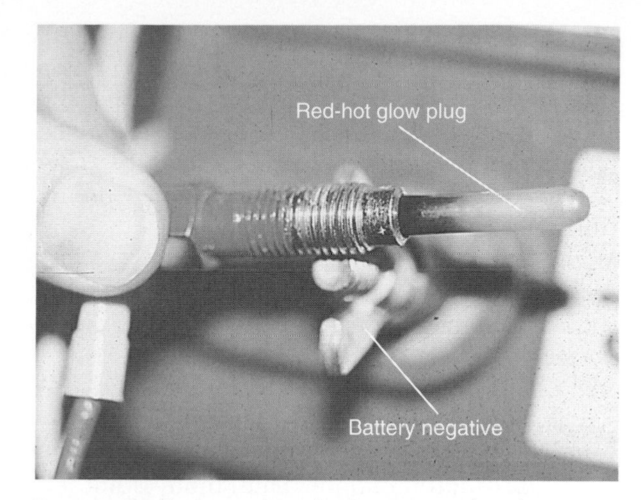

FIGURE 4-26. *Testing a glow plug by wiring it directly across a battery. The plug can be held at the top, but the probe should not be touched!*

Any safe means used to boost the temperature of the engine, battery, and inlet air will help with difficult starting. This includes using a hair dryer, lightbulb, or kerosene lantern to warm fuel lines, filters, manifolds, and incoming air; removing oil and water, warming them on the galley stove, and returning them; heating battery compartments with a lightbulb; or removing the battery, putting it in a heated crew compartment, and returning it once it is warm.

A propane torch flame can also be used to boost temperatures. Gently play the flame over the inlet manifold and fuel lines, and across the air inlet when the engine is cranked so that it heats the incoming air. Never use a torch in the presence of gasoline or propane vapors. Do not play the torch flame over electrical harnesses, plastic fuel lines and fittings, or other combustibles!

Raise temperatures slowly and evenly, playing the heat source over a broad area. Concentrated heat may crack an engine's castings. Boiling water or very hot oil may do the same.

COMPRESSION

When an engine is operating, the lubrication system maintains a fine film of oil on the cylinder walls and the sides of the pistons. This oil plays an important part in maintaining the seal of the piston rings on the cylinder walls. After an engine is shut down, the oil slowly drains back to the crankcase. An engine that has been shut down for a long period of time may suffer a considerable amount of *blowby* when an attempt is made to restart it because the lack of oil on the cylinder walls and piston rings reduces the seal, and therefore the compression.

An engine that grows harder to start over time is probably also losing compression, but this time because of poorly seated valves and piston-ring wear. (Note that a Detroit Diesel two-cycle exhibits the same symptoms if the blower is defective.) The air in the cylinders is not compressed enough to produce ignition temperatures.

Short of a top-end overhaul, there is not a lot that can be done about valve blowby. Piston blowby and a loss of cylinder lubrication, however, can be cured temporarily by adding a little oil to the cylinders. The oil dribbles down and settles on the piston rings, sealing them against the cylinder walls.

If you have plenty of battery reserve, set the throttle wide open and crank for a few seconds. Then let the engine rest for a minute. This allows three things to happen: (1) the injected diesel will dribble down onto the piston rings; (2) the initial heat of compression will take the chill off the cylinders; and (3) the battery will catch its breath. Then try cranking again.

If the engine still won't start, or if little battery reserve remains, introduce a small amount of oil directly into the engine cylinders.

Adding Oil to the Cylinders

On engines with custom-fitted oil cups on the inlet manifold (for example, many older Sabbs—see Figure 4-27), fill the cups with oil, then crank the engine. On others, remove the air filter and squirt oil into the inlet manifold as close to the cylinders as possible, while cranking the engine (see Figure 4-28). The oil will be sucked in when the engine cranks. Let the engine sit for a minute or two to allow the oil to settle on the piston rings. After starting, the engine will smoke abominably for a few seconds as it burns off the oil—this is OK. Put the air filter back in place as soon as the engine fires.

When applying oil to the cylinders, use only a couple of squirts in each cylinder. Oil is noncompressible so too much will damage the piston rings and connecting rods. Also, keep the oil can clear of turbocharger blades—a touch of the can will result in expensive damage.

Oil used in this fashion is often a magic—albeit a temporary—cure for poor starting, but the engine needs attention to solve the compression problem. To isolate the problem further, inject some oil into each cylinder in turn, then crank the engine with the throttle closed so that it does not start. If you notice a marked improvement in compression on any cylinder, then you know that one is suffering blowby around the piston and rings. If compression does not improve, suspect the valves.

FIGURE 4-27. *Putting oil into an oil cup fitted to the air-inlet manifold of a Sabb 2JZ diesel.*

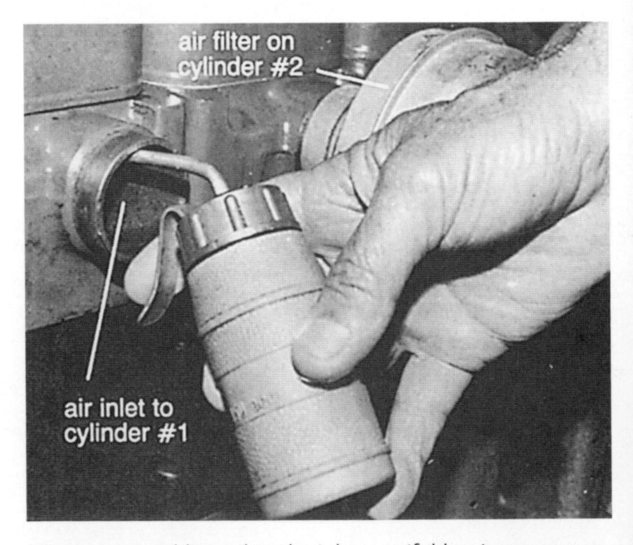

FIGURE 4-28. *Adding oil to the inlet manifold to increase combustion chamber pressure.*

A hand crank on the engine (or socket wrench on the crankshaft nut) helps in performing these tests because you can slowly turn the crankshaft and rock each piston against compression. As each piston comes up to compression, you will feel the crank handle try to bounce back. Without the use of decompression levers, you should not be able to hand crank a healthy engine through compression at slow speeds. If you can crank it through, you have considerable blowby. You can frequently tell if the valves are the culprits by listening for a hiss of escaping air.

Additional Compression Problems

Carbon buildup on valve stems, especially from prolonged low-load and low-temperature running (such as occurs when battery charging at anchor) will occasionally cause a valve to stick in its guide in the open position. This should not happen if you follow proper oil change procedures, but if tests indicate valve blowby, remove the valve cover (rocker cover) and observe the valve stems for correct position. If work has recently been performed on the cylinder head or valves, it is also possible that one of the valve clearances has been improperly set (see Chapter 7), which can hold that valve in a permanently open position.

A loss of compression can also be caused by a blown, or leaking, cylinder head gasket. The symptoms will resemble those of leaking valves, and because the head must be removed to sort out the valves, the gasket problem will become evident right away. Unless the head has been recently removed, there is little reason to suspect the gasket.

Engine wear, especially in the piston pin and rod end bearings, eventually may increase the cylinder head clearance to the point at which adequate compression cannot be achieved. Before this point is ever reached, the engine will give advance warning by knocking (see Chapter 5). A major engine rebuild will be needed.

Finally, it is possible for the operator of an engine fitted with decompression levers to leave these levers in the decompressed position, which ensures that the cylinder has no compression—silly, but worth checking before you take more drastic action. If an engine is shut down by using the decompression levers instead of closing the fuel rack on the injection pump, serious damage may occur to the valves and push rods. If you suspect this might have happened, remove the valve cover and check for bent valve stems or push rods.

Cranking Speed

No diesel will start without a brisk cranking speed (at least 60 to 80 rpm; most small diesels will crank at 200 to 300 rpm; see Figure 4-29). The engine, especially when it is cold, just will not attain a high enough compression temperature to ignite the injected diesel. If a motor turns over sluggishly, stop cranking and save the battery.

Check the battery's state of charge. If it is fully charged, check for voltage drop (which may be robbing the starter motor of power) between the battery and the starter motor. Assuming a good battery and a properly functioning starting circuit, the techniques in the Cold-Start Devices and Compression sections (see above) will help generate that first vital power stroke. Sometimes the following tricks will also boost cranking speeds:

- If fitted with decompression levers and a hand crank, turn the engine over a few times by hand to break the grip of the cold oil on the bearings. Assist the starter motor by hand cranking until the engine gains momentum, then knock down the decompression levers.
- Disconnect all belt-driven auxiliary equipment (refrigeration compressor, pumps, etc.) to reduce the starting load. Note that if a small engine is loaded up with heavy-duty auxiliary equipment (e.g., a high-output alternator), it is important to be able to switch off this equipment while

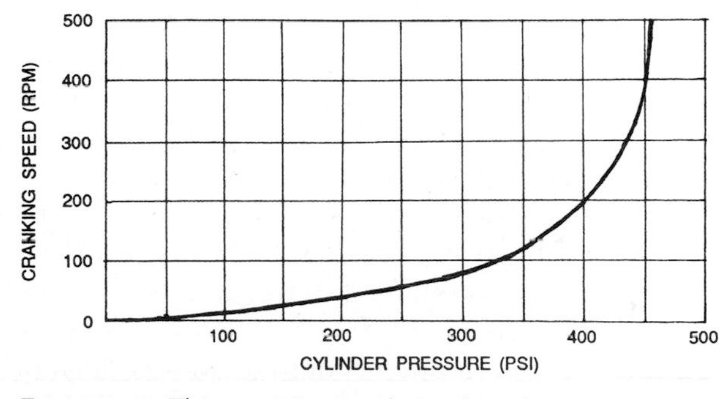

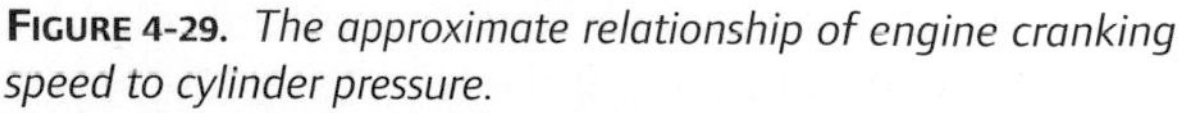
FIGURE 4-29. *The approximate relationship of engine cranking speed to cylinder pressure.*

cranking. Some multistep voltage regulators do this by putting in a time delay before the alternator is switched on, and then slowly ramping up the alternator output. But if the engine ignition switch has been left on between several cranking attempts, the time delay will have run out and the full alternator load will be present. You can reset the time delay by turning off the ignition switch.

- Place a hand over the air inlet and then crank. Restricting the airflow will reduce compression and help the engine build up speed. Once the engine is cranking smartly, remove your hand—the motor should fire. *Never block the air inlet on an operating engine*—the high suction pressures generated may damage the engine and cause injury.
- An additional trick for sailboats with a manual transmission and fixed-blade propeller is to sail the boat hard in neutral with the propeller freewheeling, then start cranking and throw the transmission into forward. The additional momentum of the propeller may bump-start the engine.

Fuel Problems

If an engine is cranking smartly, with sufficient compression to produce ignition temperatures, but still won't start, the culprit is almost certainly the fuel system. This has the potential for causing a considerable number of problems. Some are easy to check, but others can only be guessed at.

In the sections that follow, I address conventional fuel injection systems with mechanically controlled jerk or distributor injection pumps. *Electronic unit injectors, common rail systems, and most electronic control devices are basically off-limits to all but professionals.*

Check the Obvious

Diesel engines are shut down by closing off the fuel supply. On some engines this occurs when the throttle is closed; others continue to idle at minimum throttle settings and have a separate mechanical Stop control to shut off the remaining fuel supply. Has the Stop control inadvertently been left pulled out? Perhaps there is a kink in the Stop control cable so that even when the knob is pushed all the way in, the fuel supply is still closed off. Be sure to check the operation of the cable *at the injection pump*.

Is the throttle open to the position specified for starting by the engine manufacturer? If in doubt, open the throttle wide—many diesel engines will never start with the throttle closed. Trace the throttle cable from the throttle lever in the cockpit to the engine and make sure that it is actually advancing the throttle lever on the engine. On a fly-by-wire boat, have someone operate the throttle and make sure the actuator at the engine is moving the appropriate lever or rod.

Has an emergency shutdown device, such as the air flap on a Detroit Diesel two-cycle, been inadvertently tripped? Note that a Detroit Diesel with a hydraulic governor must be cranked for a few seconds to generate enough oil pressure in the governor to open the fuel rack so the engine can start. On the outside of the governor, you will see a rod with a knob on the end of it. If this is pushed in by hand and held while cranking, it will bypass the governor, opening the fuel rack.

Is there plenty of fuel in the tank? The fuel suction line is probably set an inch or two off the bottom of the tank; if the boat is heeling, air can be sucked in even when the fuel level appears to be adequate. Is the fuel valve (if fitted) open? If there is fuel and the valve is open, but no fuel is reaching the engine, is there a small filter screen inside the tank on the suction line? If so, it may have become plugged. If such a screen is fitted, throw it away; this is not the proper place to be filtering your fuel—that's why you have fuel filters. (Sometimes it is extremely difficult to get at this filter screen. You may be able to force it off the end of the pickup line, or punch big holes in it, by poking a piece of doweling or a sturdy piece of wire down the suction line into the tank.)

On many boats, the fuel tank is below the level of the engine. If air can get into the fuel system, the fuel may drain back down to the tank, especially during a long layup. To restore the fuel supply, you may need to bleed the fuel system (see below), and you'll need to locate and correct the source of the leak (see the Persistent Air in the Fuel Supply section below).

Most older engines have mechanical lift pumps whereas newer ones tend to have electric ones. With the latter, you should hear a quiet clicking when the ignition is first turned on. If you are in doubt about an electric pump's operation, loosen the discharge line and see if fuel flows.

SOLENOID VALVE

Many newer engines have a solenoid-operated fuel shutdown valve that is held in the closed position by a spring when the ignition is turned off. When the ignition is turned on, it energizes an electromagnet that opens the valve. Anytime the electrical supply to the solenoid is interrupted, the magnet is deenergized and the spring closes the valve. Any failure in the electrical circuit to the valve will automatically shut off the fuel supply to the engine.

Some solenoid fuel valves are built into the back of the fuel injection pump—they are identified by a couple of wires coming off the pump close to the fuel inlet line. Others are mounted separately but close to the pump (see Figure 4-30). A rod coming from the back of the valve actuates a lever on the pump. You can check the operation of a solenoid valve by connecting it directly to the battery with a jumper wire. Take care to get the positive and negative leads the right way around. If the valve has only one wire, this is the positive lead; if it has two, one will run to ground—you want the other one. You can also test from the solenoid positive cable to ground with a multimeter set in the ohms mode. However, first ensure the solenoid is electrically isolated, or else you will blow the fuse in your meter. A healthy solenoid will show some resistance, generally around 20 ohms. No resistance indicates a shorted solenoid; infinite resistance a burned-out solenoid.

FUEL FILTERS

The primary fuel filter should have a see-through bowl that should be checked for water and sediment (see Figure 4-31). If the bowl is opaque, open the drain on its base and take a sample. On many fuel systems, this will let air into the system, so you may also need to bleed the system (see below). It is not uncommon for a primary fuel filter element to be completely plugged. If this is the case, don't take chances—replace it along with the secondary filter element and drain the tank or pump it down until all traces of contamination are removed. If filters repeatedly clog, it is likely that sediment has built

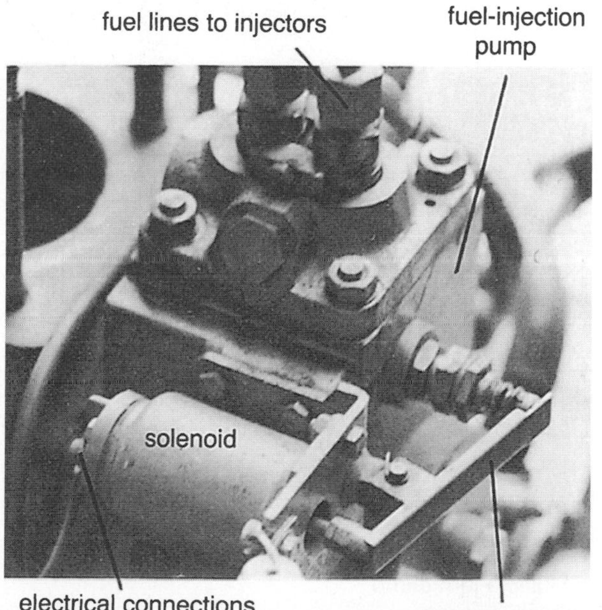

FIGURE 4-30. *A solenoid-operated fuel shutdown valve.*

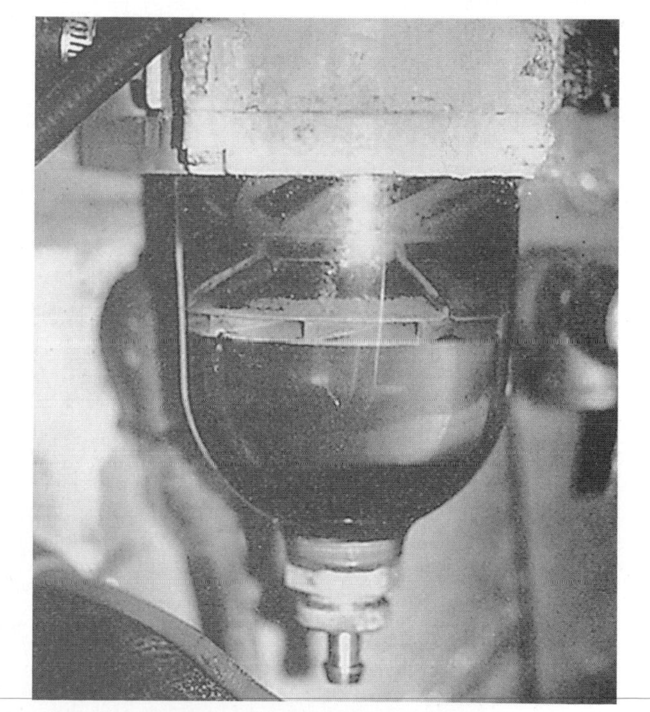

FIGURE 4-31. *A primary fuel filter with a see-through bowl. The sediment is a sure sign that the tank needs flushing.*

FIGURE 4-32. *The fuel filter inside the top of a lift pump.*

up inside the tank to the level of the fuel pickup. The tank needs opening and flushing.

If there is no primary filter, or there is any sign of contamination making its way past the primary filter, check the filter screen that is generally found inside diaphragm-type lift pumps. You can reach it by undoing the center bolt and removing the cover (see Figure 4-32).

Assuming the tank has fuel and the filters are clean, the next step is to find out if the fuel lines have air in them.

BLEEDING (PURGING) A FUEL SYSTEM

Air trapped in the fuel system can bring some diesels to a halt, although the extent to which this is true varies markedly from one engine to another. Many newer engines, together with all engines with unit injectors (including Detroit Diesel two-cycles) and those with common rail systems, can be purged of air simply by opening the throttle wide and cranking for a long enough time. Many older diesels, especially those with jerk-type fuel injection pumps, cannot—they will be completely disabled by even tiny amounts of air. Engines with distributor-type fuel injection pumps tend to fall somewhere between these two extremes.

When air has to be purged from a fuel system by hand, the process is known as *bleeding*.

Tracing the Fuel Lines

With the exception of unit injectors and common rail systems, typical fuel systems are as shown in Figures 4-33 and 4-34. Fuel is drawn from the tank by a lift pump (sometimes called a feed pump) and passes through the primary filter. The lift pump pushes the fuel on at low pressure through the secondary filter to the injection pump. (On some engines, the lift pump is incorporated into the back of the fuel injection pump rather than being a separate component.) The injection pump meters the

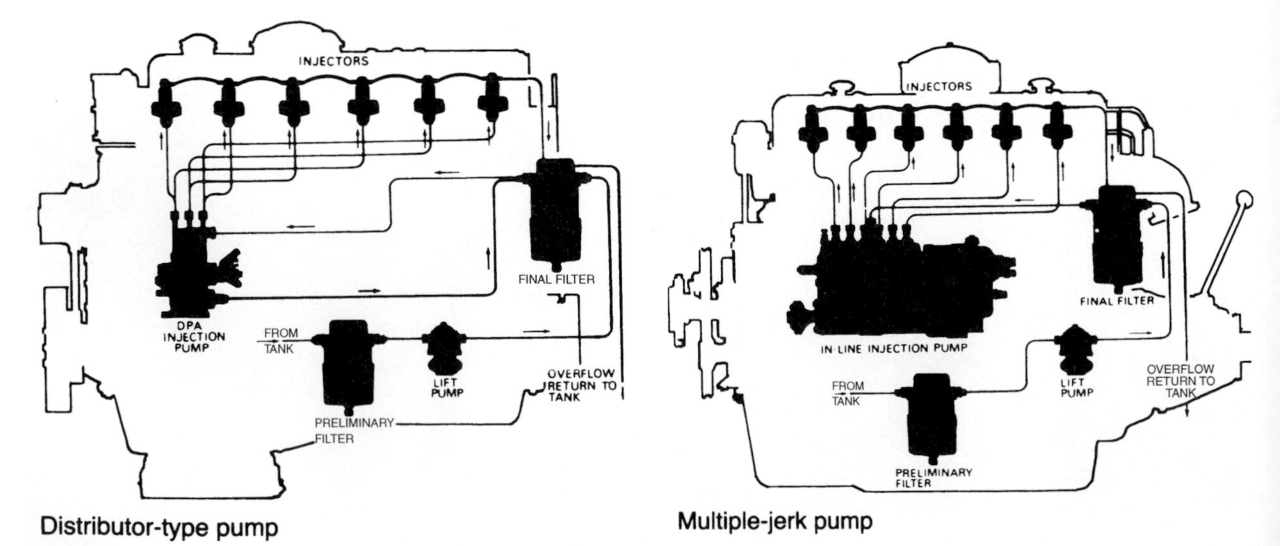

FIGURES 4-33 and 4-34. *Fuel system schematics. Note the two different types of fuel injection pumps in common use on traditional four-cycle diesel engines—distributor type and jerk type. They are easily distinguished since the fuel lines are arranged in a circle on the former and in a straight line on the latter. (Courtesy Lucas/CAV)*

fuel and pumps exact amounts at precise times and at very high pressures down the injection (delivery) lines to the injectors and into the cylinders. Any surplus fuel at the injectors returns to the secondary filter or tank via leak-off, or return, pipes. There may also be a leak-off line from the injection pump to the secondary filter (see Figure 4-35).

The more cylinders an engine has, the greater the number of fuel lines. The various filters and lines can sometimes become a little confusing. Just remember that the secondary fuel filter is generally mounted on the engine close to the fuel injection pump, whereas the primary filter is generally mounted off the engine, or on the engine bed, closer to the fuel tank. The filters should have an arrow on them to indicate the direction of fuel flow; sometimes the ports will be marked IN and OUT.

An injection line (delivery pipe) runs from the injection pump to every injector, but the leak-off pipes go from one injector to the next, then down a common pipe to the secondary filter or fuel tank. This makes it easy to distinguish delivery pipes and leak-off pipes on most engines. A few engines, however (notably many Caterpillars), have internal fuel lines and injectors that are hidden by the valve cover. In this case, each delivery pipe runs from the fuel injection pump to a fitting on the side of the cylinder head, and from there all is hidden (see Figures 4-36 and 4-37).

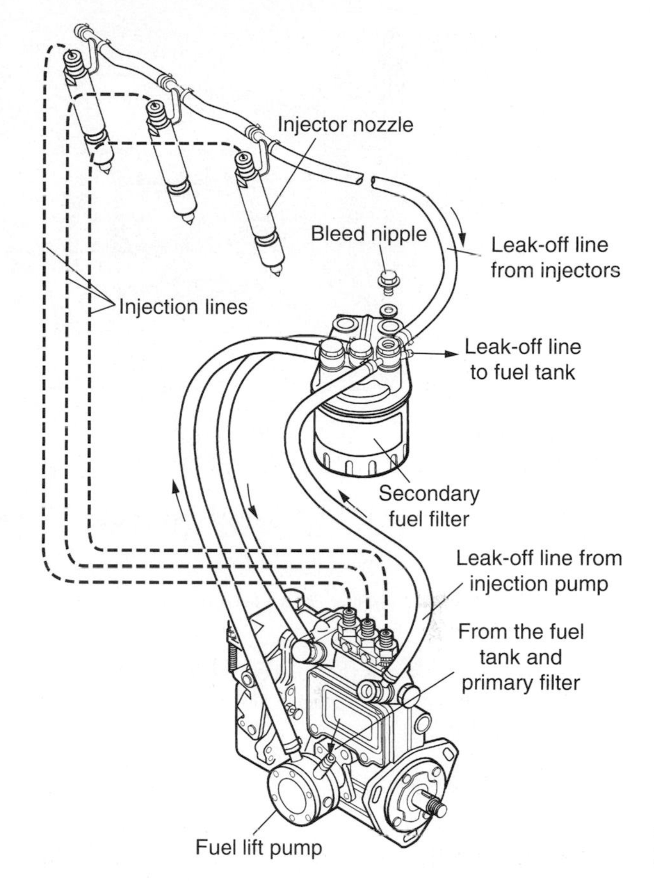

FIGURE 4-35. *Fuel lines on a three-cylinder diesel engine using a jerk-type injection pump with an attached lift pump. (Fritz Seegers)*

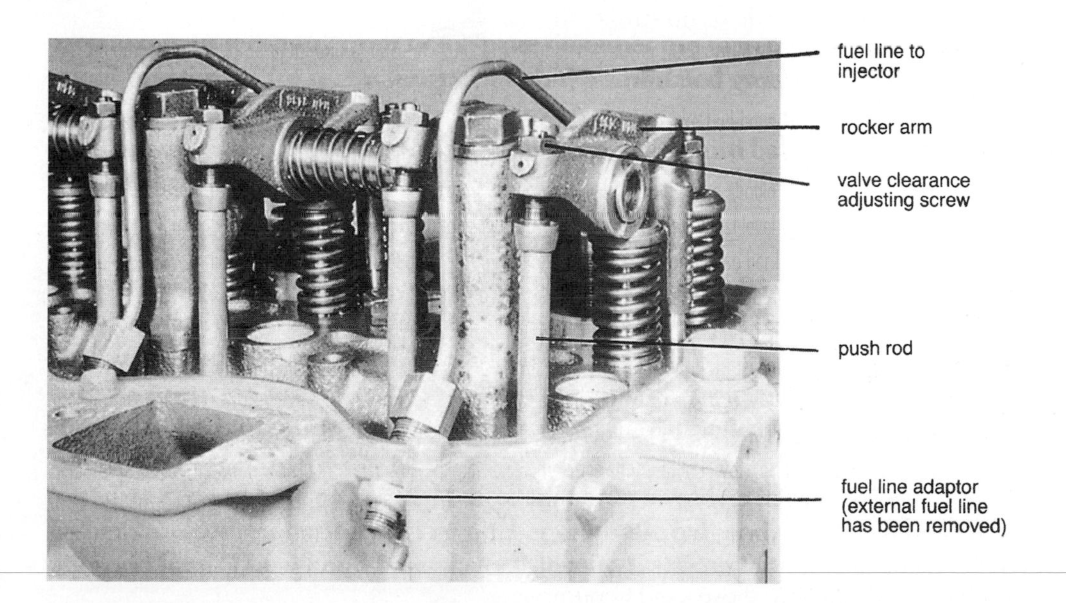

FIGURE 4-36. *Internal fuel lines. Shown here are the fuel lines inside a valve cover and their point of entry into the engine. (Courtesy Caterpillar)*

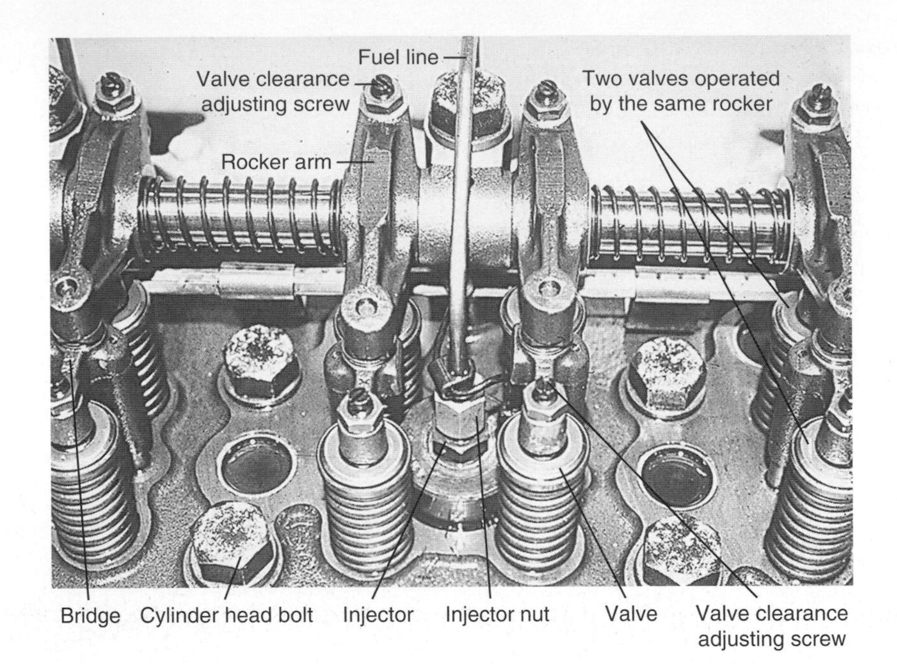

FIGURE 4-37. *Shown here is the point at which internal fuel lines connect to the injectors. This high-performance diesel has two inlet and two exhaust valves per cylinder. The rocker arm opens each set of valves via a bridge. (Courtesy Caterpillar)*

At various points in the system are bleed nipples—normally on the filters and the injection pump; there should be one on the top of the secondary filter. On the base of a mechanical lift pump is usually a small handle, enabling it to be operated manually (see Figures 4-38 and 4-39). Pump this handle up and down. If it has little or no stroke, the engine has stopped with the lift pump drive cam at or near the full stroke position. Spin the engine a half turn or so to free the manual action. (See the Lift [Feed] Pump Failure section below for a more complete explanation of this.) Electric lift pumps are activated by turning on the ignition.

Engines that do not have an external mechanical lift pump generally have a manual pump attached to the injection pump, to one of the filters, or at some other convenient point in the system (see Figure 4-40). Some boatowners like to add an electric fuel pump in the fuel line between the tank and the primary filter to make bleeding easy. However, if this is done, it is important to ensure that it does not create undue friction when the engine is running (the engine lift pump will now have to suck through whatever resistance this pump creates). As noted earlier, manual pumps similar to those used to pump fuel through outboard motor lines are also available, specially constructed for use with diesel fuel.

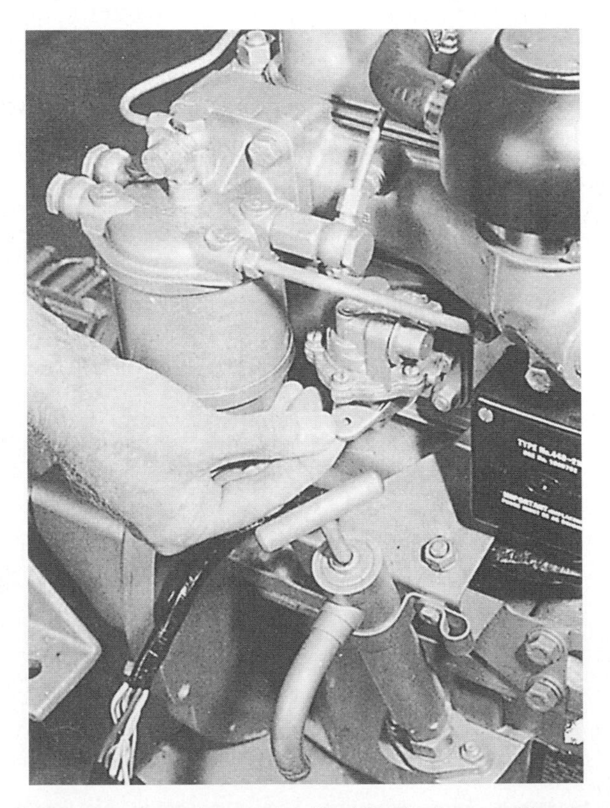

FIGURE 4-38. *Operating a manual fuel lift (feed) pump. (Courtesy Perkins Engines Ltd.)*

FIGURE 4-39. *Manual operation of a lift pump on a Yanmar.*

FIGURE 4-40. *A manual fuel pump mounted on a fuel filter.*

In all cases, bleeding follows the same procedure as that used with a lift pump.

Bleeding Low-Pressure Lines

Open the bleed nipple on the secondary filter and operate the lift pump (refer back to Figure 4-35 and see Figure 4-41). Some bleed nipples have a hose barb; if this is the case, push a piece of hose on the barb and place the other end of the hose in a container to catch the vented diesel. If the filter has no bleed nipple, loosen the connection on the fuel line coming out of the filter. Fuel should flow out of the bleed nipple or loosened connection completely free of air bubbles.

If bubbles are present, operate the lift pump until they clear, then close the nipple or tighten the connection. This process should purge the air from the suction lines all the way back to the tank, including both filters. If any of the fuel lines have a high spot, however, a bubble of air may remain at this point and be extremely hard to dislodge. (Note: If the manual lever strokes but then fails to return to its original position, its internal spring is broken. This is the cause of the fuel supply problems; see pages 103–5 for disassembly and repair procedures.)

Take the trouble to catch or mop up all vented fuel. Diesel will soften, and eventually destroy, most electric cable insulation and also the rubber feet on flexibly mounted engines.

The next step is to bleed the fuel injection pump. Somewhere on the pump body you will find one or perhaps two bleed nipples. (Some modern pumps are self-bleeding and have no nipples.) If the pump has more than one nipple, open the low one first and operate the lift pump until the fuel that flows out is free of air bubbles (see Figures 4-42, 4-43, and 4-44). Close the nipple and repeat the procedure with the higher one. The injection pump is now bled.

Bleeding High-Pressure Lines

Bleeding the fuel lines from the injection pump to the injectors is the final step. To do this, set the governor control (throttle) wide open (this is essential) and

Figure 4-41. *Bleeding a secondary fuel filter. (Courtesy Perkins Engines Ltd.)*

Figure 4-43. *Bleeding the lower nipple on a distributor-type fuel injection pump. (Courtesy Perkins Engines Ltd.)*

crank the engine so that the injection pump can move the fuel up to the injectors. This should take no more than 15 seconds. In any event, do not crank the starter motor for longer than 15 seconds at any one time because serious damage can result from internal overheating. If the engine has decompression levers and a hand crank, turn it over by hand to avoid running down the battery.

If the engine has no hand crank, you must properly bleed the system to the injection pump before

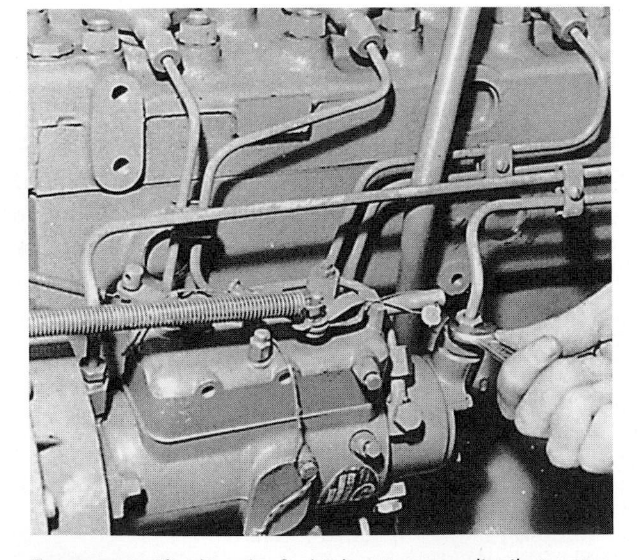

Figure 4-42. *Bleeding the fuel inlet pipe to a distributor-type injection pump. (Courtesy Perkins Engines Ltd.)*

Figure 4-44. *Bleeding the upper nipple on a distributor-type fuel injection pump. (Courtesy Perkins Engines Ltd.)*

bleeding the high-pressure lines. Frequently the battery is already low because of earlier cranking attempts; therefore, pumping up the injectors one time, let alone having to come back to try again if rebleeding is necessary, will be touch and go.

When the fuel eventually reaches the injectors, provided the engine is not running, you can hear a distinct *creak* at the moment of injection. Familiarize yourself with this noise. If it is present when you attempt to crank the engine, at least you know that fuel is reaching the engine and you won't have to bleed it (unless you are injecting water!).

If the engine still will not fire, loosen one of the injector nuts (these hold the delivery lines to the injectors—see Figures 4-45, 4-46, and 4-47) and crank again. A tiny dribble of fuel, free of air, should spurt out of this connection at every injection stroke for this cylinder (every second engine revolution on a four-cycle engine). If there is no fuel, or there is air in the fuel, the bleeding process has not been done adequately and needs repeating.

Note: DO NOT LOOSEN THE INJECTOR NUTS ON A COMMON RAIL SYSTEM. It is, in any case, unnecessary since such systems are self-bleeding (see below).

On engines with internal fuel lines, loosen the nuts on the delivery pipes where they enter the cylinder head, not the nuts on the injectors, so that none of the vented diesel gets into the engine oil (refer back to Figure 4-36).

FIGURE 4-45. *The bleed points on a Volvo MD 17C.*

Do not overtighten injector nuts because you may collapse the fitting that seals the delivery pipe to the injector (tighten the nuts to 15 foot-pounds if you don't know the manufacturer's recommended setting.) Anytime you loosen an injector nut, you must check for leaks once the engine is running. Fuel leaks on engines with internal fuel lines will drain into the crankcase, diluting the engine oil and possibly leading to engine seizure.

BLEEDING UNIT INJECTORS (INCLUDING TWO-CYCLE DETROIT DIESELS) AND COMMON RAIL SYSTEMS

These systems are self-purging. As long as the tank has fuel in it, the suction line is free of breaks, and

FIGURE 4-46. *Injectors on a Volvo MD 17C.*

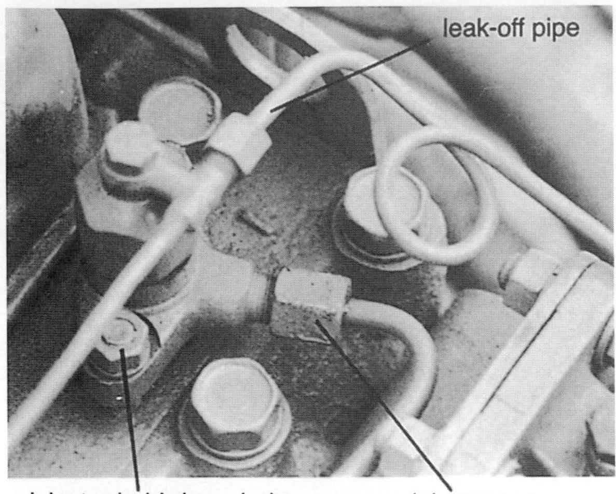

FIGURE 4-47. *The location of the injector nut.*

the fuel pump works, the diesel flowing through the system will drive out any air. To check the fuel flow, undo the return line from the cylinder head to the tank, then crank the engine; a steady flow should come out (not the little dribbles that come from jerk pumps and distributor pumps).

PERSISTENT AIR IN THE FUEL SUPPLY

One of the more aggravating problems on many traditional four-cycle diesels is persistent air in the fuel system. Air can come from poor connections, improperly seated filter housings (especially if the problem occurs after a filter change), and pinholes in fuel lines caused by corrosion and vibration against bulkheads or the engine block. If the boat has more than one fuel tank and a selector valve, the valve may be admitting air. Since the only part of the fuel system under suction pressure is that from the tank to the lift pump, this is the most likely problem area.

If the primary filter has a see-through bowl, loosen the bleed nipple on the secondary filter, operate the hand pump on the lift pump, and watch the bowl. Air bubbles indicate a leak between the fuel tank and the primary filter or in the filter gasket itself. No air means the leak is probably between the filter and the pump.

In the absence of a see-through bowl, locate the fuel line that runs from the lift pump to the secondary filter. Disconnect it at the filter. Place it in a jar of clean diesel fuel and pump. If there is a leak on the suction side, bubbles will appear in the jar.

Any air source on the lift pump's discharge side should reveal itself as a fuel leak when the engine is running. When the engine is shut down, the fuel may suck in air as it siphons back to the tank. Next time the engine is cranked, it will probably start and then die.

A similar problem can arise when the leak-off pipe from the injectors is teed into a fitting on the secondary fuel filter with another (overflow) line running from here back to the tank (this is a common arrangement; there may also be a leak-off line from the injection pump connected to the filter—refer back to Figure 4-35). When the engine is shut down, fuel will sometimes siphon down the overflow line and cause air to be sucked into the system. In this situation, there is no external evidence of the air source, making for some frustrating detective work! To cure the problem, either move the leak-off pipe so that it runs directly to the fuel tank (in which case, if there is more than one fuel tank, the fuel must be directed to the one in use or the pipe will slowly transfer fuel from the operating tank to the second tank until the first runs dry or the second overflows) or add a length of flexible fuel line between the injectors and the secondary filter. Loop this above the level of the filter and injectors and drill a small

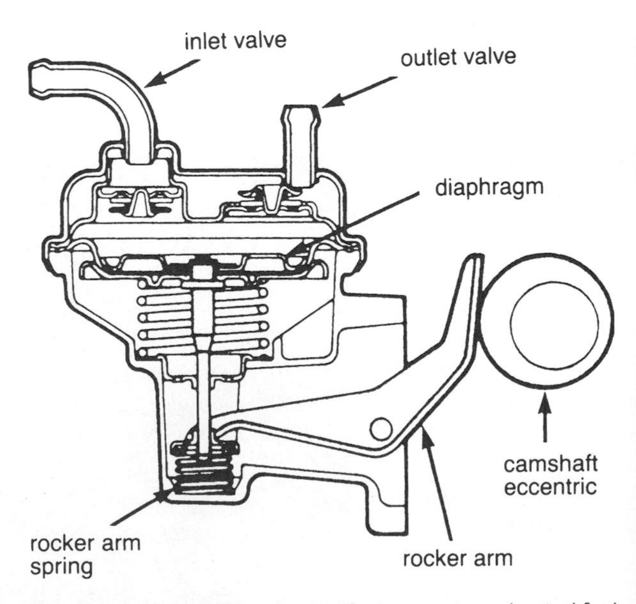

FIGURE 4-48. *A typical mechanical fuel pump. A mechanical fuel pump is mechanically actuated by a rocker arm or push rod without electrical assistance. (Courtesy AC Spark Plug Division, General Motors Corp.)*

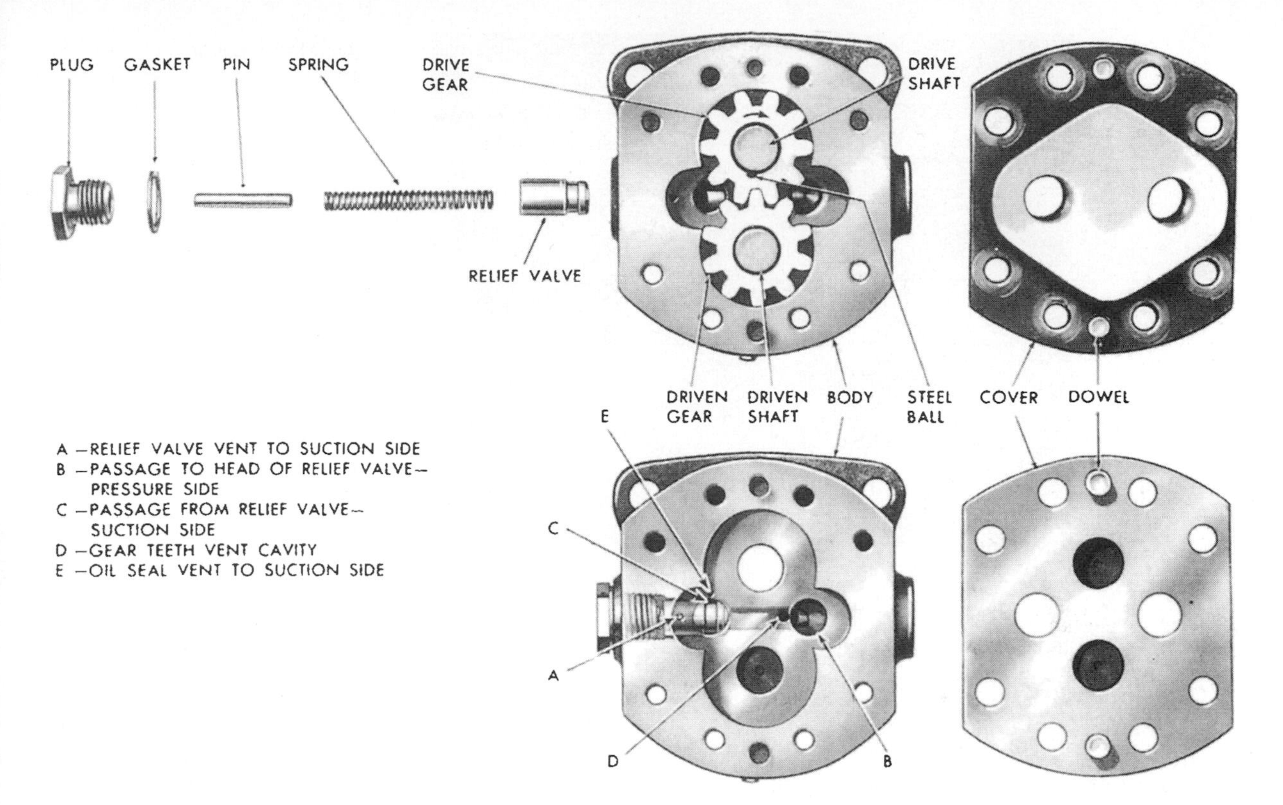

FIGURE 4-49. *A breakdown of a gear-type lift pump showing how the various parts fit together. (Courtesy Detroit Diesel)*

($^{1}/_{16}$ inch) hole in it at its highest point. This hole will act as a siphon break.

Sometimes fuel tanks are deeper than the lifting capability of the pump so the pump may fail to raise the fuel when the tank is almost empty, or it may fail to raise enough fuel (fuel starvation) at higher engine loadings. If the fuel pickup line has a filter on it inside the tank, and this becomes plugged, it will have the same effect; as previously mentioned, this is no place to filter the fuel. If the tank vent is plugged, over time the tank will develop a vacuum and once again starve the engine of fuel. When troubleshooting, the tank vent invariably gets forgotten, but if your engine starts and runs fine for some time, then begins to lose power, suspect the vent.

LIFT (FEED) PUMP FAILURE

Many small four-cycle diesels found in boats use a diaphragm-type lift pump (see Figure 4-48). Newer engines tend to use electric pumps. Larger engines with unit injectors (including Detroit Diesel two-cycles) and common rail systems use a gear-driven pump (see Figures 4-49 and 4-50).

A diaphragm pump has a housing that contains a suction and a discharge valve, plus the diaphragm. A lever, which is moved up and down by a cam on the engine's camshaft or crankshaft, pushes the

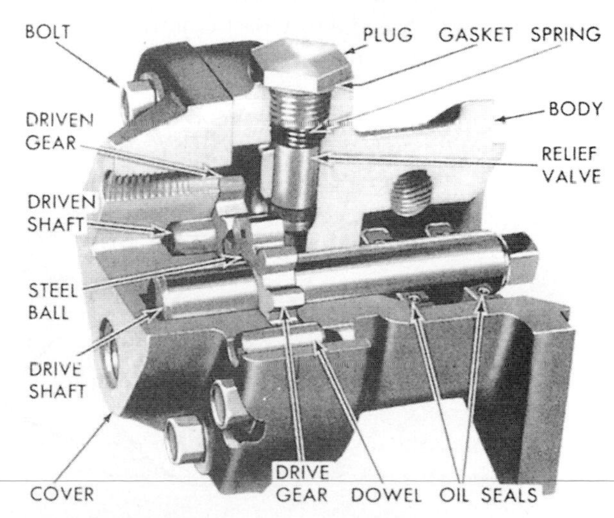

FIGURE 4-50. *A cutaway of a gear-type lift pump showing how the parts relate to one another. (Courtesy Detroit Diesel)*

FIGURE 4-51. *The diaphragm on a fuel lift (feed) pump.*

diaphragm in and out. This lever can also be activated manually, but if the engine is stopped in a position that leaves the diaphragm lever fully depressed, the manual lever will be ineffective until the engine is turned over far enough to move the cam out of contact with the lever.

Diaphragm pumps are nearly foolproof, but eventually the diaphragm will fail. When this happens, little or no fuel will be pumped out of the fuel system's bleed nipples when the lift pump is operated manually. Older pumps often have a drain hole in the base from which fuel will drip if the diaphragm has failed, but Coast Guard regulations have banned this in newer pumps.

A spare diaphragm, or better yet a complete pump unit, should be part of the spare-parts kit on boats that cruise offshore.

Diaphragms are accessible by undoing a number of screws (generally six) around the body of the pump and lifting off the top half (see Figure 4-51). The method of attaching the diaphragm to its operating lever varies; it may be necessary to remove the whole pump from the engine (two nuts or bolts) and play with, or remove, the operating lever (see Figure 4-52). When replacing the diaphragm and pump cover, keep the operating lever pressed halfway down while you tighten the cover screws.

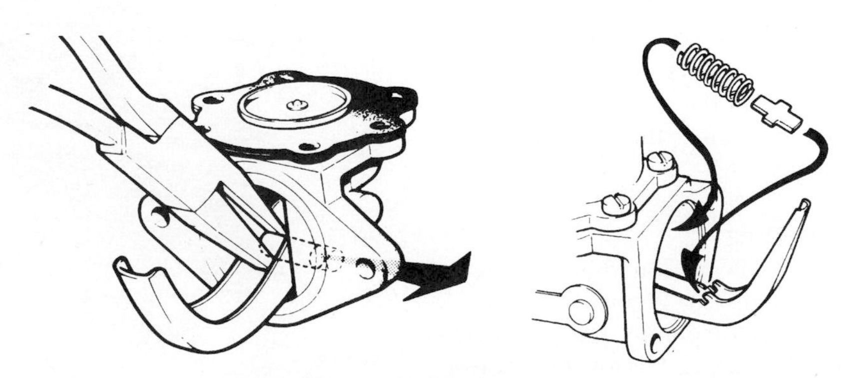

FIGURE 4-52. *Diaphragm replacement on a lift pump. (Courtesy Volvo Penta)*

Some newer engines have the lift pump built into the fuel injection pump. This may be a diaphragm pump, but two other types are also used:

1. On an in-line jerk pump, there is likely to be a piston-style pump in which a plunger moves up and down via a cam on the same camshaft that operates the individual jerk pump plungers. These pumps generally incorporate an externally operated plunger for manual priming of the fuel system.
2. On a distributor pump, a rotary vane pump is driven off the central driveshaft. These pumps cannot be operated by hand; therefore, a separate manual pump is generally included in the system at some point, usually tacked onto one of the filters.

Very Cold Weather

The diesel fuel almost universally available in the United States is known as No. 2 diesel. At very low temperatures, it congeals enough to plug up fuel filters and lines. If you suspect this to be a problem, you may only need to heat fuel lines and filters with a hair dryer or some other heat source to get things moving. If you anticipate prolonged, extra-cold weather, thin No. 2 diesel with a special low-temperature additive, kerosene, or No. 1 diesel. Please note, however, that all of these decrease the fuel's lubricating qualities; running some engines on straight No. 1 diesel, for example, can lead to injection system seizure.

Serious Fuel Supply Problems

If the air inlet and exhaust are unobstructed, compression is good, the tank has fuel, and the system is properly bled, it is time to feel nervous and check the bank balance. Not too many possibilities remain for a failure to start—basically a worn injection pump, worn or damaged injectors, or incorrect fuel injection timing.

There is just no reason for injection timing to go out unless the engine has been stripped down and incorrectly reassembled. Absent this, only some serious mechanical failure is going to throw out the timing, in which case you should have plenty of other indications of a major problem.

Worn or damaged injectors can lead to inadequate atomization of the fuel, to the extent that combustion fails to take place. Injectors are as precisely made as injection pumps and should only be disassembled as a last resort. Chapter 6 describes procedures for removing, cleaning, and checking injectors.

If the injection pump is seized or so badly worn that proper injection is no longer occurring, you can do nothing except have it rebuilt or exchange it for a new one. Changing injection pumps is covered under the section on engine timing in Chapter 7.

Let me emphasize that these problems will almost never occur in a well-maintained engine. Just about every other fault should be suspected before them.

CHAPTER 5

TROUBLESHOOTING, PART TWO: OVERHEATING, SMOKE, LOSS OF PERFORMANCE, AND OTHER PROBLEMS

OVERHEATING

Overheating can be the result of a number of things, but the primary suspect is always a loss of flow in the raw-water circuit. For this reason, as well as to prevent following waves from driving up the exhaust pipe, a water-cooled exhaust should ideally exit high enough in the stern for you to see or hear if water is coming out of it. It will then be possible for you to tell at a glance whether the raw-water side of the cooling system is functioning in some fashion, although you will not be able to tell if the flow is up to normal. (This can be done by holding a gallon container under the exhaust, timing how long it takes to fill, calculating the flow rate in gallons or liters per minute, and comparing this to the engine specifications or what is normal. It is an excellent idea to measure and log the flow rate before problems arise so as to establish a benchmark for future reference. Since raw-water flow rate is directly related to engine speed, the data must be collected at one particular speed—say, 1,000 rpm.)

It should be an iron habit to check the exhaust for proper water flow every time the engine is started. If you have overheating, you should always quickly check the gauge temperature and overboard discharge before shutting down the engine. This information will help you troubleshoot the problem, as illustrated in the following troubleshooting sections.

OVERHEATING ON START-UP

Check the raw-water side first.

Raw-Water Side

The seacock on the raw-water circuit is probably closed! If not, check the raw-water filter. If this is clear, close the seacock, disconnect the raw-water hose, and reopen the seacock to make sure water floods into the boat. If not, there's something stuck in or over the inlet, or it is plugged with marine growth such as zebra mussels. If water comes in, next check the raw-water pump (see Figure 5-1). Almost all raw-water pumps are the rubber impeller type (see Figure 5-2). If the pump runs dry, the impeller will tear.

If the pump is belt driven, check the belt. If this is OK, remove the pump cover (usually four or six screws) and check the impeller (see Figures 5-3 and 5-4). Make sure that when the pump turns, the impeller is not slipping on its shaft. If the impeller is damaged, pull it out with an impeller puller (refer back to Figure 3-25) or pliers (see Figure 5-5) or pry it out with two screwdrivers (see Figure 5-6). (Impeller manufacturers cringe at the suggestion to use screwdrivers—see Figure 5-7 from ITT/Jabsco!)

A few impellers have a locking screw (see Figure 5-8) or a retaining circlip (e.g., the Atomic 3 diesel). If the shaft comes out with the impeller, you will have to unbolt the pump from the engine to replace an adapter in the back of the pump.

Troubleshooting Chart 5-1. Overheating on Start-Up

Question	Action if YES
Is water coming from the raw-water discharge? NO ↓	YES → Check the coolant level in the freshwater circuit (if fitted). Caution: Do not remove header-tank pressure cap when hot. If the level is low, refill and find the leak.
Is the raw-water seacock closed? NO ↓	YES → Open and then check the raw-water overboard discharge. The raw-water pump may have failed from running dry.
Is the raw-water strainer plugged? NO ↓	YES → Clean and then check the raw-water discharge as above.
Has the raw-water pump failed? Inspect the drive belt and tension or replace as necessary. Make sure any clutch is operative. If the belt and clutch (if fitted) are OK, remove the pump cover and inspect the impeller vanes. Make sure the impeller turns when the engine turns. NO ↓	YES → Tighten or replace the drive belt as necessary. Replace a damaged impeller. Track down any missing vanes.
Check for collapsed or kinked raw-water hoses, an obstruction over the raw-water inlet on the outside of the hull (break a below-the-waterline hose loose as close to the raw-water seacock as possible and see if there is a good flow into the boat), or a plugged raw-water injection nozzle into the exhaust. Is the water-lift silencer frozen?	

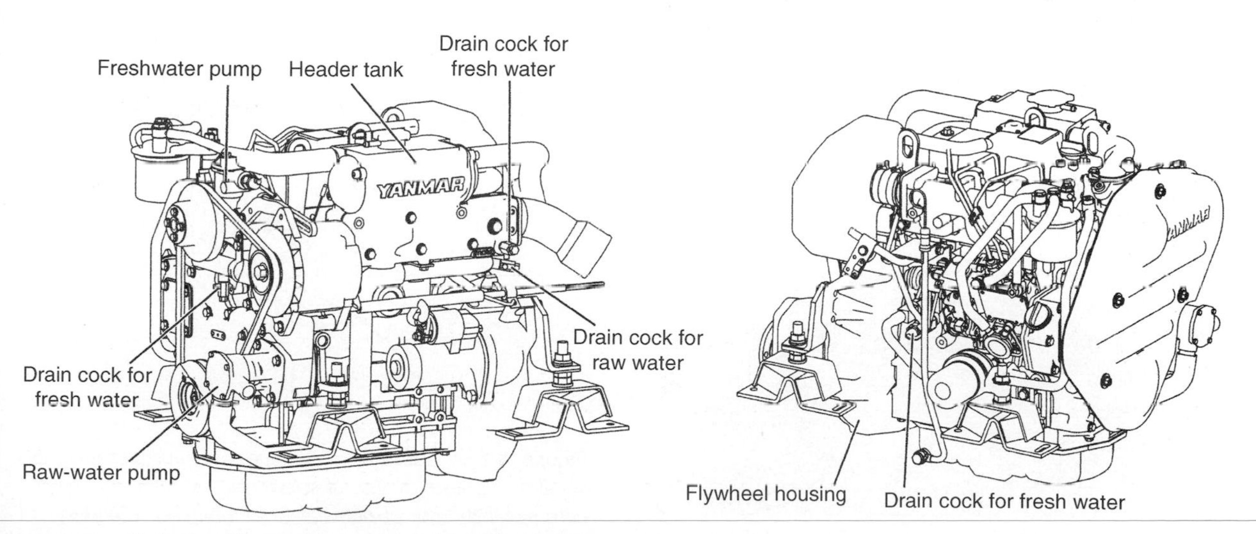

FIGURE 5-1. *Typical cooling system service points. (Courtesy Yanmar)*

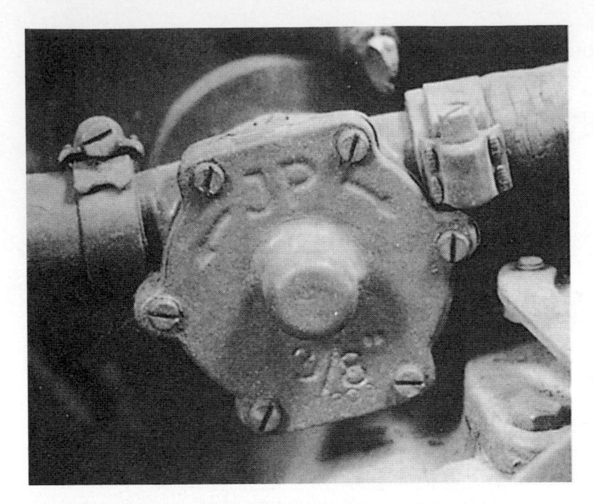

FIGURE 5-2. *A typical raw-water pump.*

FIGURE 5-3. *A raw-water pump with its cover removed. Three out of six vanes on the impeller are missing!*

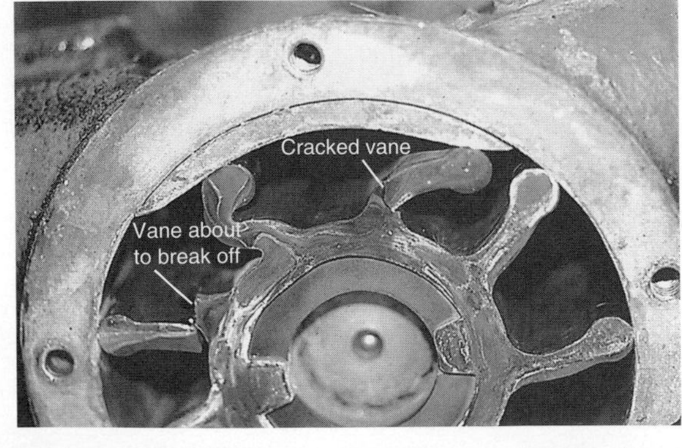

FIGURE 5-4. *Cracked vanes on a raw-water pump impeller.*

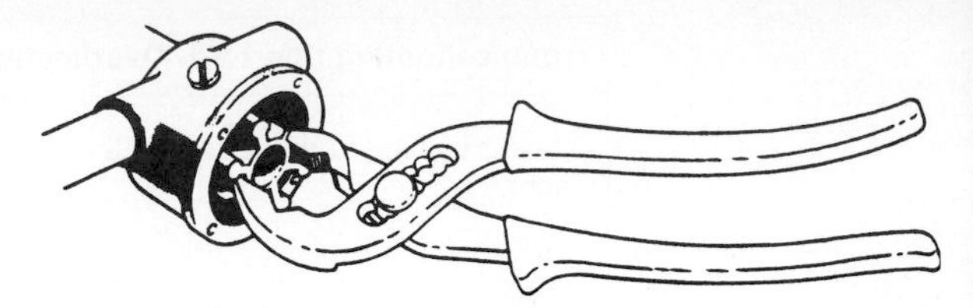

FIGURE 5-5. *Typical raw-water pump impeller removal. (Jim Sollers)*

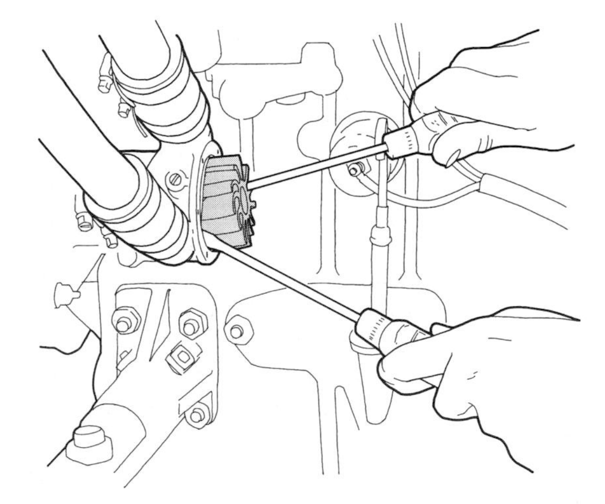

FIGURE 5-6. *Most flexible water pump impellers can be pried off or pulled out. (Jim Sollers)*

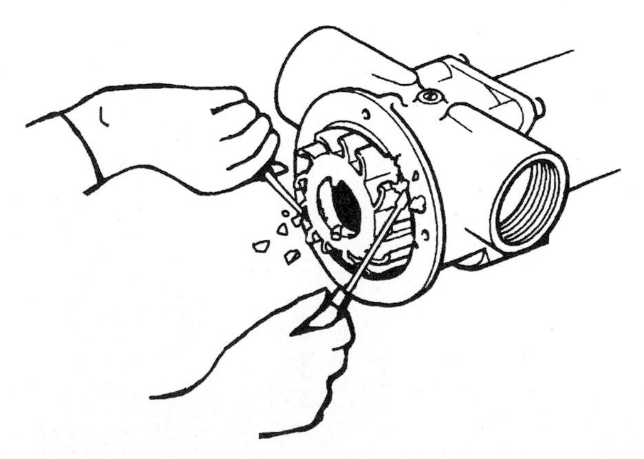

FIGURE 5-7. *ITT/Jabsco, a leading manufacturer of rubber impellers, opposes the use of screwdrivers to remove an impeller because of the risk of damage to the impeller. (Courtesy ITT/Jabsco)*

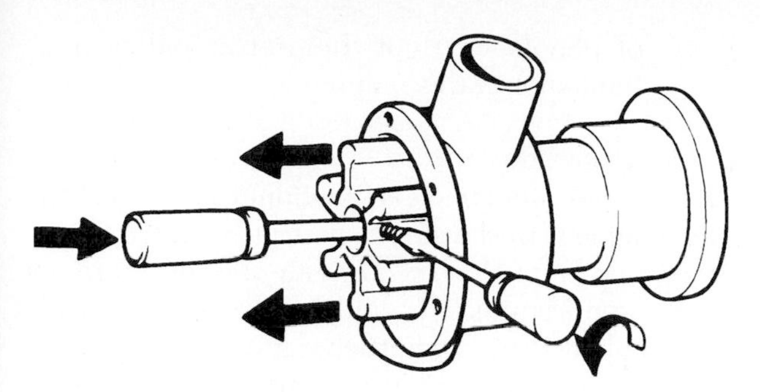

FIGURE 5-8. *Some flexible impellers are secured to the shaft by a setscrew. You must pry these out about 1/2 inch to undo the setscrew. (Courtesy Volvo Penta)*

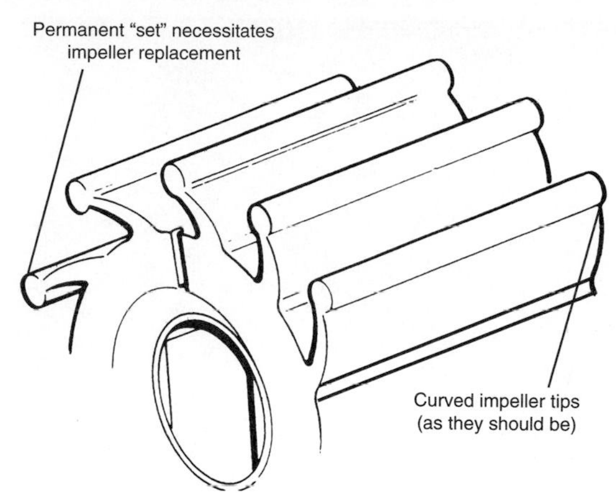

FIGURE 5-10. *Vanes with a permanent set. This impeller is ready for replacement. (Courtesy ITT/Jabsco)*

Before pulling an impeller all the way out, make a note of the way the vanes are bent down in the pump housing. Replace the impeller if the vanes have taken a permanent "set" (i.e., they do not straighten out once the impeller is removed—see Figures 5-9 and 5-10); are worn flat (the tips should be curved); or are in any other way damaged, cracked, or brittle.

Track down any pieces missing from an impeller—they will most likely be found in the heat exchanger (if fitted; see Figures 5-11 and 5-12) and sometimes at the raw-water injection elbow (see Figure 5-13). When installing a new impeller, as you push it in, rotate it so that its vanes bend down the same way as on the old one (I hope you wrote that down!). For rebuilding a pump, see Chapter 6.

If the raw-water pump is irreparably damaged and no spare parts are on board, you have a couple of options for limping home:

1. If the boat has a saltwater washdown pump, plumb this to the raw-water cooling system and use it in place of the raw-water pump. In a bind, you can also use the freshwater pump. Either pump will probably have a sufficient flow rate to enable the engine to be run at cruising speed, but you will need to keep a close eye on the temperature.

FIGURE 5-9. *Even though the impeller tips are still curved, if the vanes retain a "set" after removal, the impeller needs replacing.*

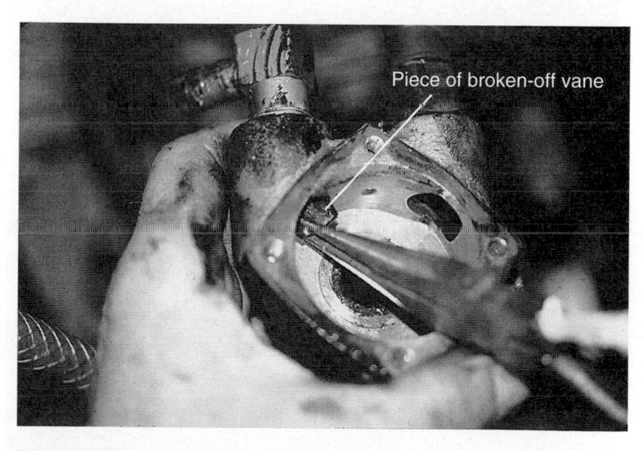

FIGURE 5-11. *Tracking down the missing vanes from the impeller shown in Figure 5-3. The first one was found in the pump discharge elbow; the second in the tubing; and the third . . .*

FIGURE 5-12. *. . . in the heat exchanger (the most likely place to find missing vanes). Note the smaller pieces of rubber lodged in two of the heat exchanger tubes.*

2. Rig a funnel well above the engine and plumb a hose from it to the heat exchanger. Fill it from a bucket. The flow rate will be slow because this is gravity fed, but that's OK because it's going to be hard work keeping the funnel filled! You can adjust the flow rate by moving the funnel up and down. Maintain the engine speed at whatever keeps the temperature from going much above normal. I know a couple of people who got themselves out of a tight spot in this manner.

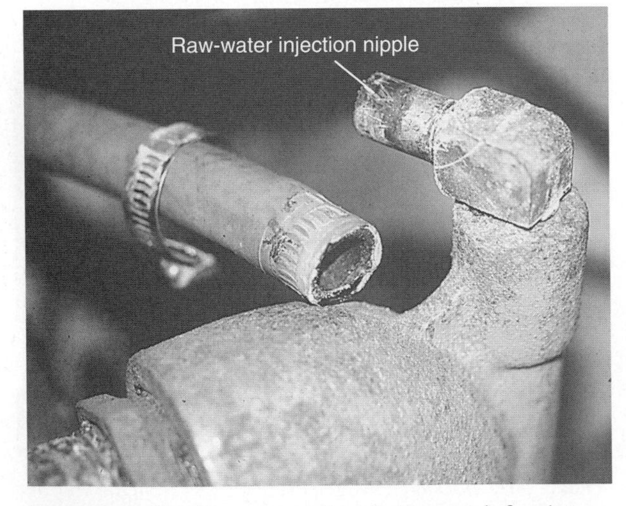

FIGURE 5-13. *Checking a raw-water injection nozzle for obstructions; this is a likely spot for debris to build up.*

Freshwater Side

If the raw-water circuit is functioning as normal and the engine is freshwater cooled, check the freshwater pump belt (which is normally also the alternator belt—see Figures 5-14 and 5-15) and the level in the coolant recovery bottle (if fitted) or expansion (header) tank (see Figure 5-16). *Warning:* Never remove the cap when it's hot. Serious burns may result.

If the level is low, find out where the water is going (adding red food dye may help in tracing leaks). Possibilities are leaking hose connections, heat exchanger or oil cooler tubes that have corroded through (see the Water in the Crankcase section later in this chapter), or a blown head gasket. A blown head gasket will likely cause air bubbles in the cooling system when the engine is running, and these will be visible in the header tank; the header tank may also smell of exhaust fumes.

In an emergency, the engine can still be run at light loads with a blown head gasket (at the risk of burning an expensive groove in the cylinder head), but remove the header tank cap to prevent a buildup of gas pressure. Note that a pressure cap on the header tank with too low a pressure rating, or a pressure cap that has failed, will allow the coolant to boil away over time (the pressure rating will be stamped on the cap; the required pressure is given in the engine specifications).

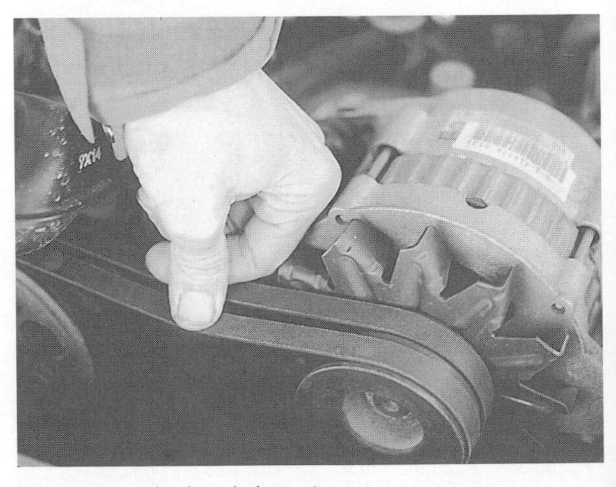

FIGURE 5-14. *Checking belt tension.*

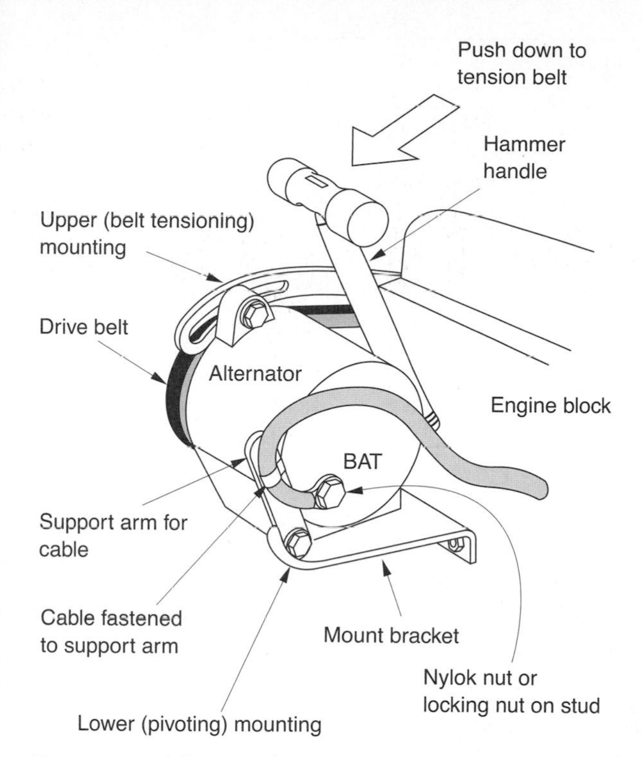

FIGURE 5-15. *Adjusting alternator/water pump belt tension. Under moderate finger pressure, you should not be able to depress the longest stretch of belt more than 3/8 to 1/2 inch (1 to 1 1/2 cm).*

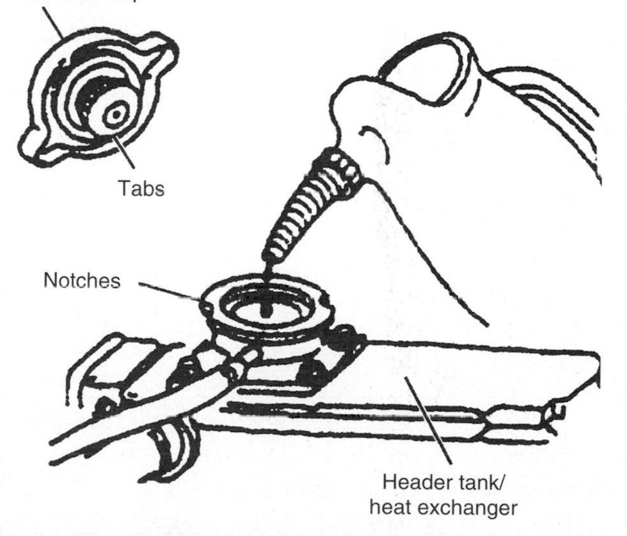

FIGURE 5-16. *A typical pressure cap (just the same as on a car's radiator). (Courtesy Yanmar)*

Finally, it's possible that the engine is, in fact, not overheating, and the gauge is reading incorrectly. If you have a meat thermometer on board, or one that reads to boiling point, you can put this in the top of the header tank to double-check the gauge reading.

OVERHEATING DURING NORMAL OPERATION

Check the Oil Level

If a low oil level is causing the engine to overheat, expensive damage may be in the making. A partial seizure of one or more pistons may already be generating a tremendous amount of heat on the cylinder walls. More likely than a lack of oil, however, is a reduction in the raw-water flow through the cooling system.

Raw-Water Side

If the raw-water intake is not set low enough in the hull, a well-heeled sailboat can suck in air. The heeling will have to be kept down until the through-hull can be moved. Note also that if a scoop-type raw-water inlet is installed facing aft (not forward), at higher boat speeds (especially on planing powerboat hulls), it will cause a reduction in pressure that may reduce or stall the water flow. On the other hand, if installed facing forward on a sailboat, when under sail, it will tend to drive water into the cooling system, which, in certain circumstances, can flood the engine with salt water. (It's best to not use a scoop fitting on a sailboat.)

The raw-water inlet screen (if fitted) on the outside of the hull may be blocked with a piece of plastic. Throttle down, put the boat in reverse, throttle up, and then shut down the engine once the boat has reverse way on it. With a little luck, the reverse propeller thrust and water flow will wash the plastic clear. To confirm this, loosen a hose below the waterline and see if water flows into the boat. If the flow is restricted, check for barnacles on the inlet screen or in the water intake.

Check the raw-water filter. If it contains a lot of silt, the heat exchanger (or the engine itself on raw-water-cooled boats) may be silted up. Feel the freshwater inlet and outlet pipes to the heat exchanger—if it is doing its job, there should be a noticeable drop in the temperature of the water leaving the heat exchanger. If not, see Chapter 6 for cleaning tips, including descaling.

Troubleshooting Chart 5-2. Overheating in Operation		
Check the raw-water overboard discharge. Is the flow less than normal? NO	YES	Check for obstructions in the raw-water circuit (see Troubleshooting Chart 5-1). In addition, check the raw-water circuit for silting, scale, and other partial obstructions (see the text).
Check the oil level. Is it low? NO	YES	Refill with the correct grade and viscosity of oil.
Is the boat overloaded? (Check for a rope around the propeller, a heavily fouled bottom, adverse conditions, excessive auxiliary equipment, or an oversized propeller.) NO	YES	Reduce the loading.
Check the coolant level in the freshwater circuit (if fitted). Caution: Do not remove the header-tank pressure cap when hot. Is the level low? NO	YES	Refill and find the leak.
Is the freshwater circuit air-locked? NO	YES	Break the hoses loose at the freshwater pump and water heater (if fitted) and bleed off any air.
Is the thermostat operating incorrectly? (Remove and test as outlined in text.) NO	YES	Replace.
There is probably a mechanical problem (e.g., faulty fuel injection; a partial seizure)—see the text.		

Raw-water-cooled engines (without a freshwater circuit) are likely to develop scale around the cylinders over time, especially if run above 160°F (71°C). I am told that if a boat has been operating in salt water and then moves into fresh water, scale formed in salt water will swell, and the engine will gradually overheat.

Note that if any other heat exchanger, or a refrigeration condenser, is fitted in line with the engine cooling circuit, an obstruction in any of these will reduce the raw-water flow through all of them.

Check for collapsed cooling hoses, a loose raw-water pump drive belt, or poor pump performance. This is another reason that a high-set overboard discharge is useful—you can readily measure the pump flow.

Where the raw water is injected into the exhaust (on both raw-water- and heat-exchanger-cooled engines), a relatively small nozzle is sometimes used to direct the water down the exhaust pipe and away from the exhaust manifold. If scale forms in the raw-water circuit, this nozzle is likely to plug, restricting the water flow. Over time, some injection nozzles will develop severe rust, plugging the nozzle; a new injection elbow is then necessary.

Freshwater Side

If the raw-water flow is normal, check any header tank as above, but only after allowing the engine to cool.

A broken alternator belt will put the freshwater pump out of commission. If you have no spare belt on board, jury-rig the pump by forming a belt out of nylon twine. Put a loop in one end, run the other end through the loop, and pull it up tight (nylon has quite a bit of stretch, so use this to tension the "belt"). If the line is small gauge, run it around several times. If no suitable line is on board, unlay a piece of three-strand anchor or mooring line to get line of a suitable size and use that. Tape the ends of the line to the belt so they don't catch in anything. You can even use panty hose.

A water pump does not create much of a load, so this belt should hold for some time. However, if it starts to slip, it will heat up and melt, so you need to inspect it regularly. You are unlikely to be able to run the alternator with it because of the higher load imposed by the alternator (especially if the batteries are down and the alternator is loaded up). If you really need the alternator, you can give it a try, but you will need to keep the engine speed down in order to keep the alternator output low enough for the belt to survive.

If the belt is OK, perhaps the engine is overloaded (e.g., by a rope around the propeller, an oversized propeller, a badly fouled bottom, or too much auxiliary equipment), causing it to generate more heat than normal. Maybe the ambient water temperature is higher than normal. For example, a boat moving into the tropics may experience a 40°F (22°C) rise in the ambient water temperature. This temperature increase will lower the temperature differential between the raw water and fresh water in the heat exchanger, causing the engine temperature to rise a little (in which case the raw-water flow rate is probably lower than it should be and/or the raw-water side of the heat exchanger needs cleaning).

Faulty injection and injection dribble can cause late burning of the fuel, which will heat up the cylinders at the end of the power stroke and would normally be accompanied by black smoke. This heat is not converted into work and must be removed by the cooling system, leading to higher temperatures. An overheated engine can generate hot spots that cause pockets of steam to build up in the cylinder or head. These can sometimes air-lock cooling passages, the cooling pump, the heat exchanger, or the expansion tank, especially if the piping runs have high spots where steam or air can gather.

Note that when a domestic water heater is plumbed into the engine cooling circuit, as most are, there is frequently a high spot in the tubing (where the hose rises above the level of the pressure cap on the header tank) that can cause an air lock (see Figure 5-17). If this is the case, you need a remote expansion tank, which must be located above the highest point in the hose runs. It will need a pressure cap rated for the recommended system pressure (see the engine manual). You should also replace the cap on the engine's header tank with one that has a higher pressure rating than the remote expansion tank to prevent the header tank cap opening before the expansion tank cap, dumping the coolant.

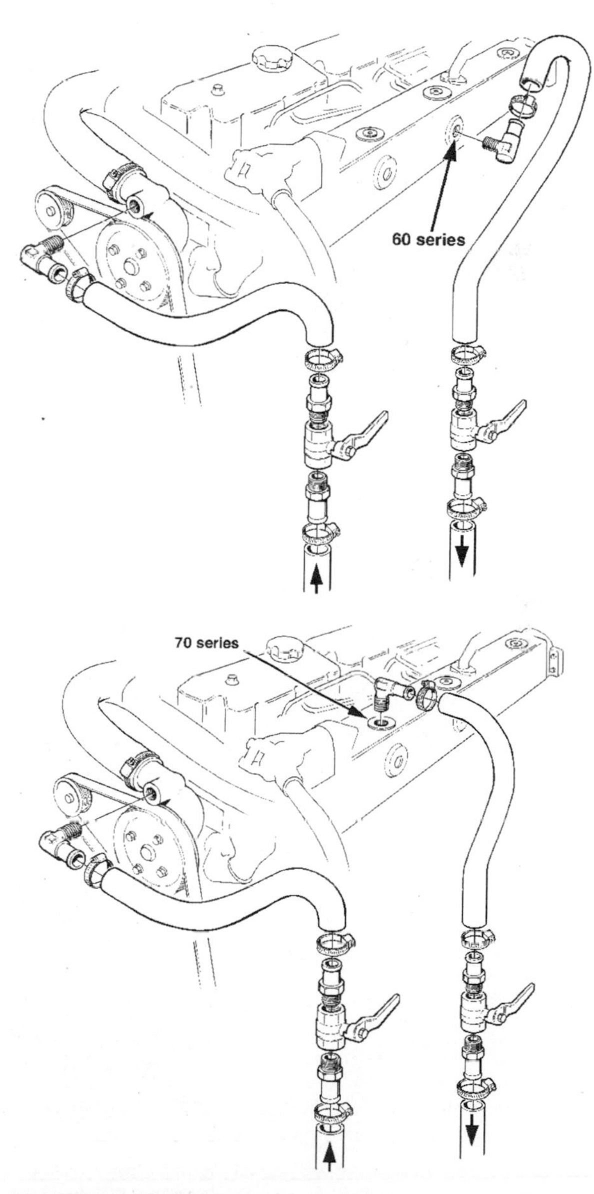

FIGURE 5-17. *Typical hot-water tank connections to a freshwater cooling system. (Courtesy Volvo Penta)*

Although thermostats are generally very reliable, they occasionally fail, usually in the closed position, which on most engines leads to overheating. You will find the thermostat under a bell housing, with one or two hoses coming out of it, near the top and front (crankshaft pulley end) of the engine (see Figures 5-18, 5-19, and 5-20). Its housing generally is held down with a couple of bolts. Remove these and gently pry up the housing. Take out the thermostat and try operating without it (see Figure 5-21). This will make most engines run cool, but will cause a few—for example, some Caterpillars—to run hot, in which case it should not be done.

To test the thermostat, put it in a pan of water and heat it (see Figure 5-22). It should open between 165°F and 185°F (74°C to 82°C), except on some Caterpillars, which open as high as 192°F (89°C), and some raw-water-cooled engines, which open between 140°F and 160°F (60°C and 71°C). Thermostats are relatively cheap, so if you have gone to the trouble of taking yours out, you may as well replace it!

Finally, problems with temperature gauges are rare. If suspected, consult the Problems with Engine Instrumentation section later in this chapter and test the actual temperature with a meat thermometer as described above.

Smoke

Aside from a little blue and maybe black smoke on start-up, especially on older engines, diesel exhaust should normally be perfectly clear. The presence of smoke often points to a problem in the making, and the color of the smoke can be a useful guide.

Black Smoke

Black smoke is the result of unburned particles of carbon from the fuel blowing out of the exhaust. On newer engines, smoke-limiting devices (see pages 21–22) and electronic engine controls generally prevent black smoke formation. If it occurs, the engine has a serious problem that needs professional attention.

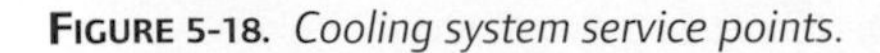
Figure 5-18. *Cooling system service points.*

Troubleshooting Chart 5-3. Smoke in Exhaust

Smoke: Black	Smoke: Blue
Obstruction in airflow: Dirty air filter Defective turbo/supercharger High exhaust back pressure	Worn or stuck piston rings Worn valve guides
Excessively high ambient air temperature	Turbo/supercharger problems: Worn oil seals Plugged oil drain
Overload: Rope around propeller Oversized propeller Heavily fouled bottom Excessive auxiliary equipment	Overfilled oil bath–type air filter High crankcase oil level/pressure
Defective fuel injection	

White
Lack of compression
Water in the fuel
Air in the fuel
Defective injector
Cracked cylinder head/leaking head gasket

Note: When an exhaust is water cooled, it is difficult to distinguish white smoke from the normal exhaust.

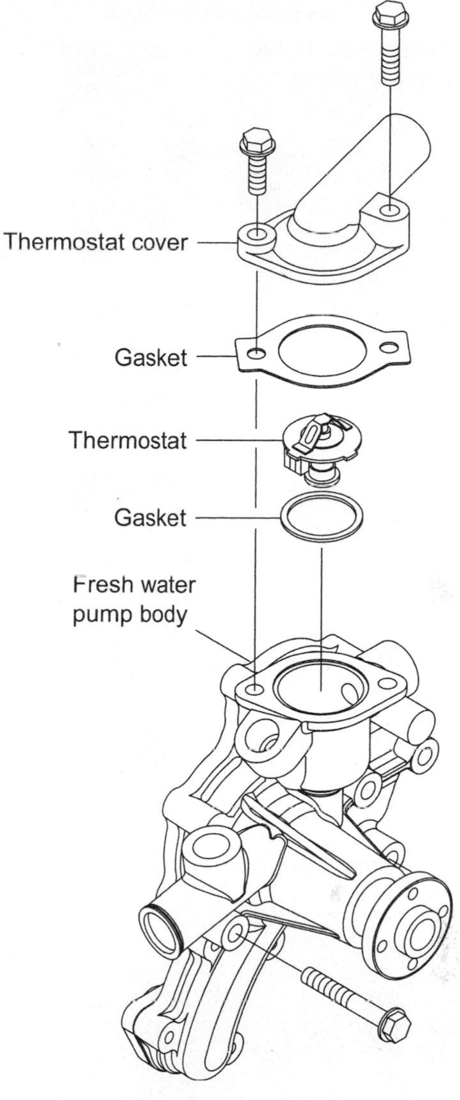

FIGURE 5-19. *Typical location of the thermostat on a modern diesel engine. (Courtesy Yanmar)*

On many older engines, any attempt to accelerate suddenly will generate a cloud of black smoke as the fuel rack opens and the engine slowly responds. Once the engine reaches the new speed setting, the governor eases off the fuel rack and the smoke immediately ceases. This smoke is indicative of a general engine deterioration: the compression is most likely falling, the injectors need cleaning, and the air filter needs changing. If the engine is otherwise performing well, you have no immediate cause for concern but the engine is giving notice that a thorough service is overdue.

If black smoke persists once the engine is up to speed, the engine is crying out for immediate attention. The following are likely causes for the smoke:

- Obstruction of the airflow through the engine. This results in an insufficient amount of air entering the cylinders for proper combustion to take place. Likely causes are a dirty air filter, restrictions in the air inlet ducting, or a high exhaust back pressure (see below). In particular, turbocharged engines and Detroit Diesel two-cycles are sensitive to high back pressure. The turbocharger slows down, and as a result, pushes less air into the engine than it requires, leaving fuel unburned. A dirty or defective turbocharger will have the same effect (see pages 126–28). Note that many engines on auxiliary sailboats are tucked away in little boxes. As often as not these are fairly well sealed to cut down on the noise levels. Unless such a box is adequately vented, this setup can strangle the

temperature-sensing pickup

thermostat housing bolts

FIGURE 5-20. *The thermostat housing on an older engine (a Volvo MD 17C).*

thermostat

FIGURE 5-21. *The thermostat removed from the Volvo MD 17C.*

FIGURE 5-22. *Checking the operation of a thermostat. (Courtesy Volvo Penta)*

engine, particularly at higher loadings and in hot climates where the air is less dense.

- Overloading of the engine. The governor reacts by opening the fuel-control lever until more fuel is being injected than can be burned with the available oxygen. This improperly burned fuel is emitted as black smoke. It is an unfortunately common practice to fit the most powerful propeller that the engine can handle in optimum conditions—a lightly loaded boat with a clean, drag-free bottom in smooth water. This practice exaggerates the performance of the boat under power, but overloading results under normal operating conditions. Adding various auxiliary loads, such as a refrigeration compressor and a high-output alternator, often compounds the problem because those loads weren't taken into account in the overall power equation. Black smoke on new boats should raise the suspicion of an overloaded engine caused by the wrong propeller or too much auxiliary equipment. (Matching engines to their loads is dealt with in more detail in Chapter 9.)

Overloading can also arise from wrapping a rope around the propeller. Apart from the likelihood of a smoky exhaust, overloading is liable to cause localized overheating in the cylinders, which will definitely shorten engine life and could lead to an engine seizure.

- Defective fuel injection. If an engine is not overloaded, and the airflow is unobstructed, poor injection is the number one suspect for black smoke. Dirty, plugged, or worn injector nozzles can cause inadequate atomization of the fuel, improper distribution of the fuel around the cylinders, or a dribble after the main injection pulse. All can lead to unburned fuel and black smoke.
- Excessively high ambient air temperatures (e.g., in a hot engine room on a boat operating in the tropics). The density, thus the weight, of the air entering the engine will be reduced, leading to an insufficient air supply, especially at high engine loadings.

FIGURE 5-23. *A heavily carbonized valve due to poor operating practices.*

BLUE SMOKE

Blue smoke comes from burning engine lubricating oil. A little blue smoke on start-up is not uncommon, but it should immediately clear. If it persists, the oil that is feeding it can only get into the combustion chambers by making it up past piston rings; down valve guides and stems; or in through the air inlet from leaking supercharger or turbocharger seals, an overfilled oil-bath air filter, or a crankcase breather. A plugged oil drain in the turbocharger will also cause oil to leak into the compressor housing and enter the air inlet.

Engines that are repeatedly operated for short periods, or are idled or run at low loads for long periods, do not become hot enough to fully expand the pistons and piston rings. They then fail to seat properly, and oil from the crankcase finds its way into the combustion chamber. In time, the cylinders become glazed (very smooth), while the piston rings get gummed into their grooves, allowing more oil through. Oil consumption rises and compression declines. Blowby down the sides of the pistons raises the pressure in the crankcase and blows an oil mist out of the crankcase breather. Carbon builds up on the valves and valve stems and plugs the exhaust system (see Figure 5-23). Valves may jam in their guides and hit pistons. *Repeated short-term operation and prolonged idling and low-load running will substantially increase maintenance costs, including major overhauls, and shorten engine life.*

WHITE SMOKE

White smoke is caused by water vapor in the exhaust or by totally unburned, but atomized, fuel. Given that almost all exhausts are water cooled, there is the potential to create steam, but in practice, the cooling water flow should be such that this situation does not happen under normal circumstances. (Note, however, that in cold weather, vaporized water in the exhaust gases is likely to condense into steam.)

To determine whether white smoke is water or fuel, hold your hand under the exhaust for a few seconds and then sniff it to see if it smells of diesel. If it does, one or more cylinders are not firing. If the weather is cold and the engine has just been started, the most likely cause is poor compression, especially if the smoke clears as the engine warms up. If the smoke persists, there may be a defective injector.

LOSS OF PERFORMANCE

HIGH BACK PRESSURE

The exhaust is an integral part of the airflow through an engine. Any restriction will generate back pressure, which will cause an engine to lose power, overheat, and probably smoke (black). The most likely causes of high exhaust back pressure are:

- A closed or partially closed sea valve on the exhaust exit pipe.
- Too small an exhaust pipe, too many bends and elbows, too great a lift from a water-lift muffler to the exhaust exit (see Chapter 9), or a kink in an exhaust hose.

- Excessive carbon formation in the exhaust system caused by long hours of operation at light loads (such as when battery-charging or refrigerating at anchor).
- In wintertime, frozen water in a water-lift-type muffler at initial start-up. This may produce a hissing immediately after cranking, or bubbles in the raw-water strainer, as the trapped exhaust looks for an escape path.

Generally, the easiest way to check for a fouled exhaust is to remove the exhaust hose from the water-lift muffler, then wipe your finger inside the muffler and use a flashlight to look up the hose (see Figure 5-24). There should be nothing more than a light carbon film. What you may find is a heavy carbon crust. In this case, the entire exhaust system, including the exhaust valves, is likely to be fouled (probably because you have been running the engine for short periods and/or for long hours at low loads). A significant overhaul is needed.

FIGURE 5-24. *Inspecting the exhaust hose from a water-lift muffler for carbon buildup. In this case, there is none.*

You should break the exhaust loose and inspect it annually until you know for sure that it is not fouling, and maybe every other year after that.

The exhaust can reveal a surprising amount of information about the operation of an engine. One of the very best methods for monitoring the performance of an engine, used on all large diesels, is an exhaust pyrometer fitted to each cylinder. These measure the temperature of the exhaust gases as they emerge from the cylinders. Variations in temperature from one cylinder to another show unequal work due to faulty injection, blowby, etc., and should never exceed ±20°F (11°C).

On occasion, exhaust pyrometers are offered as an option on smaller diesels. For those inclined to do all their own troubleshooting and maintenance, they are a worthwhile investment. This is especially so with today's higher-revving, hotter-running, and more highly stressed engines. High exhaust temperatures on any cylinder will sharply decrease engine life. The additional cost of a pyrometer installation can easily be paid for in a better-balanced and longer-lived engine.

KNOCKS

Diesel engines make a variety of interesting noises. Each of the principal components creates its own sound, and a good mechanic can often isolate a problem simply by detecting a specific knock coming out of the engine.

In addition to the injector creak already mentioned (see page 101), at any point on the valve cover you can hear the light tap, tap, tap of the rocker arms against the valve stems. (The adjusting screws that are used to set valve-lash clearances—see Chapter 7—are known as *tappets*.) With practice, anyone can pick up the note of individual tappets and get a pretty good idea if valve clearances are correct. Water pumps, camshafts, and fuel pumps have their characteristic note, as do most other major engine components.

The symphony, however, is frequently garbled by a variety of fuel and ignition knocks. Differences in the rate of combustion can cause noises that are almost indistinguishable from mechanical knocks, especially on two-cycle diesels. But if you run the engine at full speed and then shut down the governor control lever (the throttle), a fuel knock will cease at once, whereas a mechanical knock will probably still be audible, although not as loud as before because the engine is now merely coasting to a stop. Knocks that gradually get louder over the life of an engine (especially if accompanied by slowly declining oil pressure when the engine is hot) are almost certainly mechanical knocks.

Fuel-Side Knocks

Some fuel knocks are quite normal, especially on initial start-up. Most diesels are much noisier than gasoline engines and have a characteristic clatter at idle, especially when cold. The owner of a diesel will have to become accustomed to these noises in order to detect and differentiate out-of-the-ordinary fuel knocks. These can have several causes, such as:

- Poor-quality fuel (low cetane rating, dirt or water in the fuel). The fuel is slow to ignite and builds up in the cylinder. But then the heat generated by the early part of combustion causes the remaining (and now excessive) fuel to burn all at once. The sudden expansion of the gases causes a shock wave to travel through the cylinder at the speed of sound. You will hear and feel this as a distinct knock (see Figure 5-25), known as *detonation*.
- Faulty injector nozzles. These can produce a result similar to that just described. The fuel is not properly atomized, and as a result, initial combustion is delayed. Excess fuel builds up in the cylinder, and then a sudden flare-up occurs.
- Injection timing too early. This causes the fuel to start burning while the piston is still traveling up on its compression stroke. The piston is severely stressed as the initial combustion attempts to force the piston back down its cylinder before the crankshaft has come over top dead center (TDC). Timing problems should not be encountered under normal circumstances.
- Oil in the inlet manifold. On supercharged and turbocharged engines, leaking oil seals sometimes allow oil into the inlet manifold. The oil is then sucked into the cylinders and can cause detonation. In extreme cases, enough oil can be drawn in to cause engine runaway—the engine speeds up out of control and will not shut down when the fuel rack is closed (see below).

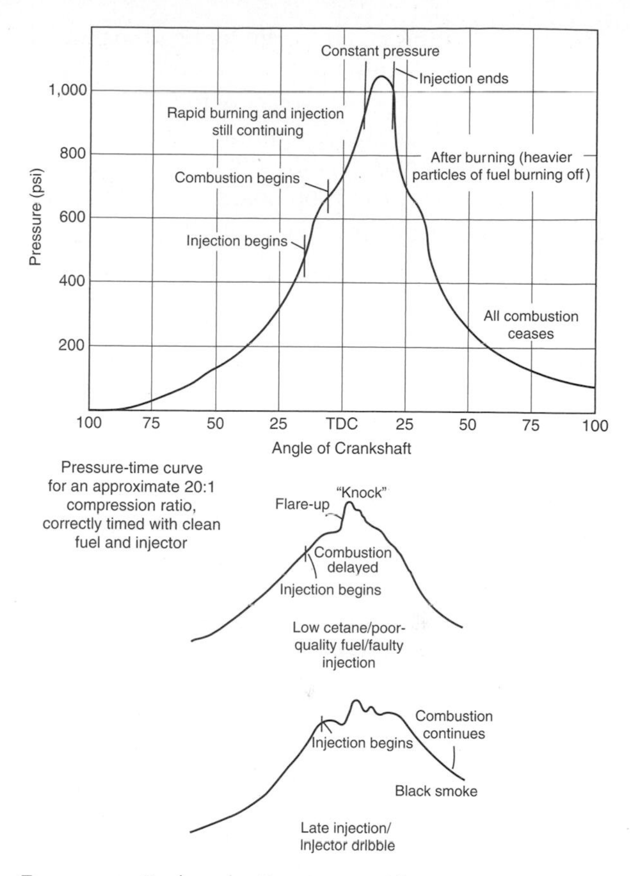

FIGURE 5-25. *Fuel combustion pressure-time curves.*

A sudden loud knocking, especially on start-up, is likely to be a stuck injector that is now dumping nonatomized diesel into the cylinder. Combustion is delayed and then there is a sudden flare-up. Damage to the piston and cylinder as well as to the fuel injection pump is possible. To determine which injector is knocking, loosen the injector lines at the injectors one at a time until fuel spurts out (don't do this with unit injectors, including Detroit Diesel two-cycles, or any engine with a common rail fuel injection system—fuel will flood out!). If the engine has internal fuel lines, loosen the delivery pipes at the external fittings on the cylinder head so that fuel doesn't run down into the crankcase.

If the knocking ceases, you have found the offending cylinder; if not, tighten the fuel line and loosen the next. If you find the offending injector but have no replacement injector on board, in an emergency you can run the engine at low loads with the injector line loose. However, you will need to collect the diesel that continually dribbles out the injector line (and you will, of course, suffer a significant loss of power). If the knocking is not stopped by loosening the injection lines, it is almost certainly a mechanical problem.

Mechanical Knocks

The more common mechanical knocks arise from the following conditions:

- Worn piston pin or connecting rod bearings. As the piston reaches TDC or bottom dead center (BDC), its momentum carries it one way while the crankshaft is moving the other. Any play in the bearings will result in a distinct noise that varies with engine speed.

  A worn connecting rod bearing knocks more loudly under load. By testing a variety of points along the crankcase and block, you can sometimes isolate the specific connecting rod at fault. A good trick is to touch the tip of a long screwdriver to the engine while placing your ear to the handle—the screwdriver will act like a stethoscope, amplifying internal sounds. Be sure to tie up long hair and keep clear of moving belts, flywheels, etc.
- Worn pistons slap or rattle in their cylinders. This is more audible at low loads and speeds, particularly when idling.
- Worn main bearings rumble rather than knock. Engine vibration increases, especially at higher engine speeds.

All these mechanical problems require professional help. Do not ignore new noises!

Misfiring

Many diesels idle unevenly due to the difficulty of accurately metering the minute quantities of fuel required at slow speeds. This is not to be mistaken for misfiring, which will be felt and heard as rough running at all speeds.

Misfiring may be rhythmic or erratic. The former indicates that the same cylinder(s) are misfiring all the time; the latter means that cylinders misfire randomly. Rhythmical misfiring is caused by a specific problem with one or more cylinders, such as low compression or faulty fuel injection. If it occurs on start-up and then clears up once the engine warms, it is almost certainly due to low compression; initially the air in the cylinder is not reaching ignition temperature, but as the engine warms, the air gets hotter until the cylinder fires.

Once again, you can track down the guilty cylinder(s) of a rhythmic misfire by loosening the injector nut on each injector in turn (with the engine running) until fuel spurts out (don't do this with unit injectors, including Detroit Diesel two-cycles, or any engine with a common rail fuel injection system—fuel will flood out!). If the engine changes its note or slows down, the cylinder was firing as it should; retighten the nut. If no change occurs, the cylinder is misfiring, and you need to investigate further (of course, if no fuel spurts out, you then know why it's not firing!).

If the engine has internal fuel lines, loosen the delivery pipes at the external fittings on the cylinder head so fuel doesn't run down into the crankcase.

Unit injectors, including Detroit Diesel two-cycles, are a little different. First of all, they have no suitable external fuel lines, and even if they did, loosening a fuel line would cause large amounts of fuel to flow out. You can, however, remove the valve cover and hold down the actuating (injector) rocker on individual injectors to disable them (see Figure 5-26; this is, however, a messy business as the rocker arms throw a fair amount of oil around). If this slows the engine and changes its note, you know that this cylinder is firing OK. If there is no change, the cylinder is missing.

Erratic misfiring on all cylinders is the result of a general engine problem, frequently contaminated fuel. If the misfiring is more pronounced at higher speeds and loads, in all probability the fuel filter is plugged. If it develops over an extended period of engine running, check for a plugged fuel tank vent or excessive lift from the fuel tank to the engine (see page 103). If it is accompanied by black smoke,

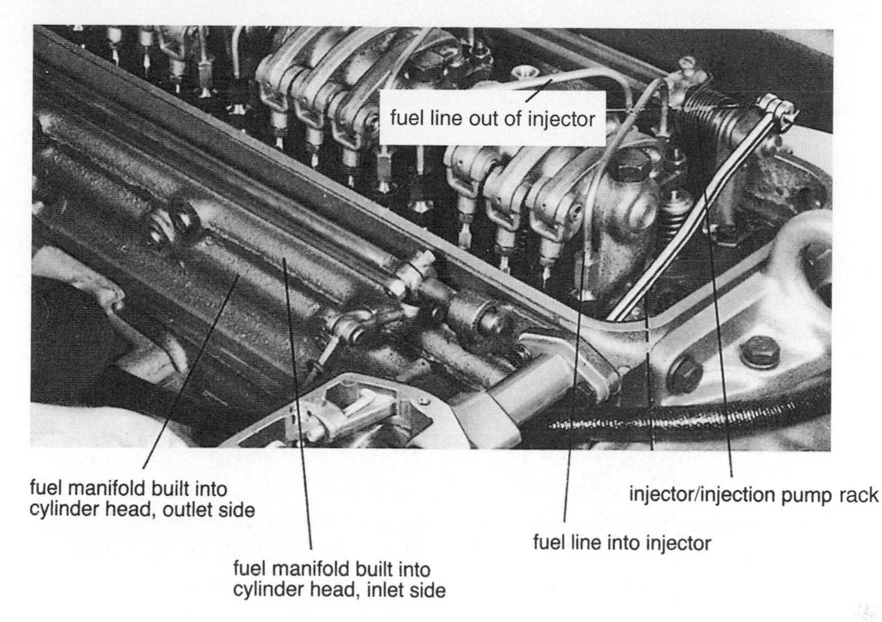

FIGURE 5-26. *Detroit Diesel fuel injection system. (Courtesy Detroit Diesel)*

the air filter is probably plugged, or there is some other problem with the airflow (carbon in the exhaust, defective turbocharger, etc.). In rare cases, poorly seated lift pump valves or a damaged diaphragm may be causing fuel starvation.

Sticking or Bent Valve Stems

Many marine diesels, particularly those fitted to auxiliary sailboats, are run at idle and at low operating temperatures for prolonged periods. As noted above (see the Blue Smoke section), this leads to carbon formation throughout the engine. Using nondetergent oils in the engine or neglecting oil and filter changes can have the same effect. Valve stems and guides become coated with carbon, and after a while, valves are likely to jam in their guides in the open position. When the piston comes up on its exhaust or compression stroke, it may hit the open valve, knocking it shut or bending the valve stem and damaging the piston crown.

As a temporary measure, you can often free sticking valves by lubricating the stem with kerosene and turning the valve in its guide to loosen it. At the earliest opportunity, the cylinder head will have to be removed, and the valves, guides, etc., thoroughly decarbonized (see Chapter 7).

Poor Pickup

Poor pickup, or a failure to come to speed, is most likely a result of one or more of the following:

- Insufficient fuel caused by a plugged filter (don't forget the filter screen in the top of the lift pump), a nearly empty tank, a plugged filter screen on the pickup tube in the tank, or a blocked tank vent causing a vacuum in the tank.
- A clogged air filter, in which case the engine is likely to be emitting black smoke. (You should have noticed that this is about the fifth time that plugged filters have been mentioned in connection with various problems! This is no accident. Routine attention to filters is neglected time and time again.)
- Supercharger or turbocharger malfunction—black smoke is likely.
- Overloading due to an improperly matched propeller, a heavily fouled bottom, too much auxiliary equipment, or perhaps a rope around the propeller.
- Excessive back pressure in the exhaust.
- Low compression, perhaps resulting in misfiring.

- Dirty or clogged injector nozzles.
- Injection pump plungers leaking due to excessive wear.
- Too much friction—a partial seizure is underway.

SEIZURE

Seizure of the pistons in their cylinders is an ever-present possibility anytime serious overheating occurs or the lubrication breaks down. Overheated pistons expand excessively and jam in their cylinders. An engine experiencing a seizure bogs down—that is to say, fails to carry the load, slows down progressively, may emit black smoke, and becomes extremely hot. If immediate steps are not taken to deal with the situation, total seizure—when the engine grinds to a halt and locks up solidly—is not far off.

Partial seizures of individual pistons can occur as a result of uneven loading, causing overheating in the overloaded cylinder. Sometimes a serious case of injector malfunction and dribble, particularly on a direct combustion engine, results in a stream of diesel washing the lubrication off a part of the cylinder wall, which leads to a partial seizure of the piston. I have even seen a new engine seized absolutely solid, while it was shut down, by the differential contraction of the pistons and cylinders in a spell of extremely cold weather.

If you detect the beginning of a partial engine seizure, the correct response is not necessarily to shut down the engine immediately because as it cools, the cylinders are likely to lock up on the pistons. Instead, instantly throw off the load and idle down the engine as far as possible for a minute or so to give it a chance to cool off. This action assumes that the seizure is not due to the loss of the lubricating oil or cooling water. In either of those situations, there is no choice but to shut down as fast as possible.

ENGINE RUNAWAY OR FAILURE TO SHUT DOWN

Diesel engines will run on oil as well as diesel. If oil finds its way into the combustion chambers, an engine can run away uncontrollably. This is a rare occurrence, but if it occurs it can be quite unnerving. Sources of such oil are:

- Leaking turbocharger oil seals.
- Too much oil (almost always from overfilling after an oil change) in the crankcase on those engines that connect the crankcase breather to the engine air inlet (most engines do this).
- Diesel dilution of the engine oil on engines with internal fuel lines.
- Seriously worn valve guides, allowing oil to work its way into the cylinders.

Propane leaks from an improperly installed propane system can also fuel a diesel engine.

In extreme cases, enough oil can be drawn into the engine to cause it to speed up out of control and not shut down when the fuel rack is closed. Runaway is more prevalent on two-cycle diesels than it is on four-cycles, and that's why some two-cycle Detroit Diesels have an emergency air flap that cuts off all air to the engine and strangles it.

In the absence of an emergency air flap, the only way to stop runaway is to cut off the oxygen supply to the engine with a CO_2 (carbon dioxide) fire extinguisher or by jamming a boat cushion or something similar in the air inlet. Don't get your hand in the way! And ensure that whatever you use is strong enough to not break up under the sudden vacuum, because the pieces may foul the engine, necessitating an expensive overhaul.

Be aware that on some turbocharged engines, the vacuum created by blocking the air inlet may suck the turbocharger oil into the cylinders, hydraulically locking them and doing expensive damage. Before resorting to this action, you need to be sure you have exhausted all other alternatives.

If not stopped, runaway can continue until the oil supply is used up. If the oil is coming through leaking supercharger/turbocharger oil seals, it will normally be coming from the engine's oil supply, and eventually the engine will seize.

Then there are engines that are stopped by pressing a button in the engine panel (e.g., most Yanmars). This closes the electrically operated fuel solenoid, shutting off the fuel supply to the engine. Sometimes the engine fails to stop when the button is pushed. The engine can always be stopped by finding the solenoid, identifying the rod it moves, and then pushing this rod in or out by hand. But before doing this, check the ignition switch—if it has been inadvertently turned off before stopping the engine, it has probably disabled the Stop button. (I can tell you from personal experience that after you've been climbing around in the engine room looking for the solenoid, and then spent some time trying to troubleshoot it, you feel a little silly when you find this out!)

High Crankcase Pressure

Smoke blowing out of the crankcase breather or dipstick hole is an indication of high crankcase pressure. This condition is likely to develop slowly, generally as a result of poorly seated or broken piston rings, which allow the combustion gases to blow by the pistons into the crankcase. Other causes may be a crack or hole in a piston crown (top), a blown head gasket, or excessive back pressure. A compression test on the cylinders will reveal if just one cylinder is blowing by, or all.

On turbocharged engines, air pressure will blow down the oil drain lines into the crankcase if the oil seals begin to give out, whereas on Detroit Diesel two-cycles, a damaged gasket between the supercharger and the block has the same effect.

Oil leaks from around the valve cover and other gaskets may be signaling excess pressure from blowby or from a plugged breather.

Water in the Crankcase

Although it should get burned off in normal operation (especially on modern diesels), a certain amount of water can find its way into the crankcase from condensation of steam formed during combustion, but appreciable quantities can come only from the cooling system. The sources are strictly limited:

- Water siphoning in through exhaust valves from a faulty water-cooled exhaust installation (see Chapter 9).
- A leak from a gear-driven (as opposed to belt-driven) raw-water pump.
- Corrosion in an oil cooler (see Figure 5-27).
- A leaking head gasket.
- Leaks around injector sleeves (where fitted).
- A cracked cylinder liner (or one with a pinhole caused by corrosion from the water-jacket side).
- A leaking O-ring seal at the base of a wet liner.

The cooling tubes on oil coolers with raw water circulating on the water side (as opposed to water from the engine's freshwater cooling circuit) are especially prone to damage. The combination of heat, salt water, and dissimilar metals is a potent one for galvanic corrosion. All too many oil coolers are made of materials unsuited to the marine environment (e.g., brass). Oil coolers are expensive and often hard to find. Before starting on a long cruise, make sure your cooler is marine grade (bronze and cupronickel), and if necessary, adequately protected with sacrificial zinc pencil anodes (see Chapter 3).

When an oil cooler tube fails during engine operation, oil is likely to be pumped into the cooling system. On a raw-water-cooled oil cooler, the oil will show itself as an overboard slick. On a freshwater-cooled oil cooler, the oil will appear in the header tank (see page 142 for suggestions on how to flush the cooling system). When the engine is at rest, water from the cooling system may find its way into the oil side of the cooler and siphon into the crankcase.

If you can identify the offending tube in the oil cooler, you may be able to plug it at both ends with a piece of doweling. However, only do this if you can open the oil cooler to expose the tube stack and you can see from which tube oil is dribbling; in reality, the necessary access to the tubes is often not available. Failing this, you can find fittings to bypass the oil and water sides of a leaking oil cooler. And while you can still run the engine until a replacement is found, do so only at low power loadings and only after changing the engine oil and filter. Keep a close eye on the oil-pressure and engine-temperature gauges.

Leaks from the freshwater side into the engine can be more damaging than saltwater leaks if the fresh water includes antifreeze. The antifreeze will cook out on bearings and other surfaces, proving just about impossible to remove. If you suspect a freshwater leak into the engine, do not run the engine until the leak has been tracked down and fixed, and the engine oil has been changed (see pages 141–42 for suggestions on how to flush the engine).

One condition may be encountered that is sometimes mistaken for a water leak into the engine but in fact is not. This is condensation in the valve cover, leading to emulsification of the oil in the valve cover (it goes gooey and turns a creamy color—see

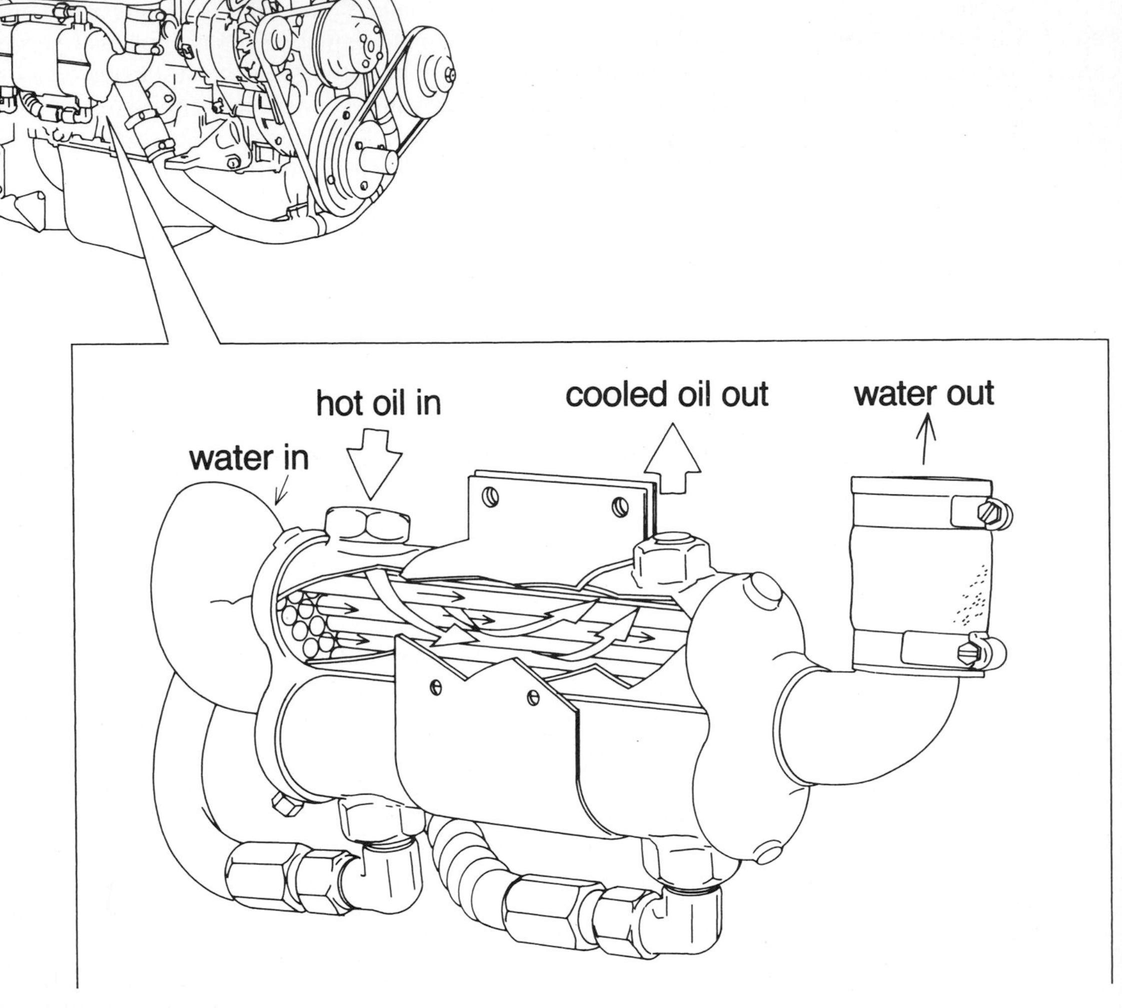

FIGURE 5-27. *Oil coolers increase engine life. Cooling water passes through small-diameter copper tubing inside the oil reservoir, drawing heat from the oil. The most likely points for corrosion are at either end of the tubes. This unit from Perkins cools engine and transmission oil. (Courtesy Perkins Engines Ltd.)*

Figure 5-28) and rusting of the valve springs and other parts in the valve train. This situation happens from time to time when an owner periodically runs the engine for relatively short periods of time to charge a battery or just to make sure it is still working. The engine never warms up properly, but it generates enough heat to create condensation in the valve cover (many old Volvo Pentas are notorious for this).

If you start the engine, run it long enough and hard enough to thoroughly warm it up and cook all the moisture out of it. If necessary, tie off the boat securely, put the engine in gear, and open the throttle a little.

OIL-RELATED PROBLEMS

LOW OIL PRESSURE

Low oil pressure is a serious problem, but it occurs infrequently. When confronted with low oil pressure many people assume that the gauge or warning light is malfunctioning and ignore the warning.

FIGURE 5-28. *Emulsified oil inside a valve cover (the cover has been removed from the engine). In this instance, the problem was caused by a blown head gasket.*

Given the massive amount of damage that can be caused by running an engine with inadequate oil pressure, this is the height of foolishness. Anytime low oil pressure is indicated, immediately shut down the engine, find the cause, and fix it.

The problem is likely to be one of the following:

- Lack of oil—the most common cause of low oil pressure and the least forgivable. If low oil pressure occurs soon after an oil and filter change, there is a strong probability that the filter was not properly tightened, or its O-ring is missing or not seating properly, resulting in the loss of the oil.
- The wrong grade of oil in the engine—with a viscosity that is too low.
- Lowering of the oil viscosity by overheating, even though the correct grade is in the engine.
- Diesel dilution of the oil from leaking internal fuel lines. (Once enough diesel has found its way into the oil to lower the pressure to a noticeable extent, you will be able to smell the fuel in the oil if you take a sample from the dipstick.)
- Worn bearings. The oil pump feeds oil under pressure through holes drilled in the crankcase (called *galleries*) and through various pipes to all the engine bearings. Oil squeezes out between the two surfaces of the bearings. The pump has a large enough capacity to circulate more oil than is needed to maintain the correct pressure in the system, and the surplus bleeds back to the sump via a pressure relief valve. In time, wear on the bearings allows the oil to flow out more freely. Eventually the rate of flow is greater than the pump can sustain at the normal operating pressure, and the pressure declines.

  Worn bearings do not, as a rule, develop overnight. A very gradual decline in oil pressure occurs, especially at low engine speeds when the engine is hot. There will also be a gradual increase in engine noise and knocks. By the time a significant loss of oil pressure occurs, many other problems are likely to be evident, and a major rebuild is called for. Any rapid loss of oil pressure accompanied by a new engine knock indicates a specific bearing failure that needs immediate attention.
- Oil pressure relief valves sometimes malfunction, venting oil directly back to the sump, with a consequent loss in pressure. Problems with pressure relief valves are rare but simple to check. Almost invariably, the pressure relief valve is screwed into the side of the block somewhere and can easily be removed, disassembled, cleaned, and put back. The spring is liable to be under some tension, so take care when taking the valve apart. After cleaning, reset the spring's tension to maintain the manufacturer's specified oil pressure. Run the engine until it is warm and check the oil pressure. If it is low, shut down the engine and tighten the relief valve spring a little (if it is adjustable). If no amount of screwing down on the spring brings the oil pressure up to the manufacturer's specifications, the problem lies elsewhere.
- A well-heeled sailboat will sometimes cause the oil pump suction line to come clear of the oil in the pan (sump), allowing the pump to suck in a slug of air. The oil pressure will drop momentarily, generally with a sudden, alarming clatter from the engine. This is especially likely to happen if the oil level is a little low. Check the level, top off as necessary, and put the boat on a more even keel.

- The failure of an external oil line or gasket (e.g., to an oil cooler) will cause a sudden, potentially catastrophic loss of oil and pressure. The engine is likely to suddenly clatter loudly. You must shut it down immediately. There will be oil all over the engine room! Less easy to spot is the loss of oil that accompanies a corroded cooling tube in an oil cooler. The oil will be pumped out of the exhaust (or into the header tank), and sometimes water will enter the crankcase (see above).
- The oil pressure gauge is unlikely to malfunction (see below for troubleshooting engine instruments). However, note that some oil pressure sending units (the piece that screws into the engine and sends a signal to the gauge) have a surge suppressor built into them (such as many Yanmars). On occasion (notably when the engine is warmed up and throttled back, causing the oil pressure to fall a little), this device will result in the gauge showing a sudden, complete loss of oil pressure, which will not recover until the engine is speeded up. It can be a little unnerving, but there is no problem with the oil pressure. To stop this happening, remove the surge suppressor from the base of the sending unit.
- Oil pumps rarely, if ever, give out as long as you keep the oil topped off and clean, and regularly change the filter. Over a long period of time, wear in an oil pump may produce a decline in pressure, but not before wear in the rest of the engine creates the need for a major rebuild. At this time, always check the oil pump.

Rising Oil Level

If the oil level in the crankcase starts to rise, the cause is likely to be water that has entered from the cooling system or diesel from a ruptured or leaking internal fuel line (if the engine has internal fuel lines). Another source is from blown seals between the engine and some hydraulic transmissions, since the oil pressure in the transmission is generally higher than in the engine. Once the problem has been identified and corrected, the oil must be changed.

High Oil Consumption

Is the engine producing a lot of blue smoke? If so, it is burning oil, most likely because of gummed-up or broken piston rings. The oil also may be entering the combustion chambers down worn valve guides, through failed seals or a plugged oil drain in a turbocharger or supercharger, or from the engine breather. (Maybe the crankcase is overfilled with oil; otherwise, there is likely to be excessive crankcase pressure as a result of blowby down the sides of pistons.)

If there is no blue smoke, check for external leaks from oil lines and gaskets. If the engine has a raw-water oil cooler, check the overboard water discharge for an oil slick as a result of a holed tube in the oil cooler. If the oil cooler is freshwater cooled, look for oil in the header tank. (Remember, do not remove the cap when hot.)

If an engine appears to stop burning oil, it may simply be due to overfilling or misreading the dipstick. However, it may also be the result of a slow water leak, or a slow diesel leak (if there are internal fuel lines), into the oil, which is keeping up with the oil consumption. It needs to be checked out.

Inadequate Turbocharger Performance

Poor turbocharger performance will cause symptoms similar to those caused by a plugged air filter—reduced power, overheating, and black smoke—so be sure to check the airflow through the engine before turning your attention to the turbocharger. This includes looking for obstructions in the exhaust, air leaks between the turbocharger and the inlet manifold, exhaust leaks between the exhaust manifold and the turbocharger, and dirt plugging the fins on any intercooler or aftercooler.

Turbochargers spin at up to 200,000 rpm; the speed of the blade tips can exceed the speed of sound, and temperatures are as high as 1,200°F (650°C)—hot enough to melt glass. The degree of precision needed to make this possible means that repairing turbochargers is strictly for specialists.

A turbocharged engine should never be raced on initial start-up—the oil needs time to be pumped up to the bearings. Similarly, never race the engine before shutting it down; the turbine and compressor wheel will continue to spin for some time but without any oil supply to the bearings, and the residual heat will turn any oil in the bearings into abrasive carbon.

Clean oil is critical to turbocharger life—the bearings will be one of the first things to suffer from poor oil change procedures. Many engines have a bypass valve fitted to the oil filter so that if the filter becomes plugged, unfiltered oil will circulate through the engine—if the filter is neglected for long, the turbocharger will soon be damaged. Note that some turbochargers also have their own oil filter, which must be changed at the same time as the engine oil filter.

Any loss of engine oil pressure, such as from a low oil level or the use of the wrong grade of oil in the engine, will also threaten the turbocharger. When a turbocharger is under load, insufficient oil for as little as 5 seconds can cause damage. Damage to bearings will allow motion in the shaft, permitting the turbines to rub against their housings.

A dirty or damaged air filter or leaks in the air inlet ducting will allow dirt particles into the turbocharger that will erode the compressor wheel and turbine. The resulting imbalance and loss of performance will lead to other problems.

Before condemning a turbocharger, make the following tests:

1. Start the engine and listen to it. If a turbocharger is cycling up and down in pitch, there is probably a restriction in the air inlet (most likely a plugged filter). A whistling sound is quite likely produced by a leak in the inlet or exhaust piping.
2. Stop the engine, let the turbocharger cool, and remove the inlet and exhaust pipes from the turbine and compressor housings (these are the pipes going into the center of the housings; see Figure 5-29). This will give you a view of the turbine and compressor wheels. With a flashlight, check for

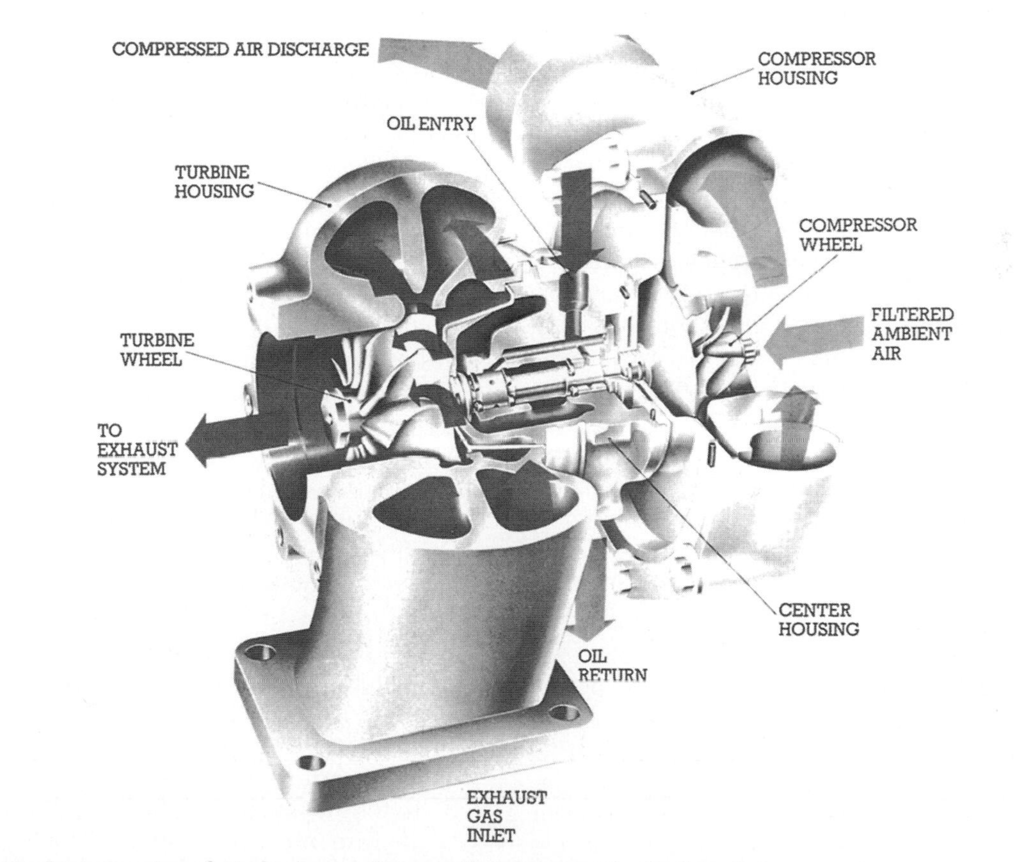

FIGURE 5-29. *A cutaway view of a turbocharger. (Courtesy Garrett Automotive Products Co.)*

bent or chipped blades, eroded blades, rub marks on the wheels or housings, excessive dirt on the wheels, and oil in the housings. The latter may indicate oil seal failure, but first check for other possible sources, such as oil coming up a crankcase breather into the air inlet, oil from an overfilled oil-bath-type air cleaner, or a plugged oil drain in the turbocharger that is causing oil to leak into the turbine or compressor housings.
3. Push in the wheels and turn them to feel for any rubbing or binding. Do this from both sides.

If these tests reveal no problems, the turbocharger is probably OK. If it failed on any count (except dirty turbine or compressor wheels—see Chapter 6 for cleaning), remove it as a unit and send it in for repair.

Problems with Engine Instrumentation

Some old engines are still found with thermometer-type temperature gauges and mechanical pressure gauges and tachometers. All will have some kind of a metal tube from the engine block to the back of the gauge. Gauge failure is normally self-evident—the gauge sticks in one position. Temperature gauges and their sensing bulbs have to be replaced as a complete unit, oil pressure gauges may just have a kinked sensing line, and tachometers may perhaps have a broken inner cable.

All new engines use electronic instruments comprising a sending unit on the engine block either connected to a gauge, warning light, or alarm (see Figure 5-30), or else wired to the engine's computer. In what follows, I ignore those tied to a computer—they require a specialist to troubleshoot—and concentrate on conventional electronic instruments.

Ignition Warning Lights

Most alternators require an external DC power source to excite the alternator before it will start generating power. An ignition warning light is installed in the excitation line to the alternator. When the ignition is turned on, current runs from the battery down this line to the alternator, causing the light to glow. When the engine fires up, and the alternator begins to put out, the light is extinguished (see Figure 5-31). If the light fails to come on when the ignition is switched on, check the bulb first (on older units; newer units have an LED, in which case test for voltage). If it is OK, the most likely problem is a break in the wire to the light or in the excitation line running to the alternator, or an electrically poor connection. If the light comes on and stays on after the engine is running, the alternator is almost certainly not putting out (for troubleshooting alternators and alternator circuits, see my *Boatowner's Mechanical and Electrical Manual*).

All Other Warning Lights and Alarms

These use a simple switch. Positive current from the battery is fed via the ignition switch to the alarm or warning light and from there down to a switch on the engine block. If the engine reaches a preset temperature, or oil pressure drops below a preset level, the switch closes and completes the circuit.

Most switches (sending units) are the earth-return type; that is, grounded through the engine block (see Figure 5-32). However, some are for use in insulated circuits, in which case they have a separate ground wire. An insulated return is preferred in marine use, especially when connected to gauges (as opposed to warning lights or alarms) since anytime the ignition is on, a small current is flowing through a gauge circuit, which may contribute to stray-current corrosion. (A warning light or alarm circuit differs in that the circuit conducts only when the light or alarm is activated; therefore, in normal circumstances, it will not contribute to stray-current corrosion.)

Many sending units incorporate both an alarm and a variable resistor that connects to a gauge. In this case, there will be two wires on an earth-return unit, and three on an insulated unit (see Figure 5-33).

If you suspect problems with an alarm or warning light (for gauges, see below), turn on the ignition switch and perform these tests:

1. Test for 12 volts between the alarm or light positive terminal and a good ground (see Figure 5-34, Test 1). No volts—the ignition circuit is faulty; 12 volts—proceed to the next step.
2. To test the alarm or light itself, disconnect the wire from the sending unit and short it

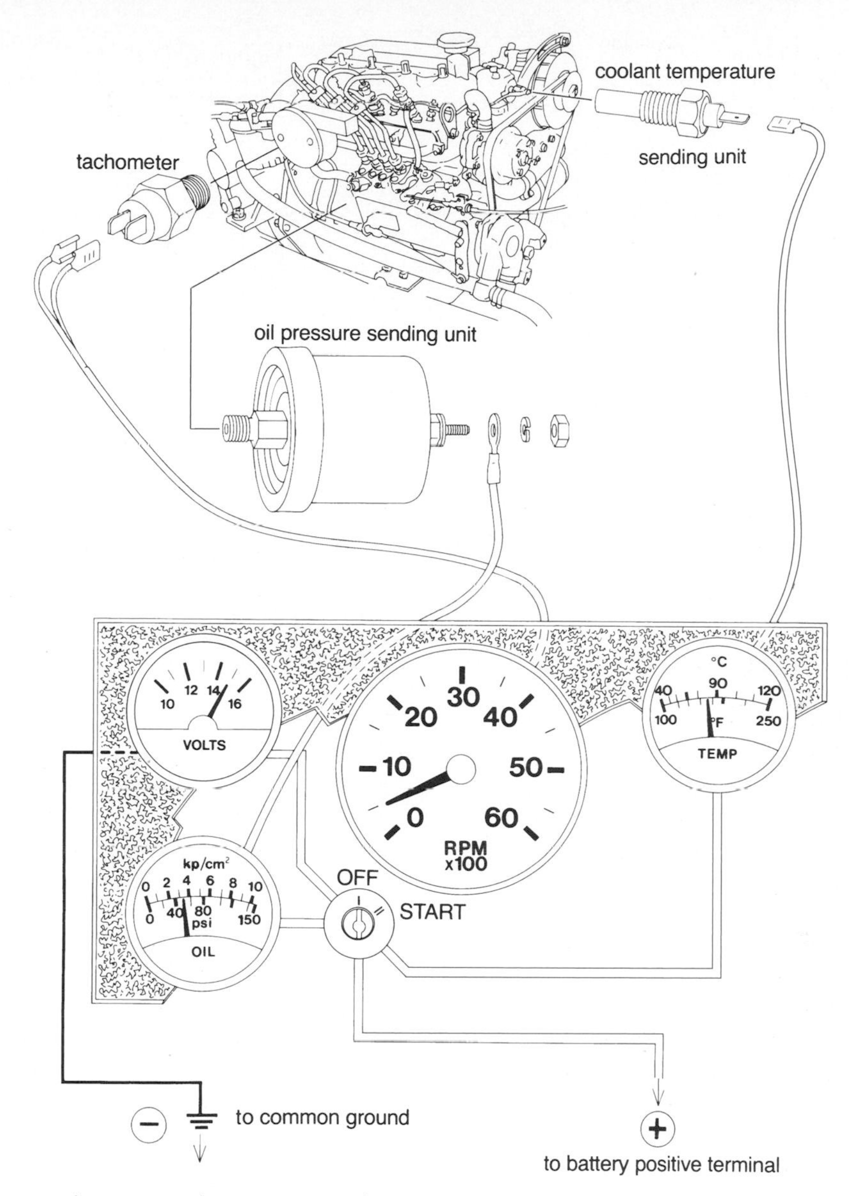

FIGURE 5-30. *How a typical conventional instrument panel receives information from the engine. (Jim Sollers)*

to a good ground (see Figure 5-34, Test 2). The alarm or light should come on. If not, conduct the same test from the second terminal on the back of the alarm or light (the one with the wire going to the sending unit) to a good ground. No response—the alarm or light is faulty; response—the wire to the sending unit is faulty.

3. If the alarm or light and its wiring are in order, the sending unit itself may be shorted (the alarm or light stays on all the time) or open-circuited (it never comes on, even when it should). Switch off the ignition, disconnect all wires from the sending unit, and test with an ohmmeter (R × 1 scale on an analog meter) from the sending unit terminal to a good ground (see Figure 5-34, Test 3). A temperature warning unit should read infinite ohms, unless the engine is overheated, in which case it

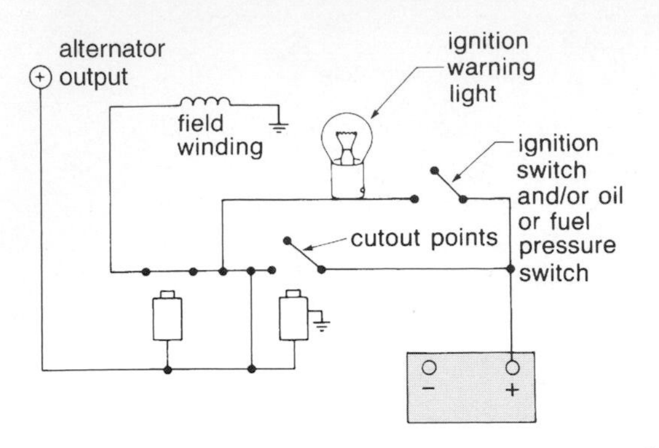

FIGURE 5-31. *A voltage regulator with excitation circuit. Closing the switch provides initial field current. The battery discharges through the ignition light into the field winding, lighting the lamp. When the alternator's output builds, the cutout points close. Now the system has equal voltage on both sides of the ignition warning light, and the light goes out.*

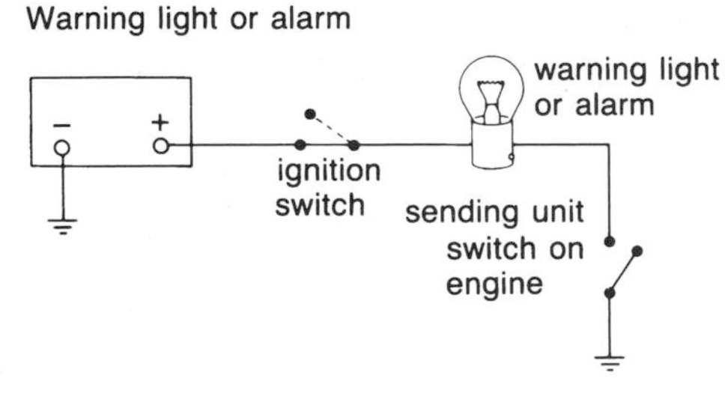

FIGURE 5-32. *A simple warning light or alarm circuit.*

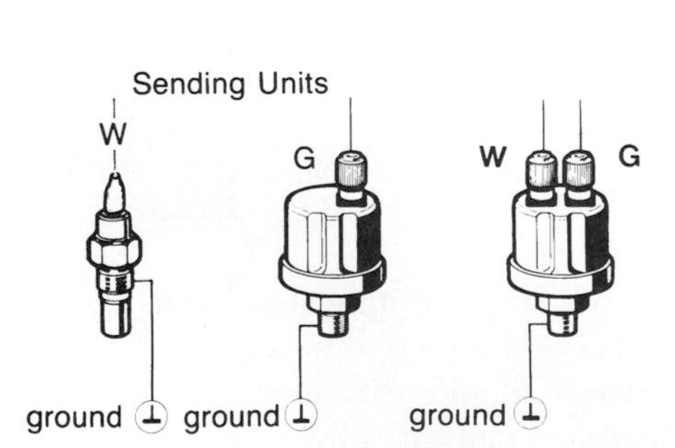

FIGURE 5-33. *Sending units for alarms and gauges.* ***Left:*** *Simple sensor with a warning contact (W), as used, for example, with an oil pressure warning light.* ***Middle:*** *Sensor with measuring contact (G) for a display instrument (gauge). This provides a continuous value for a given operational condition.* ***Right:*** *Sensor with W and G contacts for a continuous instrument display and for warning when a critical value has been reached. (Courtesy VDO)*

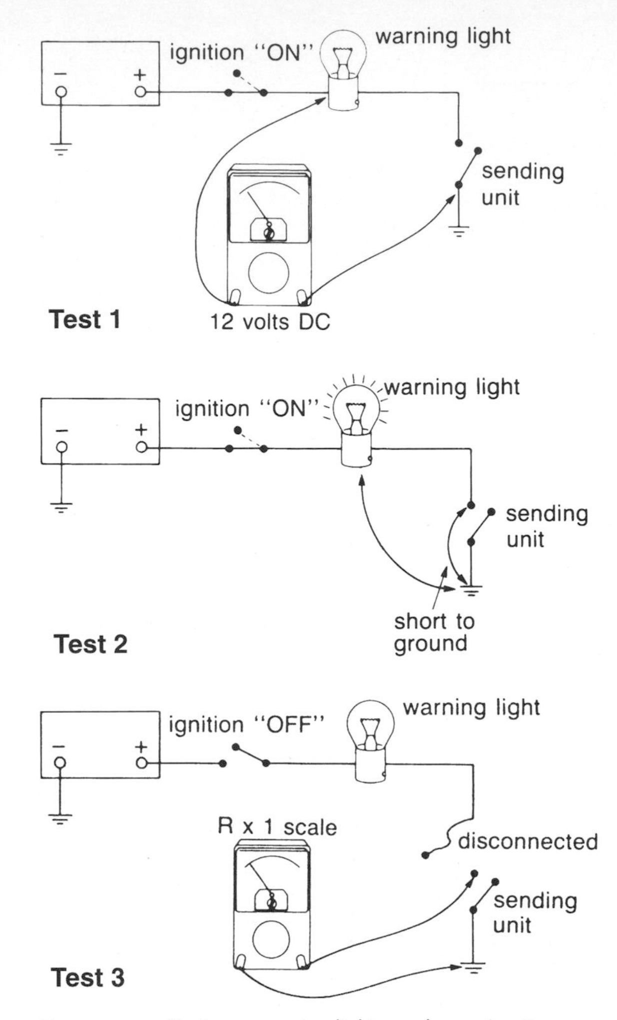

FIGURE 5-34. *Testing a warning light, or alarm, circuit.*

will read 0 ohms. An oil warning unit reads 0 ohms with the engine shut down and infinite ohms at normal operating pressures.

TEMPERATURE AND PRESSURE GAUGES

Most gauges have three terminals, although there may be as many as five. We are interested in those marked "I" or "Ign" (the power feed from the ignition switch), "G" or "Gnd" (a ground connection for the lighting circuit), and "S" or "Snd" (the connection to the sending unit on the engine block). Positive current is fed from the battery via the ignition switch to the gauge (I or Ign) and from there down to the sending unit (S or Snd) and then to ground (see Figure 5-35).

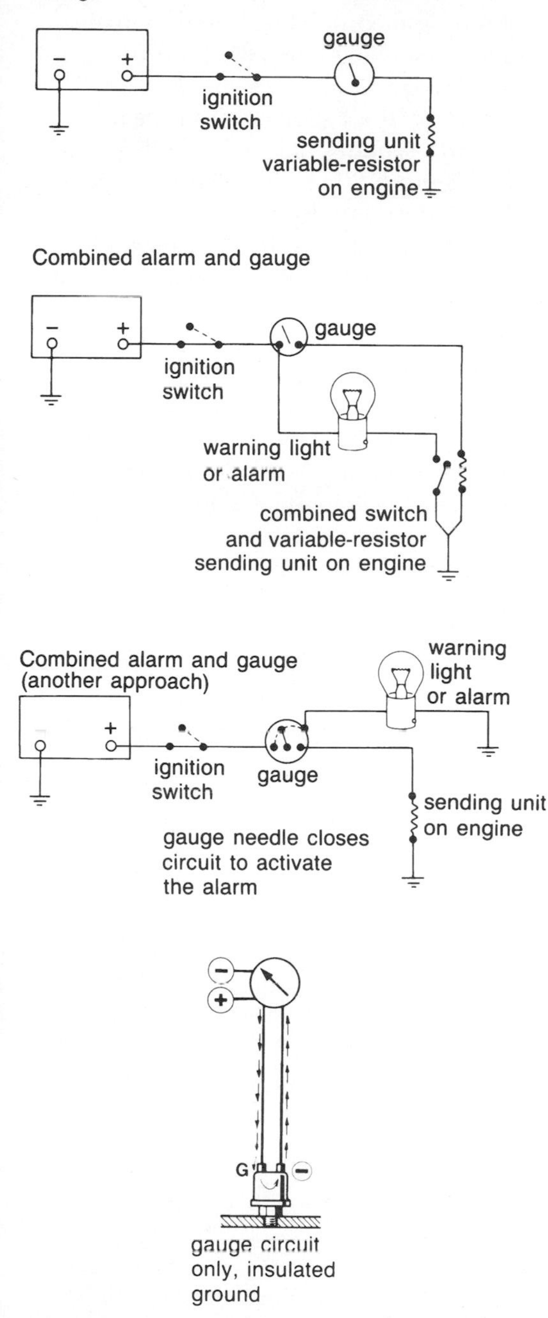

FIGURE 5-35. *Gauge circuits (top). An insulated-return sending unit and gauge circuit (bottom) with a separate ground wire (as opposed to using the engine block). This reduces the potential for stray-current corrosion. (Courtesy VDO)*

To test a gauge:

1. Test for 12 volts from the gauge positive terminal to a good ground (see Figure 5-36, Test 1). No volts—the ignition switch circuit is faulty; 12 volts—proceed to the next step.
2. Disconnect the sensing line (which goes to the sending unit) from the back of the gauge (see Figure 5-36, Test 2). A temperature gauge should go to its lowest reading; an oil pressure gauge to its highest reading, although just to confuse things, a few behave the same as a temperature gauge!
3. Connect a jumper (a screwdriver will do) from the sensing line terminal on the gauge to the negative terminal on the gauge (or a good ground on the engine block if there is no negative terminal; see Figure 5-36, Test 3). A temperature gauge should go to its highest reading; an oil pressure gauge to its lowest reading (except as noted above).

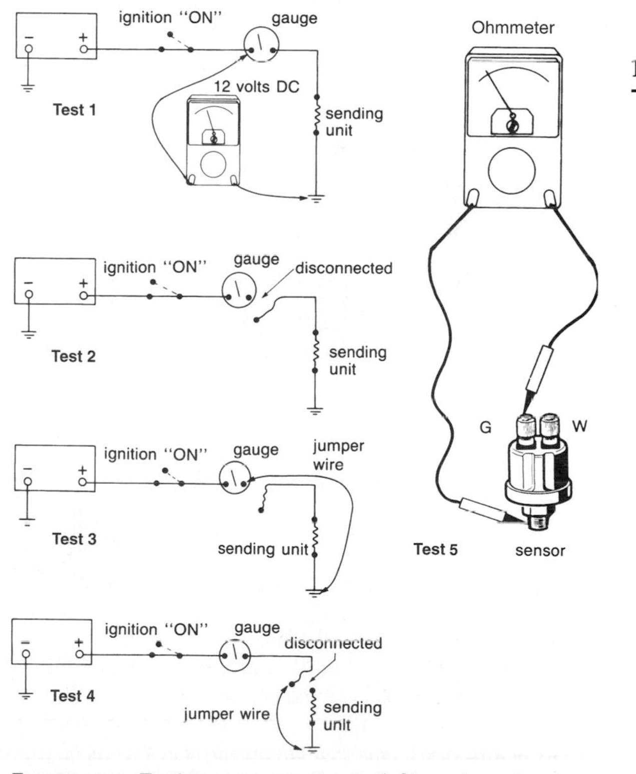

FIGURE 5-36. *Testing a gauge circuit (left) and sending unit (right). (Courtesy VDO)*

4. If the gauge passed these tests, it is OK. Reconnect the sensing line and disconnect it at the sending unit on the engine (see Figure 5-36, Test 4). A temperature gauge should go to its lowest reading; an oil pressure gauge to its highest reading (except as noted above). Short the sensing line to the engine block. A temperature gauge should go to its highest reading; an oil pressure gauge to its lowest reading (except as noted above). If not, the sensing line is faulty (shorted or open-circuited). The most likely trouble spot is in the plug connector for the main engine wiring harness (generally found toward the back of the engine). Undo the plug and check all the sockets and pins for damage or corrosion and correct as necessary. Most times, this will cure wiring problems. If not, make sure the ignition switch is turned off, then use an ohmmeter (R × 1 scale on an analog meter) to test the sending line first from the gauge to the relevant terminal on the plug, then from the other half of the plug to the sending unit. You should get a very low resistance both times. Infinity, or a high resistance, indicates resistance, or a break, in the circuit.
5. To test a sending unit, switch off the ignition, disconnect all wires, and test with an ohmmeter (R × 1 scale on an analog meter) from the sending unit to a good ground (see Figure 5-36, Test 5). Most temperature senders vary from around 700 ohms at low temperatures, to 200 to 300 ohms at around 100°F (40°C), down to almost 0 ohms at 250°F (120°C), while most oil pressure senders vary from around 0 ohms at no pressure to around 200 ohms at high pressure. However, there are differences from one manufacturer to another, and some gauges may work in the opposite direction; the important thing is to get a clear change in resistance with a change in temperature or pressure.

Tank-Level Gauges

Old-style, tank-level gauges have been electronic (conventional, rather than computer-based), but in recent years pneumatic gauges have become increasingly popular.

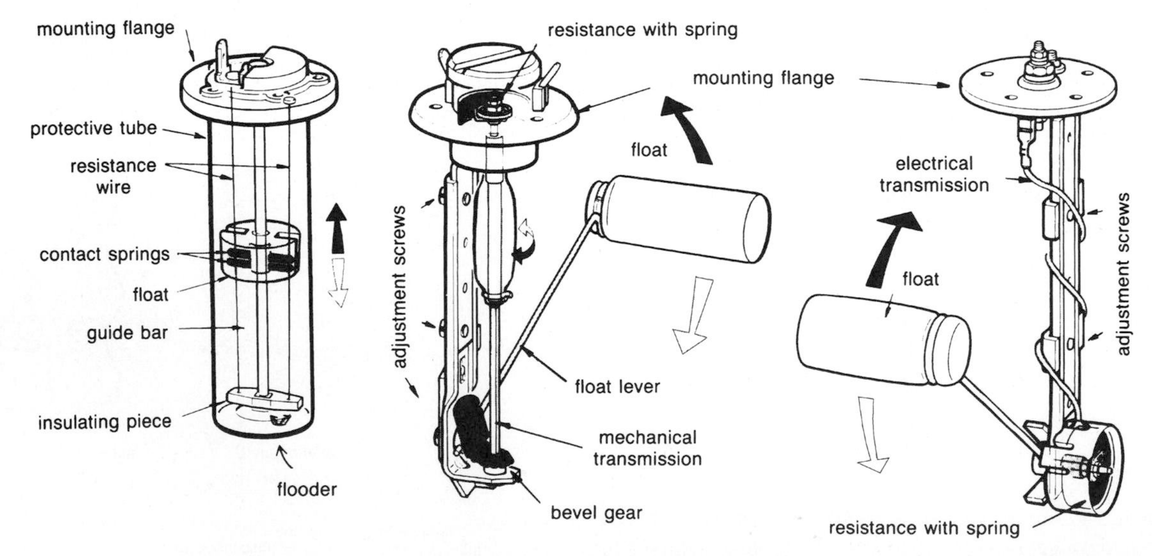

Figure 5-37. ***Left:*** *Electronic fuel tank sensor. Because of its construction, the immersion pipe sensor is suitable only for use with fuel tanks. The float unit is mounted on the guide bar, making contact with the two resistance wires via the contact springs. Resistance varies according to the height of the float, varying the gauge reading. The protective tube and the flooder provide excellent damping.* ***Middle and right:*** *Typical differences between a swinging arm–type water tank sensor (middle) and a fuel tank sensor (right). For electrically conductive media, the electrical resistance element must be positioned in the assembly flange, with the level of the liquid transmitted mechanically up to the element. (Courtesy VDO)*

Electronic Gauges

A float is put in the tank either in a tube or on a hinged arm. As the level comes up, the float or arm rises, moving a contact arm on a variable resistance (a kind of rheostat). Positive current is fed from the battery via the ignition switch to the gauge, down to the resistor, and from there to ground (see Figure 5-37). The gauge registers the changing resistance.

The same tests apply to the gauge, the sensing line, and the sending unit as to the oil pressure gauge. Sending-unit resistances are similar—near 0 ohms on an empty tank, up to around 200 ohms on a full tank.

If everything checks out OK but the unit always reads empty, the float on the sending unit is probably saturated or there is a mechanical failure—for instance, a broken or jammed arm. A saturated swinging arm–type float can be made temporarily serviceable by strapping a piece of closed-cell foam to it.

Pneumatic Sensors

Pneumatic sensors became increasingly popular in the 1990s due to their versatility and simplicity. A tube is inserted to the bottom of the tank and connected to a small hand pump mounted on the tank gauge panel. The gauge itself is teed into the tube just below the pump (see Figure 5-38).

When the pump is operated, air is forced down the tube and bubbles out into the tank. Depending on the level in the tank (and hence in the tube) more or less pressure is needed to drive all the fluid out of the tube. The gauge registers this pressure in inches of water (in the tank) or inches of diesel. The owner draws up a table converting the gauge readings to gallons—the conversion will vary from tank to tank depending on the tank size and shape. The gauges come with instructions on how to draw up this table. If the instructions are missing, or your tank is an odd shape, the best bet is simply to empty the tank, then add known quantities of fluid (e.g., 5 gallons at a time), stroking the hand pump and noting the gauge reading after each addition.

Pneumatic level gauges are more accurate than electronic gauges. What is more, apart from leaking connections or kinked tubes, there is nothing to go wrong. As many as ten tanks can be measured with one gauge simply by switching the gauge and pump into the individual tank tubes. Since there is no fluid in the tubes beyond the level of the tanks themselves, there is no possibility of cross-contamination from

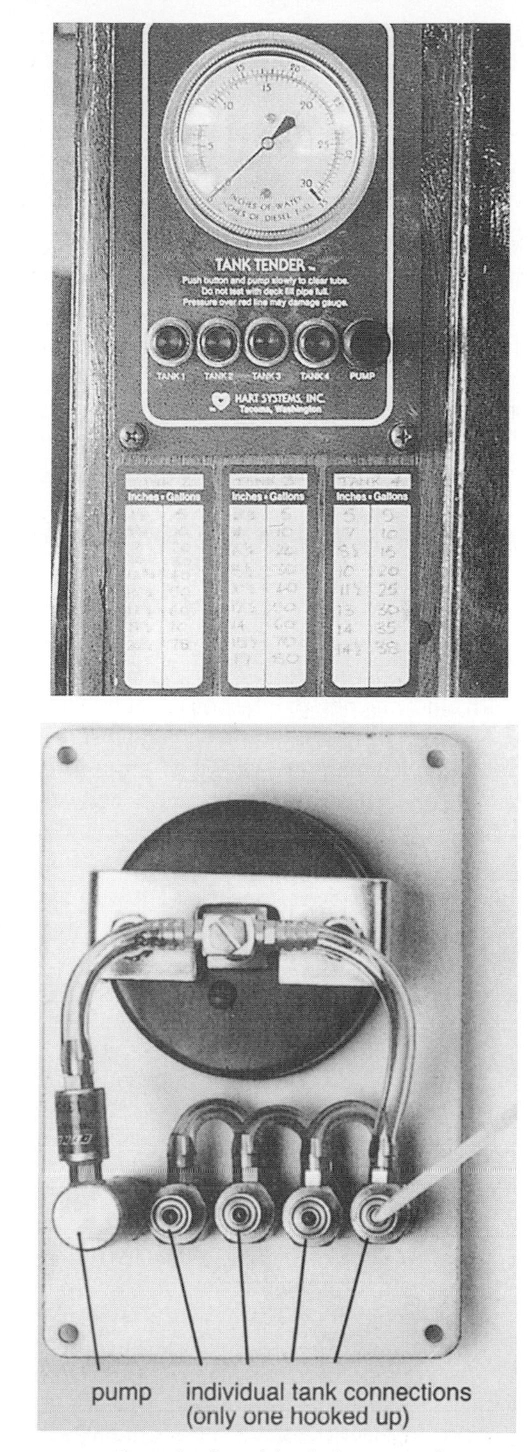

FIGURE 5-38. *Front (top) and back (bottom) of a pneumatic tank-level gauge. The gauge has been calibrated to the various tanks in the boat, with the relationship between the gauge reading and the tank volumes codified in the tables beneath the meter. (Courtesy Hart Systems Inc.)*

one tank to another. This means diesel, water, and holding tanks can all be measured with the same unit (note, however, that when inserted into a holding tank, the small air tubes have a tendency to get plugged with effluent, which then results in a false full-tank reading).

Wide, Shallow Tanks

The one problem with all conventional electronic and pneumatic gauges is that they do not work well with wide, shallow tanks because there is very little change in the fluid level (and therefore the back pressure on a pneumatic gauge) with significant changes in volume. What is more, even a little heeling will create substantial changes in fluid levels from one side of the tank to the other, resulting in false readings. An increasing number of modern boats have relatively flat bottoms and shallow bilges. No gauge will give accurate readings on tanks fitted to these spaces.

Computer-Based Electronic Gauges

The latest generation of gauges are tied into electronic engine controls with increasingly sophisticated software that can either take into account changing tank shapes, heel angles, and so on to keep an accurate track of tank volumes, or else do this by measuring consumption from a known volume. The tank readout is typically shown on a screen rather than a gauge

A Do-It-Yourself Engine Survey

With older engines, or if buying a secondhand boat, it is sometimes useful to do an engine survey. The following is a nonintrusive set of procedures that can be undertaken by just about anybody without using specialized tools, and which has no disassembly beyond loosening a hose clamp or two and undoing a few screws. Most times, it will give you a pretty good sense of impending problems. For a more detailed survey, you need to hire a professional.

1. Check for an engine hour meter and a maintenance logbook showing regular oil and filter changes and other periodic maintenance (see Figures 5-39A and 5-39B). Without this, you have no way of knowing if the engine has been reasonably cared for.
2. The fuel system is the single most expensive system on the engine:
    a. Make sure there is a primary (off the engine) as well as a secondary (on the engine) fuel filter.
    b. Inspect the primary filter for contamination. If contaminated, inspect the secondary filter.
    c. If the secondary filter is contaminated, you have reason to feel a little nervous. In any event, check any filter in the lift pump.

A

2. Fresh water engine driven pump pulls coolant from heat exchanger through engine block and thermostat back through exhaust manifold and into heat exchanger to be cooled.

07-01-82 Changed lub oil & filters T.J.D.
20.1 HRS
08-24-82 Changed lub oil & filters T.J.D.
91.0 HRS
10-16-82 Changed lub oil & filters TJD
119.5 HRS
08-02-83 Changed lub oil & filters TJD.
187.7 HRS
10-22-83 Changed lub oil & filters TJD
215.3 HRS
08-06-84 Changed lub. oil & filter TJD.
280.8 HRS
10-21-84 Changed lub oil & filters TJD
313.1 HRS
07-14-85 46 HRS CHANGED BELT
07-23-85 Changed lub oil & filter TJD
59.6 HRS Changed antifreeze
10-21-85 Changed lube oil & filters (ALL) T.J.D.
97.5 HRS
07-30-86 Changed lube oil & filter TJD
156.6 HRS
10-18-86 Changed lube oil & filters TJD
175.1 HRS
07-10-87 Changed lube oil & filter TJD
221.4 HRS
10-31-87 Changed lub oil & filters TJD
306 HRS

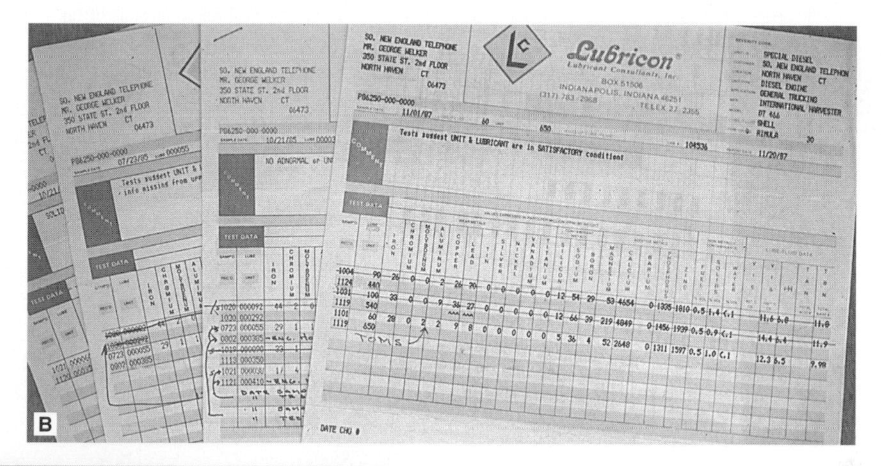

d. In all cases, draw a sample of fuel from the low spot of the fuel tank. If contaminated, pump down until clean (see Figure 5-39C).

3. Pull out the oil dipstick and wipe the oil onto your fingertips (see Figures 5-39D and 5-39E). If the oil has not been changed recently, it will be black. Up to 50 hours

Figures 5-39A to 5-39R. *A do-it-yourself engine survey. Any kind of an engine log (in this case, written in the back of the engine manual) is important (A). Regular oil sampling is an excellent indication of attention to maintenance issues (B). Fuel tank sample showing significant water contamination. If any of this has made its way through to the injectors, there may be serious damage (C). An oil sample after 50 hours on a new engine; note that it is almost translucent (D). The same engine after 150 hours—the oil is black from carbon in suspension (E). This is normal, but any kind of sludge is a sign of trouble. Checking for sludge and emulsified oil inside the oil fill opening (F). Note the oil immediately beneath this water pump, coming from a leaking oil seal where the pump bolts up to the engine (this is a gear-driven pump rather than belt driven), and the salt crystals lower down from a leaking water seal (G). A rebuild is needed. Cracked raw-water pump impeller vanes (H). Raw-water injection hose removed from the raw-water injection nipple (I). Checking the injection elbow for corrosion (J). The galvanized pipe and fittings used in this installation are especially prone to corrosion. Checking for carbon deposits (K). Quality hose clamps, but they're not much use with this kinked hose (L)! Checking hose clamps for corrosion (M). Inspecting the engine feet for softened or perished rubber (N). Inspecting the shaft seal for corroded hose clamps, a defective hose, or seal leaks (O). This packing nut may be seized to the propeller shaft, in which case the hose will get ripped apart when the transmission in engaged. Voltmeter leads on battery terminals (P). As soon as cranking commences, the voltage will start to fall. Measuring voltage drop on the positive side of the charging circuit (Q). This test needs to be at full alternator output, which is not the case here, resulting in a very low volt drop reading (0.030 volt). Usually it will be considerably higher than this. Checking the crankcase breather for blowby (R).*

A Do-It-Yourself Engine Survey (continued)

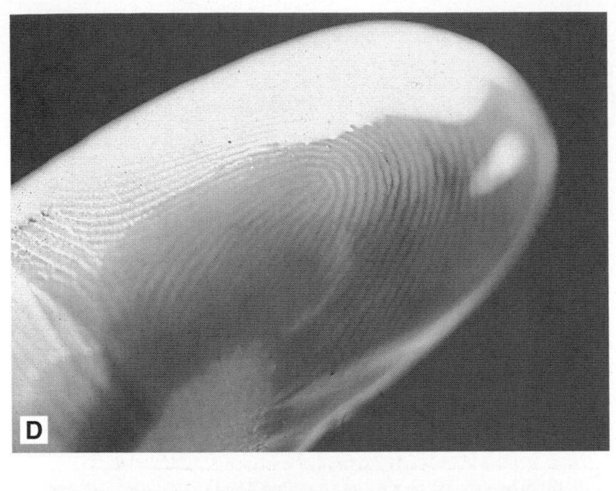
D

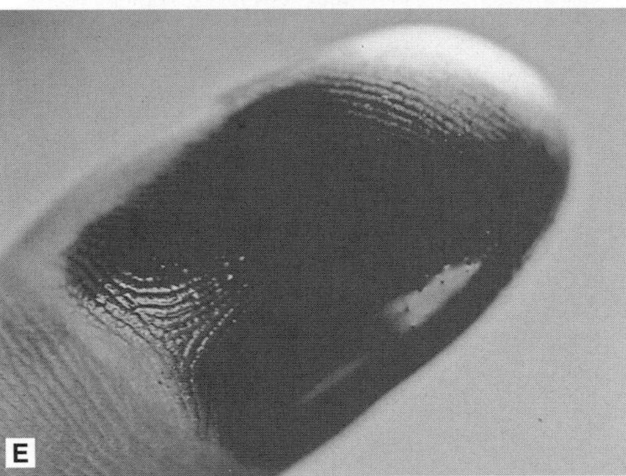
E

or more after an oil change, it should still have an element of translucency. If it is really heavily sooted, there may be blowby past the piston rings, lowering the engine compression.

4. For a more serious oil analysis, send a sample to a laboratory.
5. Remove the oil filler cap, and also any oil filler cap on top of the valve cover, and look at the underside of the cap(s). Run your finger around inside the opening (see Figure 5-39F). If there are sludgy black deposits, oil change procedures have almost certainly been seriously neglected. If there is emulsified oil (creamy in color and texture), there is water in the oil. Bad news in both cases!

F

6. Take a sample of oil from the transmission. If this is at all black, it is indicative of a slipping/burning clutch.
7. Check the raw-water circuit:
   a. Check the pump's exterior for signs of water and/or oil leaks (see Figure 5-39G). Remove the raw-water pump cover and check the impeller for missing vanes, cracks, or excessive wear (the impeller tips will be worn flat; see Figure 5-39H).
   b. Remove any zincs from the heat exchanger. If corroded to the point of nonexistence, or near nonexistence, they have been neglected and expensive corrosion problems are possible.
   c. Remove the end caps from the heat exchanger and inspect the tube stack for scaling, blockage, or corrosion.
   d. Check for the existence of a vented loop. If not present, the engine may have flooded with salt water at some point. (See Chapter 9 for details of a seaworthy exhaust installation.)
   e. Remove the valve from any vented loop and flush it in fresh water.
   f. Remove the water injection hose from the exhaust elbow and check for debris blocking the injection point (see Figure 5-39I).
   g. Inspect the elbow for corrosion (see Figure 5-39J).

8. Remove the exhaust hose from the water-lift muffler and check for anything more than a light film of carbon (see Figure 5-39K). If heavier carbon deposits are present, the valves and engine may be seriously carboned up. After cleaning the exhaust, run a back-pressure test and reduce the back pressure if it is excessive.
9. While working around the cooling and exhaust systems, inspect the hoses and squeeze them, looking for soft spots, cracking, or other problems (see Figure 5-39L).
10. When undoing any hose clamps for the previous items, inspect the screws for rusting, as well as the bands (where they are in contact with the screws; see Figure 5-39M). If rust is present, inferior hose clamps have been used, in which case the screws are not 300-series stainless steel. You should consider replacing the hose clamps (and any others like them, especially those on the exhaust system).

A Do-It-Yourself Engine Survey (continued)

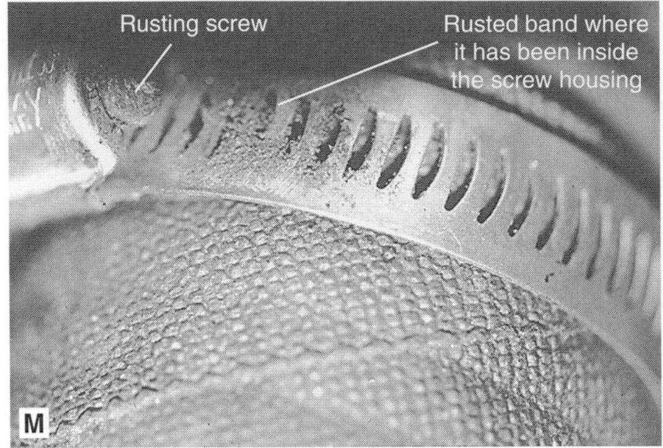

11. Inspect the rubber engine feet for signs of oil and diesel spills and for softening of the rubber (which will be caused by these spills; see Figure 5-39N).
12. Check the propeller shaft seal (stuffing box or other type of shaft seal) for signs of leaks (see Figure 5-39O), particularly for rusting of any components that may have been subjected to salt water thrown out by the propeller shaft (many engine control brackets and cables are vulnerable). Make sure there is adequate access to maintain the shaft seal (some require the engine to be removed!).
13. Operate the engine controls to make sure they are smooth and free from undue resistance and that the gearshift clicks into neutral at the appropriate point.
14. Notice that the engine has not yet been started! Make sure it is cold before trying to crank it, then activate any cold-start devices and crank. It should fire pretty well immediately. If it does not, and there is no problem with the fuel supply, the engine is probably suffering from a loss of compression, which may necessitate a major rebuild (it may be a good idea to have a mechanic run a compression test). It's best to run this test on a colder day (the colder, the better; the warmer the weather, the less likely it is that compression problems will be revealed).

15. Immediately after cranking, check the exhaust for smoke. An initial thin blue haze is OK, but if it persists, the engine is burning oil. White vapor may be unburned fuel, in which case the engine will almost certainly be misfiring and quite likely has a compression problem (especially if it was slow to start).
16. Shut down the engine, close the throttle, and pull out any stop lever or close the fuel rack so the engine will not start. Place a DC voltmeter across the hot terminal on the starter motor and the battery positive terminal, and crank. You will be measuring the volt drop on the starting circuit (see the Starter Motors section in Chapter 4). If the volt drop is more than 1.0 volt on a 12-volt system (2.0 volts on a 24-volt system), the starting circuit needs attention. Repeat this from the engine block to battery negative.
17. Now move the meter leads to the battery posts (positive and negative) and crank for 15 seconds. The voltage will stabilize and then slowly decline (see Figure 5-39P). If it starts to drop rapidly, the cranking battery is either not charged or is reaching the end of its life.
18. By now the battery will be somewhat discharged. Start the engine, speed it up, and immediately (while the alternator is at full output) test with a DC voltmeter from the alternator's positive (output) terminal to the battery's positive terminal (see Figure 5-39Q). You will be measuring the volt drop in the charging circuit. If it is more than 0.5 volt on a 12-volt system, or 1.0 volt

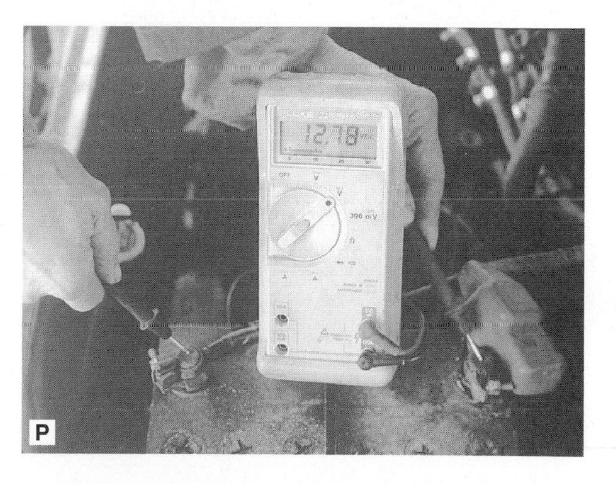

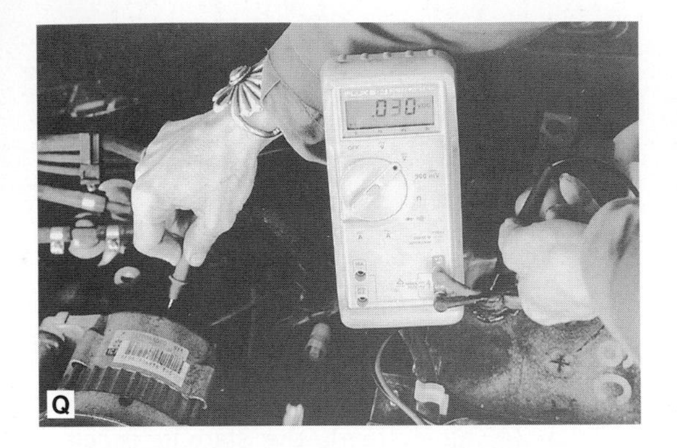
Q

R

on a 24-volt system, the cables are undersized, there is other unwanted resistance, or the circuit has isolation diodes that need removing (see my *Boatowner's Mechanical and Electrical Manual* for a discussion of this issue).

19. Load up the engine (including turning on any auxiliary equipment driven by the engine) and make sure it will come to full rated rpm. If not, the propeller may be incorrectly sized. The engine may emit some black smoke as you come up to speed. If this continues when at speed, there may be an obstruction in the inlet air or the exhaust, or a fuel injection problem.
20. While the engine is at full load, find the crankcase breather (generally from the crankcase to the air inlet) and see if you can get your finger over the end of it (see Figure 5-39R). If it is blowing out a steady stream of gas, there is serious blowby and a major overhaul is in the cards.
21. Run the engine for a while, monitoring the voltage across the battery posts until it stabilizes (this may take some time). This stabilized voltage is the absorption voltage for the charging system. If it is below 14.2 volts on a 12-volt system (28.4 for 24 volts), the batteries are probably being perennially undercharged (see my *Boatowner's Mechanical and Electrical Manual*).

CHAPTER 6

REPAIR PROCEDURES, PART ONE: COOLING, EXHAUST, AND INJECTION SYSTEMS

So far, I've covered routine maintenance and the most common diesel engine problems encountered by boaters. In this chapter, and especially the next, I get deeper into repairs than most people will want to go, although in a bind in some far-off corner of the world, you may be forced to try some of these procedures.

However, before tearing into an engine, take the time to carefully analyze problems (see Figure 6-1). Quite often an overenthusiastic mechanic will take the wrong thing apart, in the process spending a lot of unnecessary money and destroying the evidence needed to make a correct analysis of the problem. For example, consider the case of a boatowner who had been experiencing problems with very uneven thrust. A mechanic told him the transmission was defective and replaced it. The problem was still there with the new transmission, and the mechanic gave up. It turned out to be nothing more than water in the fuel system!

Remember also that if you pull an engine to pieces but fail to find your problem, you'll be forced to consult a mechanic. The mechanic may have to reassemble the engine and run it to find out what the problem was in the first place, then disassemble it to fix it! This can get very expensive.

The Cooling System

Routine maintenance and common problems were covered in Chapters 3 and 5. What follows are a few procedures less likely to be needed.

Cleaning Heat Exchangers and Oil Coolers

In time, corrosion, silt, and sand may block the cooling tubes in heat exchangers. The engine will show a slow but steady rise in temperature. If the heat exchanger or oil cooler has removable end caps, the tubes can be cleaned out. You may need to push a rod through them to clear blockages. In this case, use a wooden dowel (not metal), or a copper welding rod, and do not use too much force—the relatively soft tubes can be easily damaged.

Note that some tube stacks are a sliding fit in the heat exchanger barrel, sealed with a large O-ring at either end. These can be pushed out for further cleaning and servicing. To do this, set the heat exchanger body on end, find a block of wood that covers much of the other end of the tube stack, and tap this with a plastic or rubber mallet.

Flushing the Lubricating System

If an oil cooler is in an enclosed engine cooling circuit, (i.e., it has the engine's fresh water circulating through it and not raw water), and if the engine has ethylene glycol in the cooling circuit (as it should have), a failed oil cooler tube will allow the glycol into the crankcase (as may a blown head gasket). When mixed with oil, ethylene glycol forms a varnish that coats bearing surfaces and may lead to engine seizure. It also forms a soft, sticky sludge that will adhere to oil passages and coat valves, causing them to stick.

Drain the oil, change the filter, and flush the engine with a solution of two parts butyl cellosolve

FIGURE 6-1. *Before tearing into an engine, take the time to analyze problems.*

to one part SAE 10 engine oil at the earliest possible opportunity. Butyl cellosolve is a hazardous substance—make sure you follow all the safety procedures outlined on the butyl cellosolve container. Fill the engine to its normal level, start it, and run it at a fast idle (with no load) for 30 minutes. Thoroughly drain it, change the filter, refill it with SAE 10 oil, and run it for another 10 minutes. Drain it, change the filter once again, and refill with its normal oil. Even with this procedure, it is very hard to remove the varnish.

FLUSHING THE COOLING SYSTEM

If oil has entered the water side of an enclosed cooling system, it will show up in the header tank. Drain the cooling system and remove the thermostat (on most engines, but not where this produces a high temperature—e.g., some Caterpillars). Flush the system with fresh water to remove as much of the oil as possible. Refill it and run the engine at a fast idle until it is warm. Remove the pressure cap from the header tank (be careful!) and add a cup of nonfoaming detergent (e.g., automatic dishwasher soap; use approximately 2 ounces of soap per gallon of water). Run the engine another 20 minutes. If there is still undissolved oil on the surface of the water, add another cup of soap and run another 10 minutes. Drain and refill. If oil is still present, repeat the procedure. When the system is clean (see Figure 6-2), replace the thermostat (if removed), then refill with water and antifreeze.

FLUSHING RAW-WATER-COOLED ENGINES

Raw-water-cooled engines (without a freshwater circuit) are likely to develop scale around the cylinders over time, especially if run above 160°F (71°C).

Silt and other loose deposits can sometimes be flushed out of the block of a raw-water-cooled engine by doing the following: remove some item (e.g., a thermostat) that gives access to the cooling passages; open the cylinder block drain valve (or better still, remove it to create a larger hole); and

FIGURE 6-2. *The hoses have been removed from this heat exchanger in order to inspect the freshwater side. The tube stack is basically clean, although there is a small amount of easily removed debris.*

run water from a hose through the block. Use as much pressure as possible.

Scale and salt deposits can generally only be removed chemically. You will need to consult with a professional. In a pinch, if no professional help is available, an engine with a cast-iron block and cast-iron cylinder head can be descaled with phosphoric or oxalic acid. (The latter is used in a solution of 8 ounces to 2.5 gallons of water. You can get it from engine-supply houses and in boatyards that use it for cleaning hulls.) Pump the acid solution around and around the engine to dissolve the scale. When all the scale is dissolved, thoroughly flush the engine with fresh water, neutralize any remaining acid with an alkaline solution (2 ounces of sodium carbonate dissolved in 2.5 gallons of water is recommended), then thoroughly flush again. *Do not use this procedure on aluminum blocks or cylinder heads without first seeking professional advice.*

Bleeding a Cooling System

When you start an engine after any part of the cooling system has been drained, you must check the flow of coolant, particularly on the raw-water side. If a water pump is air bound, or water fails to circulate for any other reason, the vanes can strip off rubber impellers in seconds, and cylinder heads can crack in minutes.

Bleed an air-bound cooling system just as you would a fuel system, starting from the water source and working toward the pump. Once the pump is primed, it should push the coolant through the rest of the system. However on occasion, air locks can form in hoses, stalling out the water flow; the long hose runs associated with the heat exchangers in domestic water heaters are particularly prone to this. If this is a problem, progressively loosen hose connections around the system, allowing water to vent out of each connection before retightening. If you encounter a stubborn air lock on the freshwater side, you can often drive it out using one of the small pumps that fit in the chuck of an electric drill. These are available at most hardware stores for around $20.

Water Pumps

Pumps are generally of the piston, diaphragm, or flexible-impeller type on raw-water circuits; and of the centrifugal type (standard on automobiles; see Figure 6-3) on freshwater circuits. As long as you keep the drive belt tight, the latter rarely give problems, but if yours does, simply exchange it for a new one. Signs that a replacement is needed are leaking seals around the driveshaft, leaks from the drain hole on the underside of the pump body, squealing bearings, undue play in the bearings, or any roughness in the bearings when you turn the pump by hand with the belt removed. Note: If the pump is ever run dry, the seals will fail rapidly.

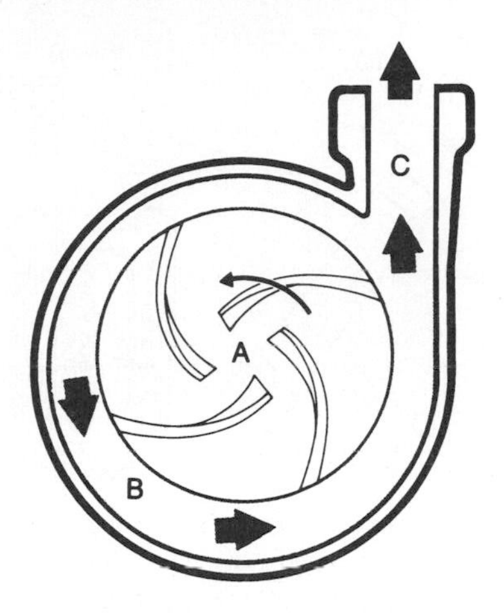

Figure 6-3. *Operation of a centrifugal pump. Liquid enters the inlet port in the center of the pump (A). The level of liquid must be high enough above the pump for gravity to push it into the pump, or the pump must receive initial priming. Centrifugal forces generated by the rotating and curved impeller forces the fluid to the periphery of the pump casing, and from there toward the discharge port (B). The velocity of fluid discharge translates into pressure in the system downstream from the pump (C). The flow rate is dependent upon restrictions in the inlet and outlet piping and the height that the liquid must be lifted. (Courtesy ITT/Jabsco)*

Piston pumps (see Figure 6-4) suck water into a cylinder through an inlet valve and push it back out through a discharge valve on the next stroke. The piston is generally sealed in its cylinder with a rubber O-ring, or on some older engines with a dished leather washer bolted to the bottom of the piston.

The piston seal and valves are the only items likely to cause trouble. The valves and their seats slowly deteriorate due to corrosion, leading to a gradual loss of pumping capability. Trash stuck in the valve seats can completely disable the pump. (A raw-water circuit must have an effective filter.)

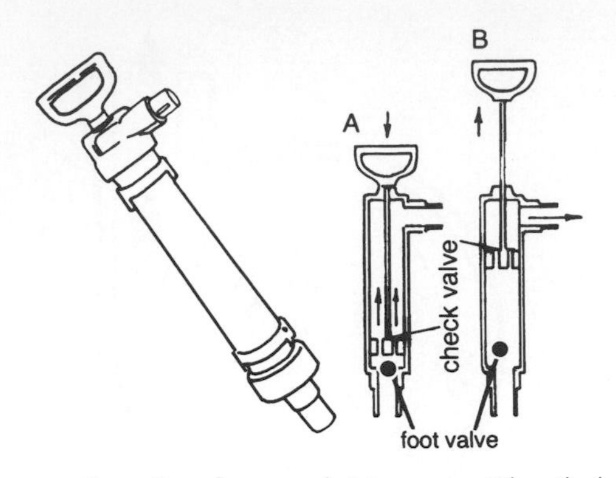

FIGURE 6-4. *Operation of a manual piston pump. When the handle is depressed, fluid is trapped between the base of the piston and the foot valve and forced up into the pump body through the check valve in the base of the piston (A). As the handle is lifted, the check valve closes and fluid is lifted to the discharge port. At the same time a vacuum is created below the check valve, which draws more fluid into the pump chamber through the foot valve (B).*

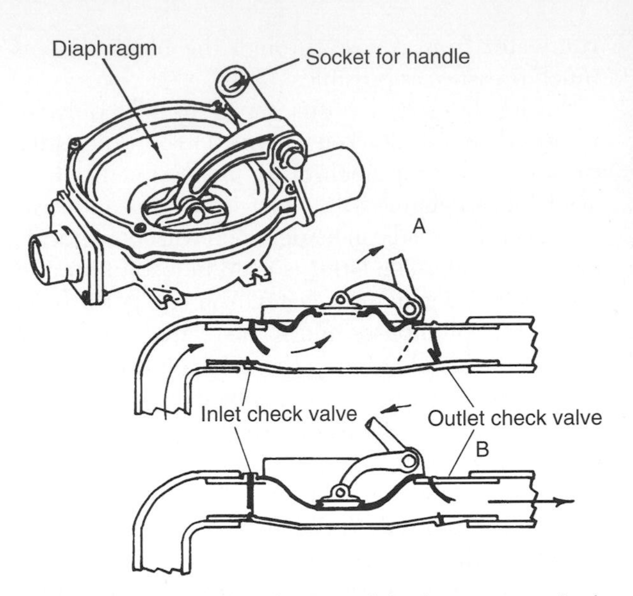

FIGURE 6-5B. *Operation of a manual diaphragm pump. As the handle is pulled back, the flexible diaphragm expands, creating a vacuum that pulls back the inlet check valve and draws fluid into the pump chamber (A). When the handle is pushed down, hydraulic pressure forces the inlet valve to close against its seat, at the same time opening the outlet check valve open and expelling the fluid (B).*

Over a winter layup, when the engine is drained and left for several months, the piston seal will sometimes stick to its cylinder wall and then tear when the engine is first started in the spring. The puzzled owner knows that everything was functioning just fine when the boat was laid up, and can't make out why there is no water flow in the spring. Always carry a spare O-ring or some leather on board. In an emergency, any thin piece of leather (e.g., from an old wallet), cut in a circle and worked up the sides of the piston with a generous smearing of grease, will temporarily replace a leather washer.

Diaphragm water pumps (see Figures 6-5A and 6-5B) operate in just the same way that diaphragm lift pumps do. Valves slowly corrode or collect trash. The only other likely point of failure is the diaphragm. The pumps generally have a hole in their base—if the diaphragm fails, water will dribble out. Diaphragms are easily replaced by removing the pump cover.

FIGURE 6-5A. *A manual diaphragm pump. (Courtesy Whale)*

FLEXIBLE-IMPELLER PUMPS

Flexible-impeller pumps form the vast majority of raw-water pumps in small boats (see Figure 6-6). Since they are a relatively common source of problems, they need to be looked at in some detail.

Modes of Failure

The principal causes of failure are running dry and swollen impellers caused by chemicals (generally oil and diesel) in the raw water. In the former case, the impeller vanes are likely to strip off; in the latter, the impeller jams in the pump's body. In either case, a new impeller is needed. Always carry a couple of spares on board. The original pump will almost certainly have a neoprene impeller, but when you buy replacements, get nitrile or (preferably) Viton impellers because they have a far greater resistance to chemicals. They are, however, a little more expensive than neoprene impellers. The Globe Rubber Company (www.globerubberworks.com) manufactures long-life replacement impellers for many popular pumps.

Flexible-impeller pumps like to be used often. If left for long periods without running (e.g., over the

Troubleshooting Chart 6-1. Flexible Impeller, Vane, and Rotary Pump Problems: No Flow	
If the pump is belt driven, is the pump pulley turning? YES ↓	NO → Tighten or replace the drive belt.
If a clutch is fitted, is it working? (The center of the pulley will be turning with the pulley itself.) YES ↓	NO → Adjust or replace the clutch.
For gear-driven pumps, and for belt-driven pumps on which the pulley is turning, proceed as follows:	
Remove the pump cover; are the impeller vanes intact? YES ↓	NO → Replace the impeller and track down any missing vanes. The pump probably ran dry; find out why. Check for a closed seacock, plugged filter, collapsed suction hose, or excessive heeling that causes the suction line on a raw-water pump to come out of the water. Less likely is a blockage on the discharge side causing the pump to overload.
Are the vanes making good contact with the pump body? YES ↓	NO → The impeller is badly worn and needs replacing. On vane and rotary pumps the vanes may be jammed in the impeller and just need cleaning.
Does the impeller turn when the pump drive gear or pulley turns? YES ↓	NO → The impeller, drive gear, or pulley is slipping on its shaft, or the clutch (if fitted) is inoperative. Repair as necessary.
If the impeller turns, the pump may just need priming. Otherwise the suction or discharge line must be blocked. Check for a closed seacock, plugged filter, collapsed suction hose, or excessive heeling that causes the suction line on a raw-water pump to come out of the water.	

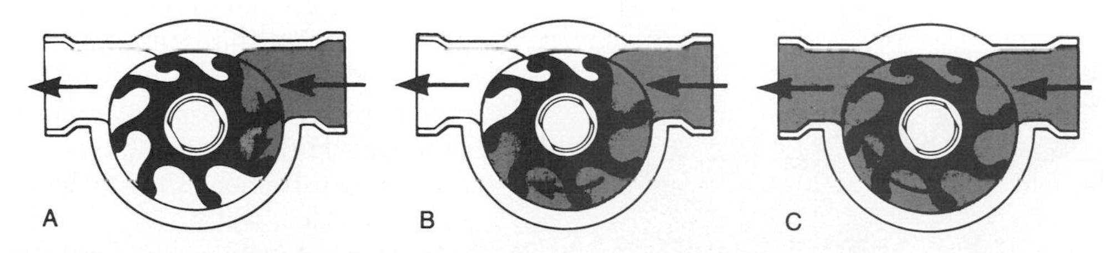

FIGURE 6-6. *Operation of a variable-volume flexible-impeller pump. The housing is flat on the top, which squeezes the impeller blades toward the shaft. As the blades expand, a vacuum is created that draws liquid into the pump body (A). As the impeller rotates, each successive blade draws in liquid and carries it to the outlet port (B). When the blades are compressed, they expel the remainder of the liquid, creating a continuous, uniform flow (C). (Courtesy ITT/Jabsco)*

winter), the vanes tend to stick to the pump housing. When the pump is restarted, it may blow fuses (if electrically driven) or strip off its vanes. If this is a constant problem, if the pump cover is sealed with a gasket (as opposed to an O-ring) try fitting an overthick cover gasket (see the Reassembly section below), although this may result in a small loss of pump efficiency. Otherwise, after a shutdown period, the pump cover will need loosening until the pump starts spinning, and the impeller is thoroughly lubricated with water.

Other common problems are leaking seals (quite likely as a result of running dry; there will be telltale signs of water dribbling out of the weep hole on the base of the pump) and worn or corroded bearings. Apart from normal wear, bearings will be damaged by water from leaking seals, an overtightened belt, and misalignment of drive pulleys or couplings. Optimum belt tension permits the longest stretch of belt to be depressed by 1/3 to 1/2 inch with moderate finger pressure. Check pulley alignment by removing the belt and placing a rod in the groove of the two pulleys; any misalignment will be clearly visible. Pumps that operate in sandy water will experience a gradual loss of performance due to wear on the pump housing, cover plate, wear plate (if fitted), and impeller.

Disassembly

Despite the variety of flexible-impeller pumps in service, most share many construction similarities (see Figures 6-7 and 6-8). Removing the end cover (four to six screws) will expose the impeller. Almost all impellers are a sliding fit on the driveshaft, either with splines, square keys, Woodruff keys, one or two flats on the shaft, or a slotted shaft. If an impeller puller is not available, use a pair of needle-nose pliers or Vise-Grips to grip the impeller and pull it out. Some impellers are sealed to their shafts with O-rings; most are not. If the impeller will not come loose, it may be one of the few locked in place with a setscrew (Allen screw; in particular, some Volvo Penta and Atomic 4 engines). On some of these, you can pull out the impeller far enough (1/2 inch) to release the screw, but if the screw is inaccessible, you must disassemble the drive side of the impeller and knock the impeller out on its shaft.

Figure 6-7. *A typical flexible-impeller pump.*

The impeller vanes should have rounded tips (not worn flat) and show no signs of swelling, distortion, or cracking of the vanes, or any kind of set (bend). If in doubt, replace the impeller. If an O-ring is fitted to the shaft, check it for damage. If the impeller has a tapered metal sleeve on its inner end (extended insert impellers), inspect the sleeve and discard the impeller if there is any sign of a step where the sleeve slides into the shaft seal.

If it is necessary to remove the cam (e.g., to replace a wear plate—see next paragraph), loosen the cam retaining screw, tap the screw until the cam breaks loose, then remove the screw and cam. Clean all surfaces of sealing compound.

Some impellers have a wear plate at the back of the pump chamber. If fitted, the plate can be hooked out with a piece of bent wire. Note the notch in its top; this aligns with a dowel in the pump body. If the wear plate is grooved or scored, replace it.

There are three types of shaft seals:

1. Lip-type seals, which press into the pump housing and have a rubber lip that grips the shaft.
2. Carbon-ceramic seals, in which a ceramic disc with a smooth face seats in a rubber boot in the pump housing. A spring-loaded carbon disc, also with a smooth face, is sealed to the pump shaft with a rubber sleeve or O-ring. The spring holds the carbon disc against the ceramic disc, and the extremely smooth faces of the two provide a seal.
3. An external stuffing box (packing gland), which is the same as a propeller-shaft stuffing box. These are not very common. (For care and maintenance, see pages 242–45.)

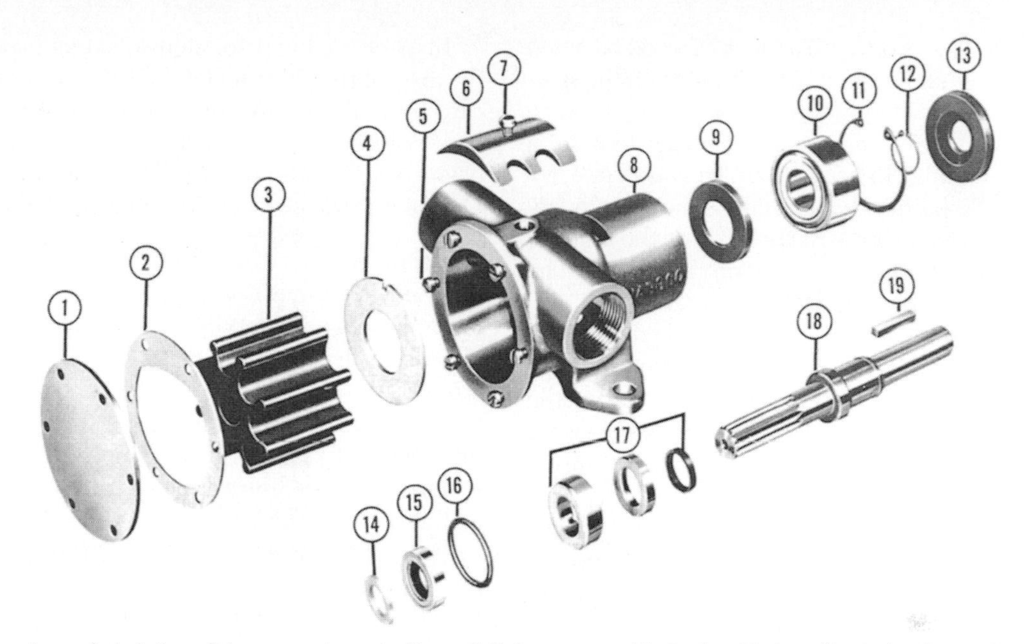

FIGURE 6-8. *An exploded view of the pump shown in Figure 6-7. Pump cover (1). Gasket (2). Impeller (splined type; 3). Wear plate (4). Pump cover retaining screws (5). Cam (mounted inside pump body; 6). Cam retaining screw (7). Pump body (8). Slinger (to deflect any leaks away from the bearing; 9). Bearing (10). Bearing-retaining circlip (11). Shaft-retaining circlip (12). Outer seal (13). Inner seal assembly, lip type (14, 15, 16), or inner seal assembly, carbon-ceramic type (17). Pump shaft (18). Drive key (19). (Courtesy ITT/Jabsco)*

Although it's possible to hook out and replace some carbon-ceramic seals with the shaft still in place (this cannot be done with lip-type seals), in most cases the shaft must be taken out. To do this, take apart the drive end of the shaft. First unbolt the pump from its engine or remove the drive pulley (if it's belt driven); then proceed as follows:

If the pump has a seal at the drive end of the shaft (number 13 in Figure 6-8), hook it out. When removing any seals, be extremely careful not to scratch the seal seat. You will usually find a bearing-retaining circlip behind the seal in the body of the pump (pumps bolted directly to an engine housing may not have one). Remove the circlip, flexing it the minimum amount necessary to get it out. If it gets bent, it should be replaced rather than straightened and reused.

Support the pump body on a couple of blocks of wood and tap out the shaft, hitting it on its impeller end (see Figure 6-9; the exception is pumps with impellers fastened to the shaft—these must be driven out the other way). Do not hit the shaft hard; be especially careful not to burr or flatten the end

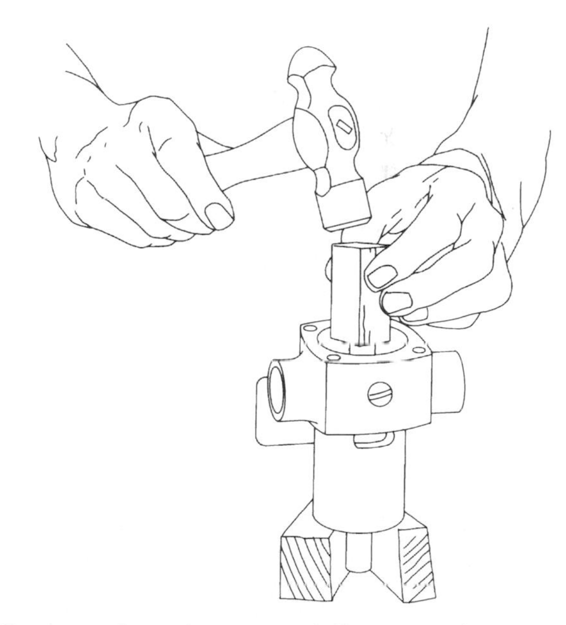

FIGURE 6-9. *Removing a pump shaft. Remove the end cover and impeller. Support the pump body with a couple of blocks of wood, impeller end up. Protect the end of the impeller shaft with a block of wood and lightly tap the shaft free. (Jim Sollers)*

of the shaft. It is best to use a block of wood between the hammer and the shaft, rather than to hit the shaft directly.

If the shaft won't move, take another look for a bearing-retaining circlip. If there truly isn't one, try hitting a little harder. If the shaft remains stuck, try heating the pump body in the area of the bearing with hot water or gentle use of a propane torch. The shaft will come out complete with bearings.

The main shaft seal can now be picked out from the impeller side of the body using a piece of bent wire. There may be another bearing seal on the drive side, and quite probably a slinger washer between the two. If the washer drops down inside the body, retrieve it through the drain slot.

To remove the bearings from the shaft, take the small bearing-retaining circlip off the shaft, support the assembly with a couple of blocks of wood placed under the inner bearing race, and tap out the shaft, hitting it on its drive end (see Figure 6-10; use a block of wood between the hammer and the shaft once again).

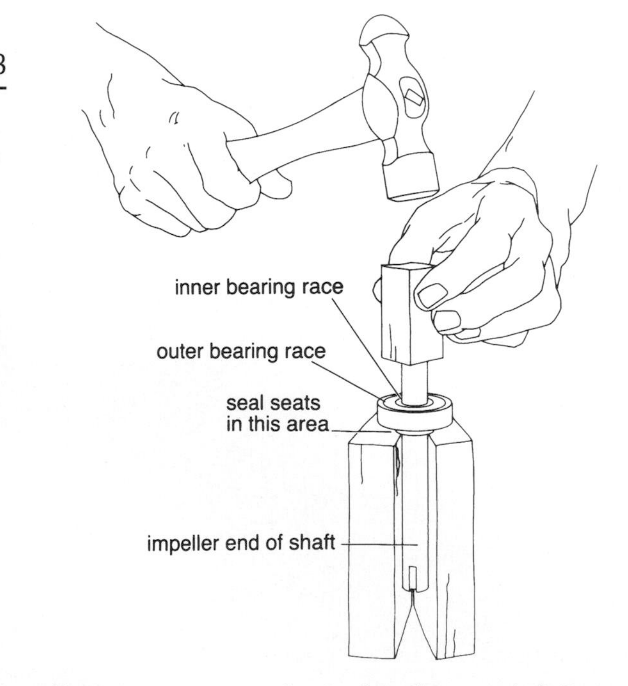

FIGURE 6-10. *Removing a bearing from the pump shaft. Remove the bearing-retaining circlip from the shaft. Support the bearing with a couple of wooden blocks placed under the inner bearing race. Protect the shaft end with a block of wood and lightly tap the shaft free of the bearing. (Jim Sollers)*

Inspect the shaft for signs of wear, especially in the area of the shaft seal. Spin the bearings and discard them if they are rough, uneven, or if the outer race is loose. Scrupulously clean the pump body, paying special attention to all seal and bearing seats. Do not scratch bearing, seal, or seating surfaces.

Reassembly

To fit new bearings to a shaft, support the inner race of the bearing and tap the shaft home. To make the job easier, first heat the bearing (e.g., in an oven to around 200°F/93°C but no more) and cool the shaft (in the icebox). The shaft should just about drop into place. Replace the bearing-retaining circlip, with the flat side toward the bearing.

Put the new shaft seal in the pump body and also the inner bearing seal (if fitted). Lip-type seals have the lip toward the impeller. Carbon-ceramic seals have the ceramic part in the pump body, set in its rubber boot with the shiny surface facing the impeller.

A lip-type seal is lightly greased (with petroleum jelly or a Teflon-based waterproof grease), but a carbon-ceramic seal must not be greased. The seal faces must be wiped spotlessly clean—even finger grease must be kept off them—and the seals lubricated with water.

All seals must be centered squarely and pushed in evenly (see Figures 6-11A and 6-11B). If the seal

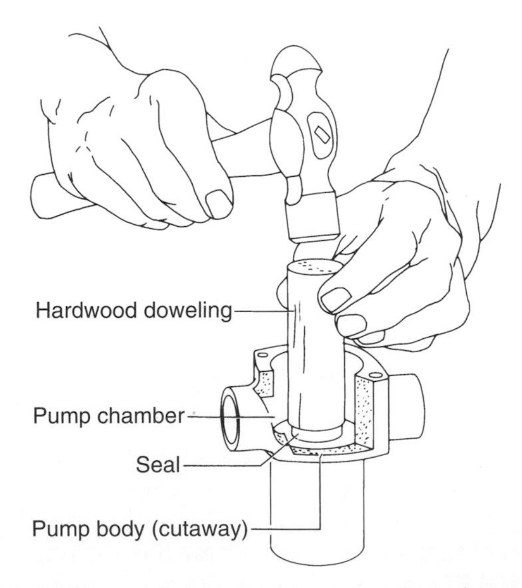

FIGURE 6-11A. *Replacing a pump seal. Seat the seal squarely in its housing. Using very soft hammer taps and a piece of hardwood doweling the same or slightly larger diameter as the seal, push the seal down until flush with the pump chamber. (Jim Sollers)*

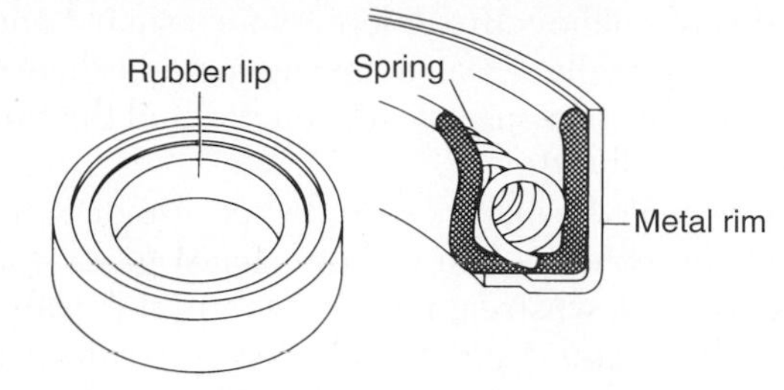

FIGURE 6-11B. *When replacing a pump seal, make sure the hardwood doweling seats on the seal's metal rim and not the rubber lip. The seal goes in with the rubber lip toward the pump chamber. (Jim Sollers)*

is bent, distorted, or cockeyed in any way, it is sure to leak. A piece of hardwood doweling the same diameter as, or a little bigger than, the seal makes a good drift. Push the seal down until it is flush with the pump chamber.

If the pump has a slinger washer, slide it up through the body drain and maneuver the shaft in from the drive side of the body, easing it through the slinger and into the shaft seal (see Figure 6-12). Pass the shaft through any seals very carefully.

Seat the bearings squarely in the pump housing, support the pump body, and drive the bearings home evenly, applying pressure to the outer race (a socket with a diameter just a little smaller than the bearing works well—see Figure 6-13). Once again, heating the pump body (but not with a propane torch this time because you may damage the seals) and cooling the bearings will help tremendously. Refit the bearing-retaining circlip with the flat side to the bearing and press home the outer bearing seal (if fitted), with the lip side toward the pump impeller.

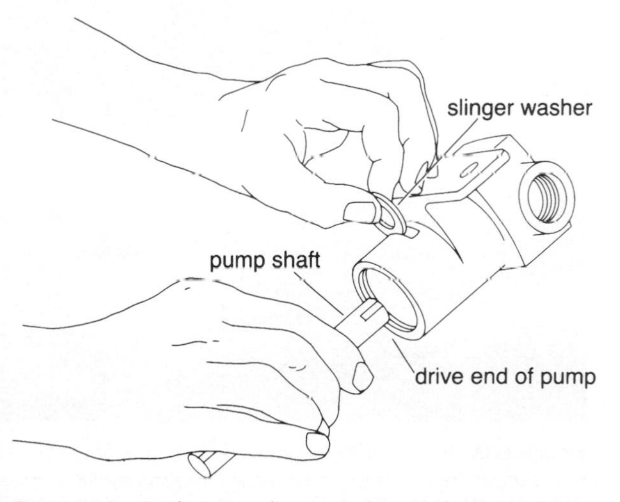

FIGURE 6-12. *Replacing a slinger washer. Push the washer up through its slot in the pump body, threading the shaft through it and into the shaft seal. (Jim Sollers)*

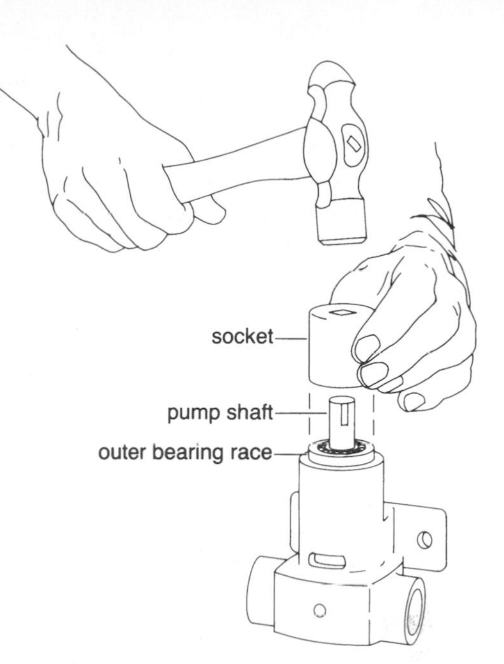

FIGURE 6-13. *Driving home a bearing. Use a ratchet-drive socket with a diameter slightly smaller than the outer bearing race. Make sure the bearing is squarely seated in the housing before you drive it home. (Jim Sollers)*

Turn now to the pump end. If the pump has a carbon-ceramic seal (see Figure 6-14), clean the seal face, lubricate it with water, and slide the carbon part up the pump shaft, with the smooth face toward the ceramic seat. Some seals use wave washers to maintain tension between the carbon seal and the seat; most use springs. (If the new seal has both, discard the wave washer.)

Replace the wear plate, locating its notch on the dowel pin. Lightly apply some sealing compound (e.g., Permatex) to the back of the cam and to its retaining screw. Loosely fit the cam. Lightly grease the impeller with petroleum jelly, a Teflon-based grease, or dishwashing liquid, then push it home, bending down the vanes in the opposite direction to pump rotation. Replace the gasket and pump cover. Tighten the cam screw.

If the pump cover has a paper gasket (many now use O-rings), the correct gasket is important—too thin, and the impeller will bind; too thick, and

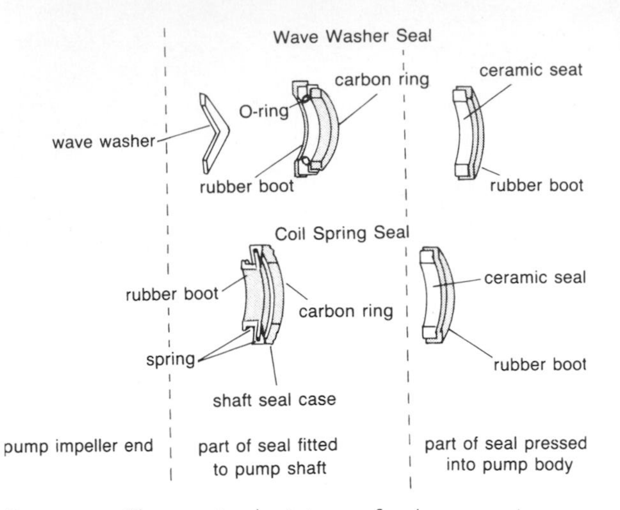

FIGURE 6-14. *There are two basic types of carbon-ceramic pump shaft seals—wave-washer seals (uncommon) and coil-spring seals. In either case, the ceramic seal and rubber boot are pressed into place in the pump body, and the carbon-ring seal and retainer are fitted to the pump shaft. (Courtesy ITT/Jabsco)*

pumping efficiency is lost. Most pump gaskets are 0.010 inch (ten thousandths of an inch) thick, but on larger pumps they may be 0.015 inch. As noted previously, some impellers on intermittently used pumps have a tendency to stick in their housings during periods of shutdown and, if electrically driven, blow fuses when the pump is started. To stop this, loosen the pump cover on initial start-up, then tighten it. (You can achieve this same result, with only a small loss in pumping efficiency and without loosening the cover screws, by fitting an overthick gasket.)

When refitting a flanged pump to an engine, be sure the slot in the pump shaft, or the drive gear, correctly engages the tang or gear on the engine. Also make sure the pump flange seats squarely without pressure. Some pumps have spacers or adapters that go between the end of the pump shaft and the engine driveshaft. Pulley-driven pumps must be properly aligned with their drive pulleys, and the belt correctly tensioned.

THE EXHAUST SYSTEM

Exhausts on many auxiliary sailboats tend to slowly plug with carbon due to prolonged hours of low-load operation, such as running a refrigerator or charging batteries at anchor. Break loose your exhaust and inspect it annually. Clean a heavily sooted exhaust, and the exhaust passages on the engine and the turbocharger (if fitted).

Water-cooled exhausts corrode because of exposure to the combination of hot exhaust gases and salt water. Downstream from where the water enters the system, good-quality, wire-reinforced steam hose is the best material for exhaust pipes (it should be labeled SAE J2006 or UL 1129—see Figure 6-15). Fire-retardant fiberglass is best for water-lift mufflers. However, if the raw-water circuit fails, both materials are likely to burn up, although any high-temperature engine alarm (if fitted—an excellent idea) should sound before this happens. Double-clamp all hose joints with all-stainless 300-series (see page 63) hose clamps. Inspect the hoses and clamps annually.

EXHAUST INJECTION ELBOW

Sooner or later, the injection elbow will corrode through. The elbow is frequently custom-made and hard to replace, so you should inspect it annually and carry a spare on board. If you have to have a new one custom fabricated, make sure the welder uses welding rods similar to the metal used in fabricating the elbow (e.g., stainless rods on stainless steel). Dissimilar metals will greatly accelerate the

FIGURE 6-15. *Poor-quality, sagging, soft-wall exhaust hose. When it gets hot, it sags some more, significantly interfering with the exhaust flow and engine performance at higher engine loads. To avoid such problems, use quality J2006 or UL 1129 marine exhaust hose.*

FIGURE 6-16. *Galvanized and plain steel fittings in an exhaust system, such as those shown, are an invitation to corrosion.*

FIGURE 6-17. *A corroded galvanized exhaust elbow (left). The hot gases and water from the exhaust have eaten through it. The same elbow patched with a rubber inner tube and hose clamps (right). The repair lasted for 200 hours of engine-running time.*

rate of corrosion (see Figure 6-16). Under no circumstances have the joint brazed.

You can temporarily repair a holed elbow by tightly wrapping a strip of rubber cut from an inner tube around the elbow and clamping it with two hose clamps (see Figure 6-17). Alternatively, use an underwater epoxy (e.g., Marine-Tex) to patch the hole.

CLEANING TURBOCHARGER WHEELS

As a rule turbochargers (see Figure 6-18) should be left to the experts. However, if removal of the inlet and exhaust ducting and the tests outlined on pages 126–28 reveal no problems other than a dirty compressor wheel or turbine, you can generally clean these without help.

Mark both housings and the center unit with scribed lines so you can put them back together in the same relationship to each other. Allow the unit to cool before removing any fasteners, or the housings may warp. If the housings are held on with large snap rings (circlips), leave them alone—they will come apart easily enough but will require a hydraulic press to put back together! Housings held with bolts and large clamps may be taken apart (see Figure 6-19).

If the housings are difficult to break loose, tap them with a soft hammer or mallet. Pull them off squarely to avoid bending any turbine blades. Only use noncaustic solutions to clean turbines (degreasers

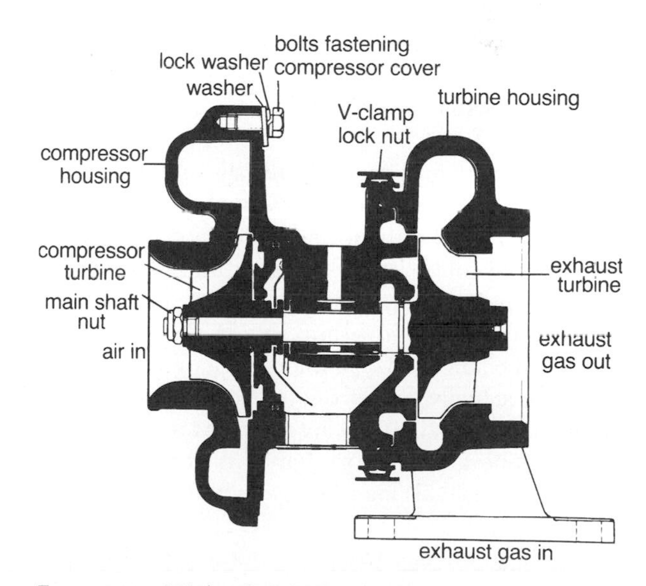

FIGURE 6-18. *A Holset 3LD/3LE turbocharger. (Courtesy Holset Engineering Company)*

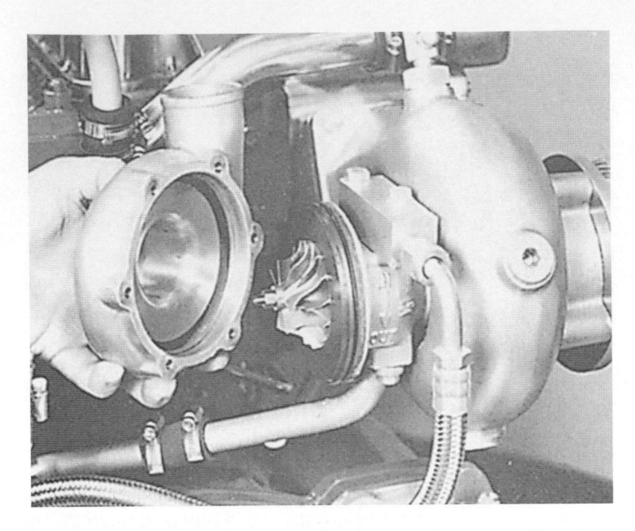

FIGURE 6-19. *Removing a turbocharger housing. (Courtesy Perkins Engines Ltd.)*

work well; I am told that various carburetor cleaners also do a good job) using soft-bristle brushes and plastic scrapers. Do not use abrasives; the resulting damage to the blades will upset the critical balance of the turbines. Make no attempt to straighten bent blades—the turbocharger demands a specialist's help.

The procedure for reassembly is the reverse of disassembly. Be sure to accurately line up all scribed marks. After reassembly, spin the turbines by hand to make sure they are turning freely. Before starting the engine, crank it for a while to get oil up to the turbocharger bearings.

GOVERNORS

Problems with governors are usually indicated by the engine's failure to hold a set speed, particularly on start-up. Hunting is the most common symptom—the engine continuously and rhythmically cycles up and down. However, not all such speed fluctuations are caused by the governor, so first check for excessive load changes (e.g., a microwave on less than full power is actually going from full power to no power at timed intervals; this will cause the governor on a small diesel generator to work hard), misfiring cylinders (see pages 120–21), and poor injection (see later in this chapter).

If these checks fail to reveal a problem, most likely some part of the governor mechanism or fuel injection pump control linkage is sticking. The governor has to overreact to any change in load to overcome resistance to movement of the fuel rod. Then the rod moves with a jerk and goes too far, causing the engine to overspeed or underspeed. The governor then overreacts once again, but in the other direction, and the engine underspeeds or overspeeds, and so on. Excessive slack or play in the fuel pump control linkage (throttle linkage) will cause similar behavior.

Many governors are inside the fuel injection pump housing. Injection pumps are items for the experts—if the problem cannot be solved by cleaning and adjusting the external fuel-control linkage, call in a professional.

If the governor is inside the engine (as opposed to the fuel injection pump), the governor mechanism itself may be causing the problem. In this case, it is quite likely that a buildup of sludge (from lack of adequate engine oil changes or from prolonged low-load operation) is interfering with correct governor operation. You will have to clean the governor.

Older Detroit Diesel two-cycles may have an externally mounted governor operating an injector control tube (see Figure 6-20). If you remove the valve cover, all the operating levers will be accessible. You should be able to move the injector control tube, and thus all the injector racks, through their full range without any rough spots or sticking. If you detect friction, disconnect the injectors one by one and test again. If the friction ceases, check the individual injector rack. If it is sticking, it may simply be because the injector hold-down clamp is improperly positioned or too tight (try loosening the clamp, repositioning it, and tightening), but more likely the injector needs an overhaul. If there is still friction in the linkage after all the injectors are disconnected, check the various support brackets on the control tube for binding and also the position of the control lever (running to the governor) on the control tube.

The only other likely maintenance item on a governor is an occasional adjustment of the idle setting on the speeder spring. There is normally no reason to alter this. If the engine will not idle correctly, it's almost certainly due to some other problem. The idle setting should only be adjusted when the

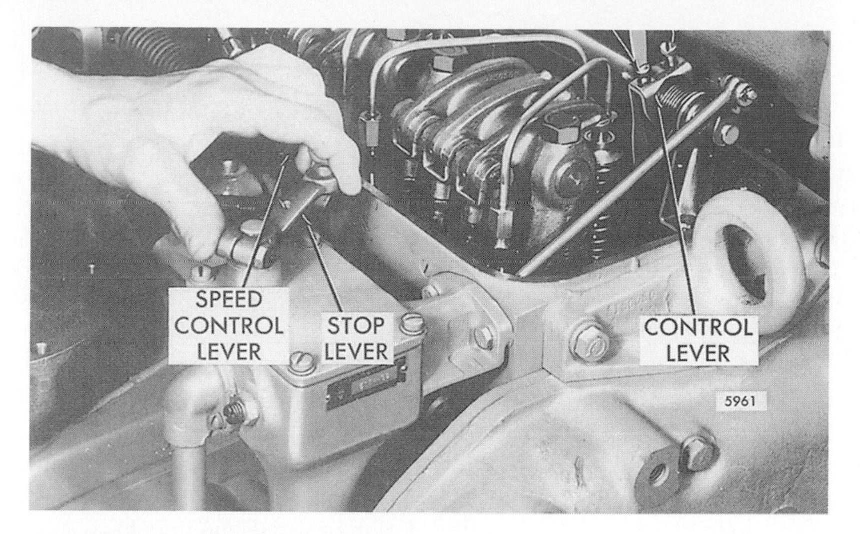

Figure 6-20. *A Detroit Diesel fuel-rack control linkage. (Courtesy Detroit Diesel)*

engine is running well. Engine-mounted governors have a screw and locknut somewhere on the outside of the block. Governors inside injection pumps generally have an external low-speed screw acting directly on the throttle control lever.

Somewhere there will also be a maximum fuel setting screw and locknut that will almost certainly be tied off with lockwire and sealed (see Figure 6-21). Do not tamper with it. Breaking the seal automatically voids any engine warranty. If it's not sealed and if the engine appears to be overloaded at full throttle (produces black smoke, overheats), reduce the maximum fuel setting. Increasing maximum fuel settings above manufacturer's set points can lead to engine seizure.

Fuel Injection Pumps

Before jumping to the conclusion that you have a defective fuel injection system, always check every other likely cause of your problem, particularly an obstructed airflow through the engine (plugged air filters, collapsed ducting, poor turbocharger or blower performance, a carboned-up exhaust pipe, etc.).

The only user-serviceable parts of a fuel injection system are the fuel strainer and rubber diaphragm in the lift pump and the diaphragm found in some

Figure 6-21. *A maximum fuel setting screw on a Volvo MD 17C.*

fuel injection pumps. See pages 103–5 for how to service lift pumps.

Diaphragm Fuel Injection Pumps

A few older engines have vacuum-type governors (see page 28). In this case, the injection pump has a diaphragm in the back of it. This diaphragm pushes against the fuel-control rack on one side and is controlled by a vacuum line to the air-inlet manifold on the other side. A holed diaphragm can lead to a loss of power, excessive black smoke, a very rough idle, and overspeeding.

Test an injection pump diaphragm with the engine off. The diaphragm is contained in a round housing at the back of the pump (the end not attached to the engine). Coming out of the top of this housing is a vacuum-sensing line that leads to the air-inlet manifold. Disconnect this line at the pump.

At the opposite end of the pump, just above the flange that holds it to the engine, is a protective cap, inside of which is the fuel-control rack. Remove this cap. On the side of the pump, below and just forward of the vacuum line, is the pump-control lever (throttle). Hold this in the Stop position.

If you place a finger tightly over the vacuum connection on the pump and let go of the control rod lever, the fuel-control rack at the engine end of the pump should move a short distance and then stop, as long as you block the vacuum fitting with your finger. If there are any leaks in the diaphragm or around the seal to its housing, the control rack will keep moving (perhaps slowly) until it is as far out as it can go. In this case, the diaphragm needs inspecting, and probably replacing.

You can get to the diaphragm by removing its cover (four screws) and undoing the bolt holding it to the fuel rack.

Injection Pump Oil Reservoirs

Although most pumps are lubricated with diesel fuel, some in-line jerk pumps have an oil sump that contains regular engine oil. This becomes diluted with diesel fuel over time and occasionally needs changing. The sump will have an oil drain plug and a dipstick or level plug. In the latter case, take out the level plug and fill the sump with oil until it runs out of the hole. Replace the plug. No other area of the fuel pump is serviceable in the field, and it should be left strictly alone.

Injectors

The following information applies to "conventional" injectors. It does not apply to many unit injectors or to any electronic injectors (which should be left alone).

Lucas/CAV recommends: "In the absence of specific data, a figure of 900 hours of operating between servicing is a useful guide for the boatowner." Aside from pulling the injectors to have them serviced, leave injectors alone. Individual injector needle valves are matched to their bodies to within 0.00004 inch (four one-hundred thousandths of an inch). Equipment of this degree of precision needs to be disassembled by specialists. "It is not possible for the owner or crew to recondition or service an injector without the essential nozzle setting outfit, special tools, technical data, and service training," states the Lucas/CAV handbook. "Any tampering or attempts at servicing without these essentials will always make matters worse." The following information is therefore given only for those in a dire emergency, and after all other procedures have failed to solve a problem.

Servicing a fuel injection system requires extreme cleanliness. Before you attempt to break loose fuel lines or remove injectors, thoroughly clean the area around them. The instant any fuel lines are disconnected, you must cap both loose connections. Once you remove an injector from a cylinder head, plug its hole to prevent dirt from falling into the cylinder.

Injection pipes are pre-formed from steel tubing. When moving them out of the way, undo them from both ends and lift them out. DO NOT BEND THEM! Normally, they are all clamped together. Rather than undo the clamps, it is often easier to remove them as a set.

Injectors are held in place by a metal plate bolted to the cylinder head, or they are screwed directly into the head (see Figure 6-22). The former can sometimes be hard to break loose—dribble a little penetrating oil down the side of the injector 1 to 2 hours before you pull it out.

On some engines (for example, many Volvo Pentas), the injector is in a sleeve that is directly cooled by the engine's cooling circuit. Occasionally the sleeve sticks to the injector and comes out with it. To prevent this, the injectors are normally pulled using a special tool, which you almost certainly will not have! Therefore drain the coolant from the

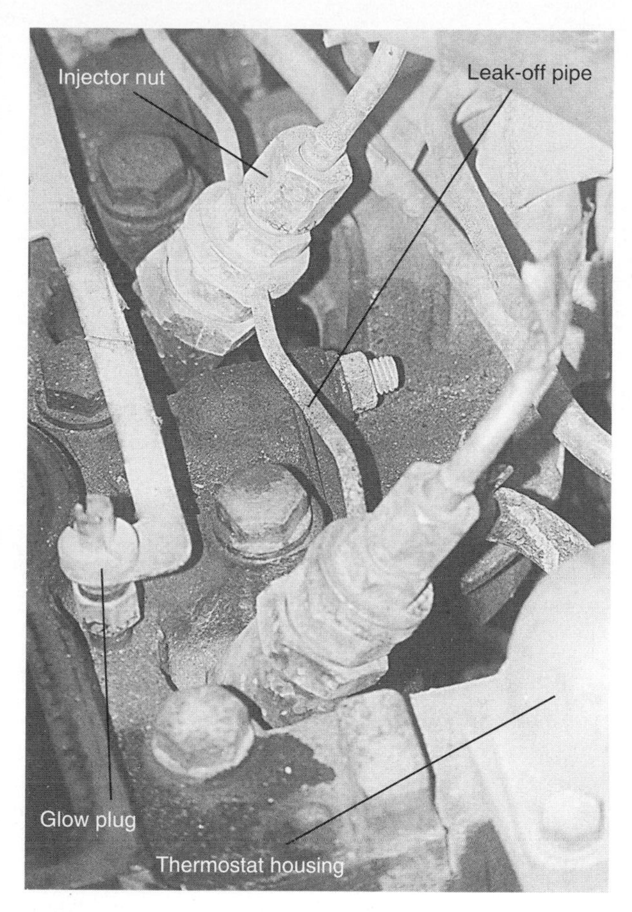

FIGURE 6-22. *Screw-in-type injectors.*

block before attempting to pull any injectors. This eliminates the risk of coolant running into a cylinder. Once again, dribble penetrating oil down the sides of the injector, let it sit for a while, then loosen the injector by working it from side to side. If the sleeve comes out, you will have to call in a trained mechanic because installation of a new sleeve requires special tools.

Injectors, and many fuel lines, are sealed with copper washers. Do not lose any of these, and be sure that they go back in the right place on reassembly.

TESTING AN INJECTOR

Now that the injector is out of the cylinder head, you can check its operation by reconnecting it to its delivery pipe, bleeding the line, opening the throttle, cranking the engine over, and observing the spray pattern (see Figures 6-23 to 6-26). Do this only after you have loosened the delivery pipes to

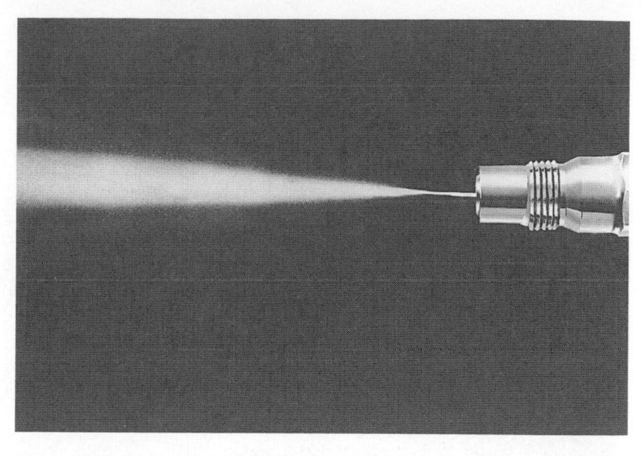

FIGURE 6-23. *Correct pintle spray pattern. (Courtesy Lucas/CAV)*

the other injectors (if these pipes have not been disconnected) to prevent the engine from starting. Rotate the engine at a speed of at least 60 rpm (normal cranking speed on most small diesels is more than adequate). Each type of injector will produce a distinct spray pattern, but all should have certain features in common:

1. A high degree of atomization of the diesel.
2. Strong, straight-line projection of the spray from the nozzle as a fine mist with no visible streaks of unvaporized fuel.
3. No dribbling or drops of fuel (the nozzle should remain dry after injection is complete).
4. Fuel should come out of all the holes in the nozzle in even proportions.

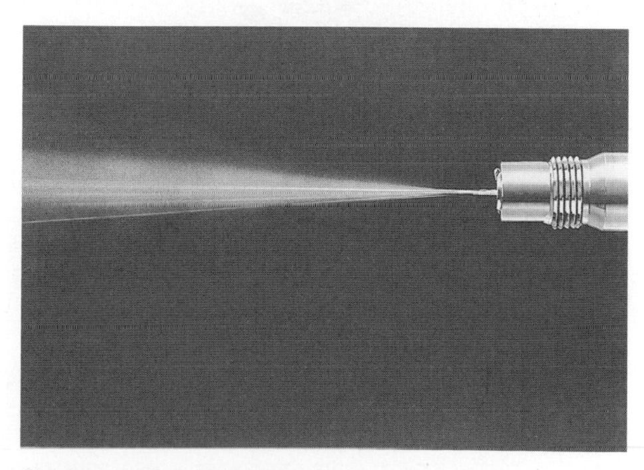

FIGURE 6-24. *Incorrect pintle spray pattern (excessive streaking). (Courtesy Lucas/CAV)*

FIGURE 6-25. *Correct multi-hole spray pattern. (Courtesy Lucas/CAV)*

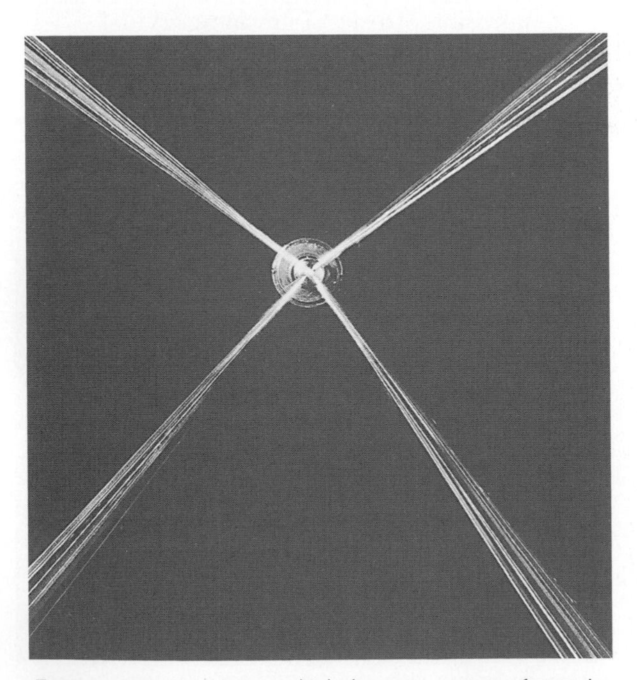

FIGURE 6-26. *Incorrect multi-hole spray pattern (excessive streaking). (Courtesy Lucas/CAV)*

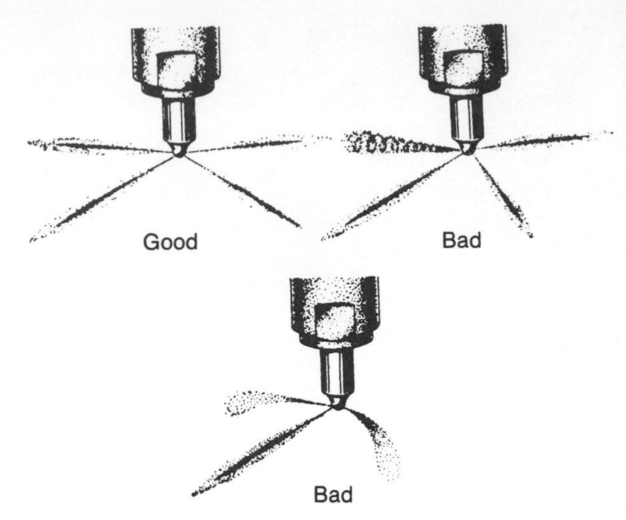

FIGURE 6-27. *Pressure-testing the injectors. Check the spray pattern and also that the fuel jets cease simultaneously at all the holes and do not drip afterward. (Courtesy Volvo Penta)*

Figure 6-27 illustrates different spray patterns. If you have an injector that you know is good, test this one first to familiarize yourself with a correct spray pattern before going on to test suspect injectors. If you are in doubt about the spray patterns, take the injectors to a fuel injection shop to have them tested. Testing (though not rebuilding!) is cheap.

Whenever an injector is tested, you must keep well out of the way. The diesel fuel is fine enough, and has more than enough force, to penetrate skin and blood vessels. It can cause severe blisters and blood poisoning.

If the spray pattern is defective or the nozzle drips, the injector nozzle needs cleaning. Cleaning the carbon off the outside of the nozzle and using a very fine wire to clear the holes in the nozzle may be sufficient. Use a brush with brass bristles and diesel fuel on the nozzle—never use steel brushes because they can damage the holes. If a proper set of injector-cleaning prickers is not available, a strand of copper wire may work. The pricker off a Primus stove or kerosene lantern is another alternative. Do not enlarge the holes or break off a pricker in an injector hole.

DISASSEMBLY

Do not dissemble two-spring and needle lift injectors (see pages 25–26)—send them back for a replacement. Unit injectors, such as those found on

Detroit Diesel two-cycles, incorporate the fuel injection pump. They are more complex than those illustrated below and no attempt should be made to disassemble them. Electronic injectors are also off-limits.

If you have to disassemble a conventional injector, first soak it for several hours in Gunk or diesel fuel to loosen everything up. Bosch stipulates a minimum of 4 hours. Two ounces of caustic soda dissolved in 1 pint of water with ½ ounce of detergent will go to work on carbon deposits. This volume of caustic soda must not be exceeded or corrosion is possible. Boil the parts to be decarbonized in this solution for an hour, continually topping off the water to compensate for evaporation. Before you reassemble the injector, thoroughly flush and dry it to remove all traces of the caustic soda.

To disassemble an injector, hold the injector body firmly. A special vise is recommended, although it will probably not be available. The injector must not be directly clamped in a steel vise because the vise may distort it. Remember how precise everything is! Place protective wooden blocks around the injector body and the vise, using only the minimum necessary pressure.

Within the injector is a powerful spring (see Figure 6-28). Sometimes the spring pressure is externally adjustable (as illustrated in Figure 6-29) by removing a cap nut on the top of the injector. If this is the case, back off the locknut and screw, carefully counting an exact number of turns, until the spring is no longer under tension. This spring determines the pressure at which the injector opens and is set at the factory on a special testing device. In the field without the proper equipment, the spring pressure cannot be accurately reset, so the best you can do is put it back where it was.

On other injectors, the opening-spring pressure is set by fitting a number of shims (spacers) under the spring (see Figure 6-30). The more shims, the higher the opening pressure. Every 0.001 inch (one thousandth of an inch) of increase in shim thickness raises the opening pressure by about 55 psi. These injectors have no external spring adjustment, and you can move directly to removal of the nozzle assembly.

Unscrew the injector nozzle nut. Some injectors are designed to project a spray in a specific direction. In this case, the nozzle and injector body are held in the correct relationship with a small dowel. If the nozzle nut is particularly hard to break loose, soak the whole assembly again because excessive force is likely to damage the dowel pin. Sometimes a sharp tap on the end of the wrench is necessary to break the grip of carbon in the injector and get the nut moving.

Once the nut is off, you can remove the nozzle and its components. Be sure to carefully note the order of all the parts. On no account must injectors be mixed up—nozzles and needle valves are machined as matching sets and must always go together.

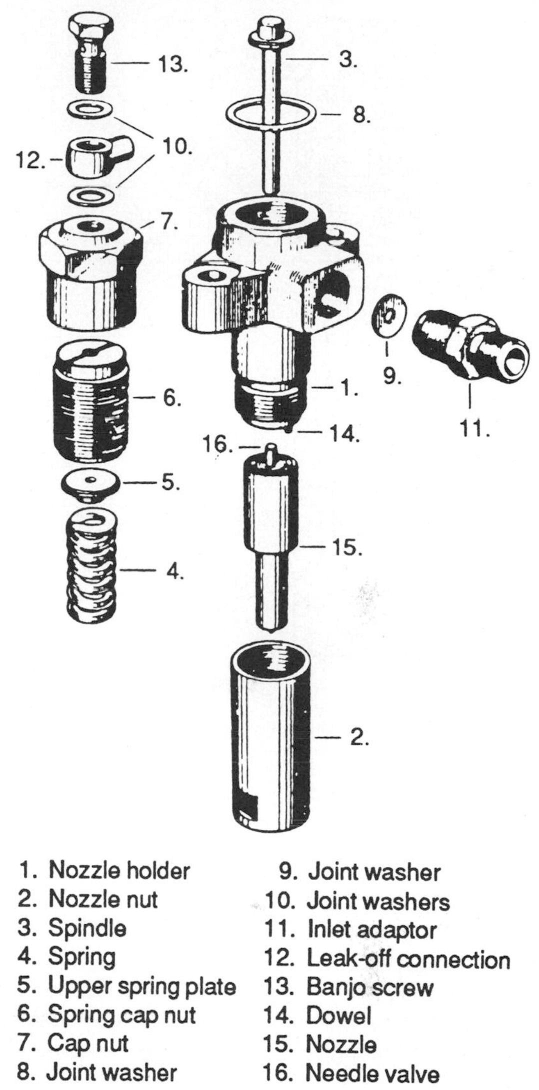

1. Nozzle holder
2. Nozzle nut
3. Spindle
4. Spring
5. Upper spring plate
6. Spring cap nut
7. Cap nut
8. Joint washer
9. Joint washer
10. Joint washers
11. Inlet adaptor
12. Leak-off connection
13. Banjo screw
14. Dowel
15. Nozzle
16. Needle valve

FIGURE 6-28. *An exploded view of a multi-hole injector. (Courtesy Lucas/CAV)*

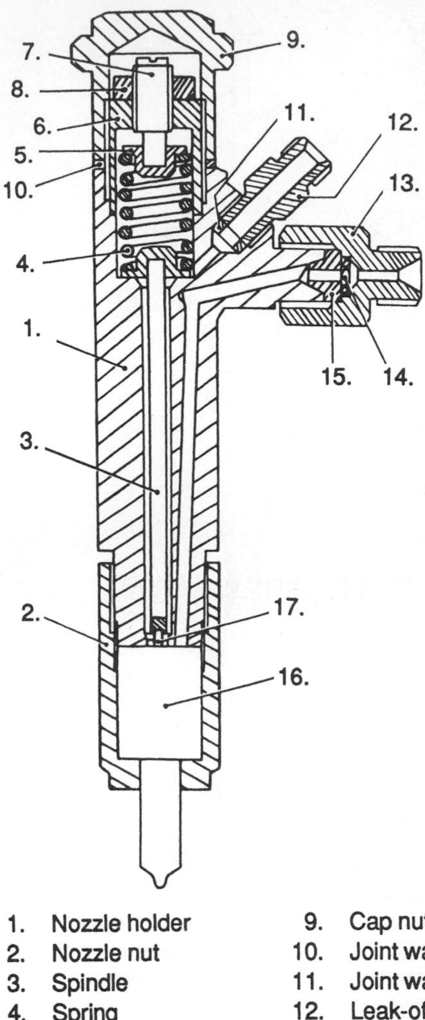

1. Nozzle holder
2. Nozzle nut
3. Spindle
4. Spring
5. Upper spring plate
6. Spring cap nut
7. Spring adjusting screw
8. Locknut
9. Cap nut
10. Joint washer
11. Joint washer
12. Leak-off adaptor
13. Inlet adaptor
14. Filter
15. Nipple
16. Nozzle
17. Needle

FIGURE 6-29. *An injector with a spring adjusting screw. (Courtesy Lucas/CAV)*

CLEANING

All contact surfaces within the injector should be clean and bright. Use the caustic soda solution, appropriate scrapers, or both to clean up carbon. Do not scratch the needle valve, its seat, or the nozzle bore. Never use abrasive cleaning or grinding compounds on the needle or its seat in the nozzle. These are machined at different angles to ensure a tight line contact (see Figure 6-31). Any grinding or lapping will destroy this fit.

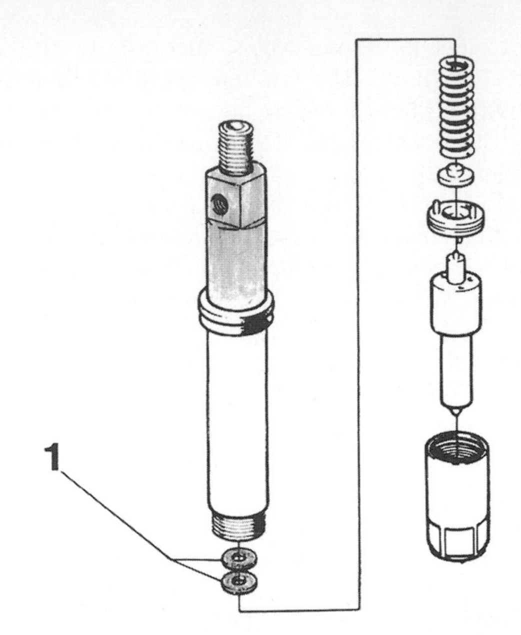

FIGURE 6-30. *An injector with internal shims (1) to set the opening pressure of the spring. (Courtesy Volvo Penta)*

Injector manufacturers sell nozzle-cleaning kits (see Figure 6-32). Lucas/CAV issues the following instructions for cleaning the nozzle of a multi-hole injector with the tool kit:

1. Remove all traces of carbon deposit from the exterior of the nozzle [refer to Figure 6-33] and from the needle with the wire brush. Polish the needle with a piece of soft wood; do not use an abrasive cleaning compound.
2. Clean the gallery with the scraper.
3. Clean the nozzle seat with the scraper, using the appropriate end of the scraper.
4. Similarly, clean the cavity with the scraper.
5. Use the pin vise with the appropriate size of pricker wire to clean the spray holes in the nozzle tip. The vise must be used carefully to avoid the risk of breaking the pricker wire in a spray hole [see Figure 6-34].

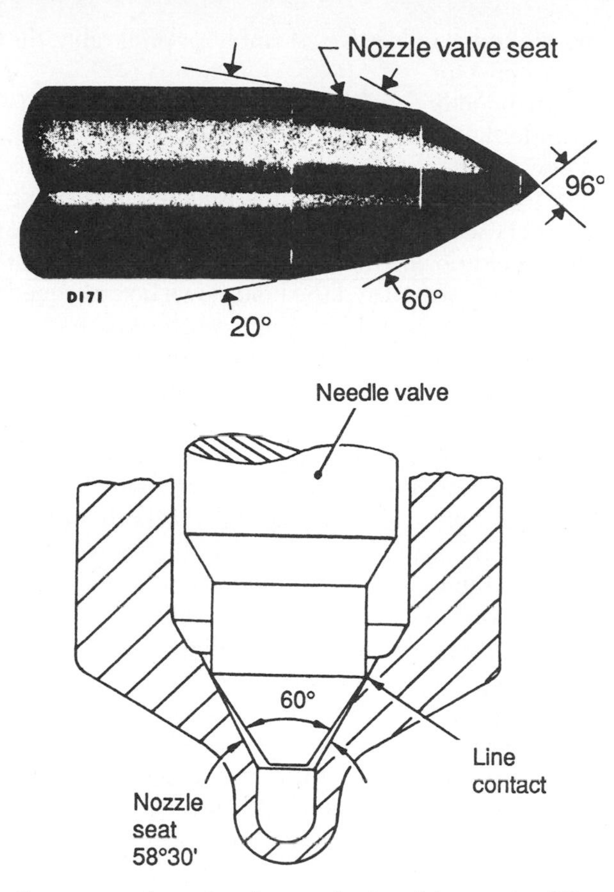

FIGURE 6-31. *A nozzle valve seat (top) and the angular difference between the nozzle seat and the needle valve (multi-hole injector; bottom). (Courtesy United Technologies Diesel Systems; Lucas/CAV)*

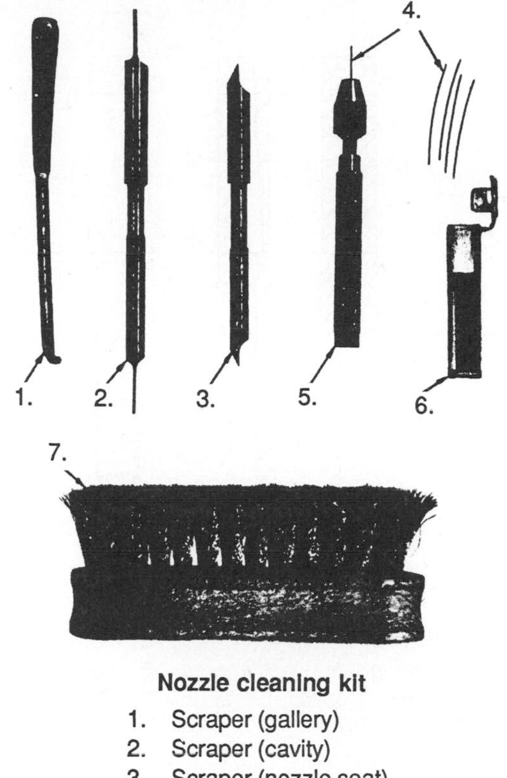

FIGURE 6-32. *A nozzle-cleaning kit. (Courtesy Lucas/CAV)*

A clean needle valve should drop easily into its seat and fall back out when the injector is inverted. Injector parts must be spotlessly clean (only use lint-free rags or paper towels) and dipped in clean diesel before reassembly.

REASSEMBLY

This process is the reverse of the disassembly procedure. The things that need to be checked, such as the nozzle lift and spring pressure, are beyond the scope of an amateur. This is why, if at all possible, injectors should be left alone. As already noted, screw down the opening pressure-adjusting screw (if fitted) to its previous position. All shims must be put back exactly as before, in the same order, with a thick spacer on the top and the bottom of the thin spacers. Now you can check the spray pattern again.

If you have lost the spring setting of an adjustable spring, turn the adjusting screw until it is finger-tight.

Thereafter, every turn represents 900 to 1,000 psi opening-spring pressure. Multi-hole injectors are generally set to around 2,200 psi (2 to 2½ more turns), and pintle nozzles to around 1,500 psi (1½ more turns). If the spray pattern is still poor, it may be improved by tightening the spring pressure by up to ¾ of a turn more, but certainly not beyond this. If after all this the needle valve fails to seat properly and the injector refuses to operate correctly, you can do nothing, short of replacing the

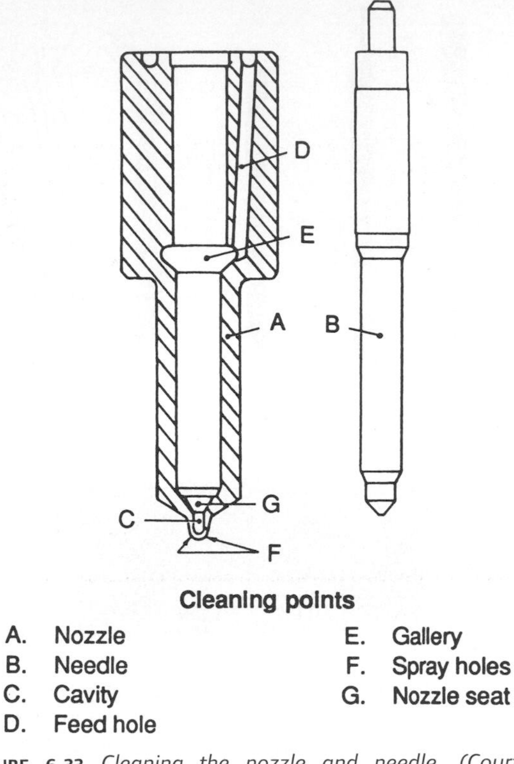

FIGURE 6-33. *Cleaning the nozzle and needle. (Courtesy Lucas/CAV)*

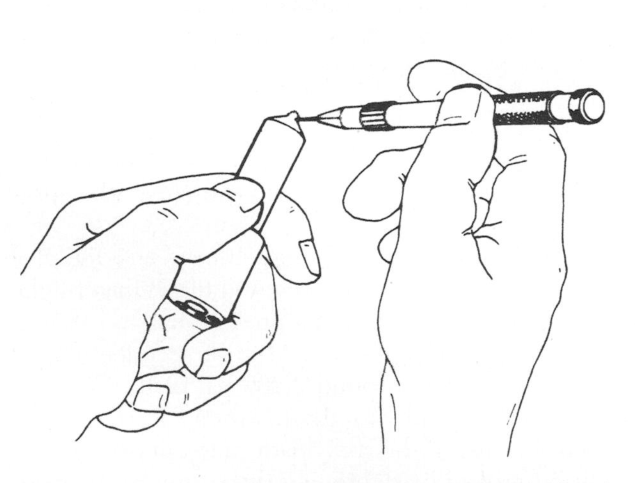

FIGURE 6-34. *Clear spray holes by using the probing tool fitted with the appropriate cleaning wire. Fit the wire in the chuck so it protrudes only about 1.5 mm, giving maximum resistance to bending. Place the wire into each hole, pushing and rotating gently until each hole is cleared. Note: Different nozzles will need different cleaning-wire sizes. (Courtesy Lucas/CAV)*

nozzle and needle valve assembly, or preferably, the whole injector.

An injector must make a gas-tight seal in its cylinder head. The hole in the head must be clean; a new sealing washer (if one is used) should be fitted if at all possible (if not, see the Annealing Washers section below). If the injector is the type that's held down with a steel plate, the plate must be squarely seated and the hold-down bolts torqued evenly. A smear of high-temperature grease (e.g., Never-Seez) around the injector barrel will prevent corrosion from locking the injector in the cylinder head.

Fuel lines are specifically made for individual cylinders, and must be returned to these cylinders. The lines will make an exact fit. Put both ends in place at the same time, then hand tighten before the final tightening. You should never have to bend the line or force it to fit. Be especially careful with internal fuel lines since any bending may cause the line to rupture at a later date, allowing diesel fuel to dilute the engine oil. If a fuel line needs replacing, you must buy the correct individual line from the engine manufacturer. Fuel line nuts must not be overtightened, because this distorts and fractures flare fittings—tighten most nuts to 12 to 15 foot-pounds (16 to 20 Nm).

Always check a fuel system for leaks after you have removed and replaced an injector or fuel pipe. Where there are internal fuel lines, run the engine for 20 to 30 minutes, then remove the valve cover. Check the cylinder head and the various lubricating oil puddles for any signs of diesel leakage.

If there is blowby past an injector, either it did not seat properly, or the retaining nuts were done up unevenly. Remove it and put it back again (with a freshly annealed washer if it uses one).

ANNEALING WASHERS

Injectors are often sealed in cylinder heads with soft copper washers. These may also be used on other parts of the fuel system. After being subjected to high temperatures, or simply over time, these washers become hard, and if you reuse them, they lose their sealing properties. Hardened copper can be readily softened by heating it to a cherry-red color with a propane torch (or over a propane stove or Primus), then dropping it into cold water. This is known as annealing. (Copper-based metals are annealed by rapid cooling, whereas iron-based metals are annealed by slow cooling. Rapid cooling of iron induces hardness.)

GASKETS

Sometimes you may have to improvise a gasket in the field. Your repair kit should contain a roll of high-temperature gasket material, plus some cork or rubber-based material for valve cover and pan (sump) gaskets. Most other gaskets can be made from brown paper, if necessary using several layers. Old paper charts also work well.

Even complicated gaskets are relatively simple to make, but first you have to scrupulously clean the two mating surfaces (see Figure 6-35). The trick is to lay a sheet of the gasket material over the piece that needs the gasket. Use the end of a ball-peen hammer (see Figure 6-36) to lightly tap the gasket into all the bolt holes. The relatively sharp edges of most castings will cut the gasket material, making a perfect hole. Now slip a bolt through the gasket to hold it in place. Repeat this procedure at a few more widely spaced bolt holes to hold the sheet of gasket paper securely. Tap out the remaining holes or other areas until you complete the gasket. The keys to success are striking the gasket in a way that forces it against the sharpest edge of the piece being gasketed, while using the minimum amount of force necessary (especially on aluminum) to avoid burred edges, cracked castings, and other damage. A flurry of light taps is far better than a heavy blow.

FIGURE 6-35. *All surfaces must be scrupulously cleaned before installing new gaskets.*

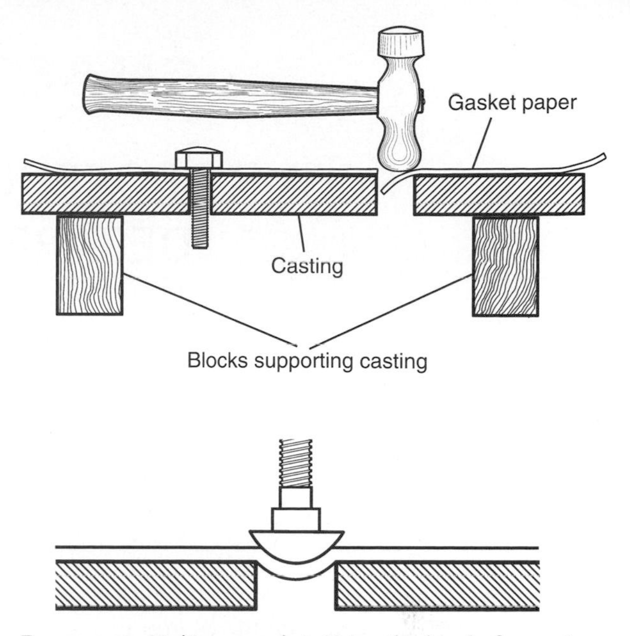

FIGURE 6-36. *Making a gasket. Using the head of a carriage bolt works well (bottom). (Fritz Seegers)*

If you don't have a ball-peen hammer, you can use a box-end wrench, an upside-down carriage head bolt, or any other curved metal object. Where the outline of the piece being gasketed is not sharp enough to cut the gasket paper, rub an oily finger over it. This leaves a clear enough line to follow with a knife or pair of scissors.

CHAPTER 7

REPAIR PROCEDURES, PART TWO: DECARBONIZING

Sooner or later carbon buildup in the cylinders and cylinder head will necessitate decarbonizing and doing a valve job. Decarbonizing consists of removing the cylinder head, and perhaps the pistons, and removing baked-on carbon from them and all associated parts (valves, etc.).

In the case of engines whose valves are operated by push rods, this procedure is well within the capability of an amateur mechanic and should give no cause for alarm. As in all other areas of maintenance, as long as you keep the work area and engine clean and approach the job calmly and methodically, you shouldn't encounter any insurmountable problems.

On engines with an overhead camshaft (a camshaft in the cylinder head eliminates the need for push rods), removing the cylinder head to decarbonize will disturb the valve (and possibly the injection pump) timing. At the end of this chapter, I describe the general principles that govern pump and valve timing, but you may need the manufacturer's manual to learn the specific procedure for timing a particular engine.

When should you decarbonize? There are two schools of thought: (1) perform all maintenance at preset intervals, or (2) wait until you have problems. The logic of the first position is that preset maintenance intervals will deal with conditions likely to cause problems before they get out of hand. The logic of the second position is that no two engines operate under the same conditions, and one engine may run five times longer than another before it needs decarbonizing, or any other substantial overhaul. I lean toward this position.

"If it ain't broke, don't fix it" is not such a bad idea, as long as you are religious about performing routine maintenance, especially oil and filter changes, and ensuring clean fuel. This alone will go a long way toward lengthening the intervals between substantial overhauls. The next most important thing is to avoid prolonged running at low loads.

Once a problem becomes evident (smoky exhaust, loss of power, difficult starting, etc.), you must deal with it immediately. If these conditions are caused by valve or piston blowby, for example, the hot gases won't take long to do some serious damage.

PREPARATORY STEPS

The following must be done before the cylinder head can be removed:

1. Ensure the engine is spotlessly clean. Anytime an engine is opened up, all kinds of damaging dirt will fall into it if you haven't cleaned it.
2. As a general rule, everything you take off an engine should go back in the same place and the same way around. This is especially important for pistons, valves, push rods, rockers, and other moving parts. All these parts wear where they rub on mating surfaces. If they are switched around on reassembly, a high spot on one part may now rub against a high spot on another part,

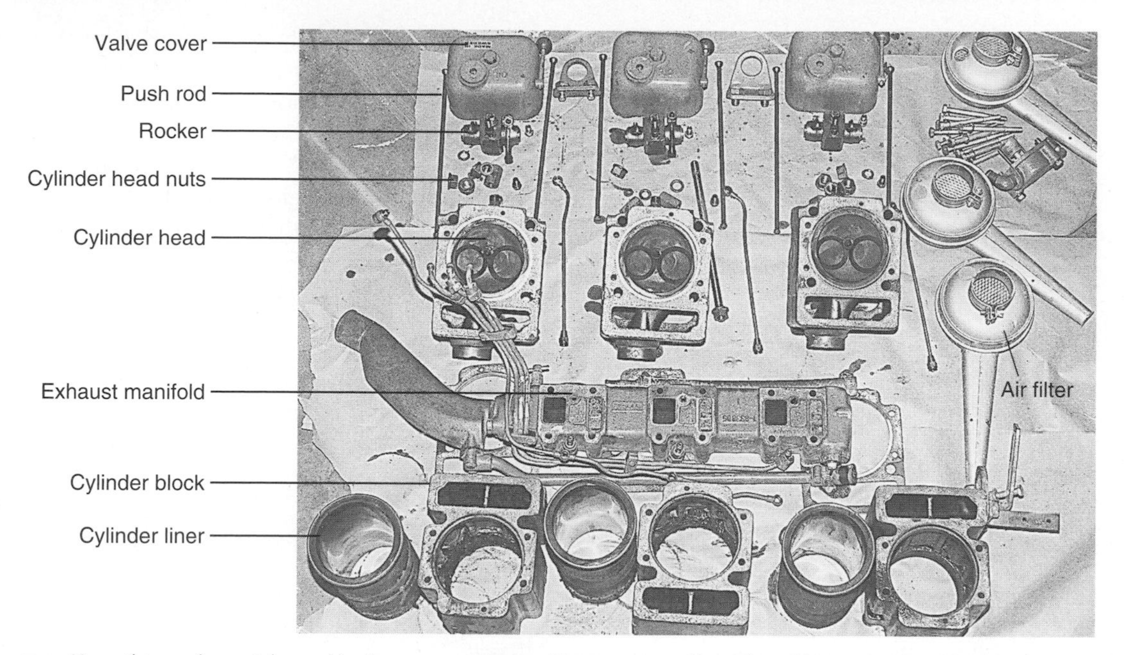

Figure 7-1. *Keep things clean, tidy, and in the correct relationship to one another. This older engine has separate cylinder blocks and cylinder heads for each cylinder. All modern small engines combine these into a single block and head.*

and overall engine wear will greatly accelerate. Set aside a clear area and protect it with newspapers or a clean cloth. As you remove pieces from the engine, spread them out in the correct relationship to one another so that you won't be confused on reassembly (see Figure 7-1). Finding a suitable space on a boat is sometimes hard, but it should be done.

3. Drain the engine of coolant to at least a level below the cylinder head. A drain valve or plug should be located somewhere at the base of the block (see Figure 7-2). Since most sailboat engines are below the waterline, you must first close the engine water seacock on a raw-water-cooled engine. On engines with a heat exchanger and a header tank, loosen the radiator cap

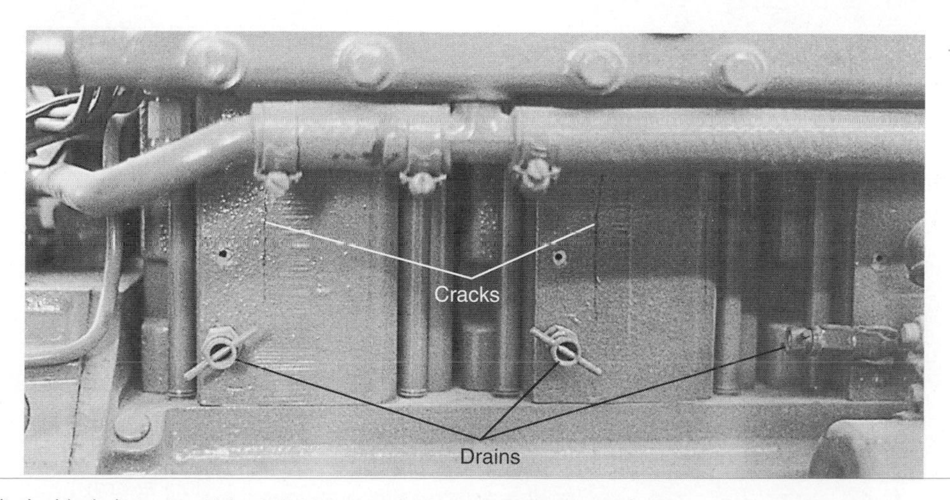

Figure 7-2. *Cylinder block drains in a Volvo MD 17C. Note the individual cylinder block drains; most modern single-block engines have a single drain. When winterizing, the owner of this engine neglected to drain the cylinders, which froze and cracked (the faint vertical lines).*

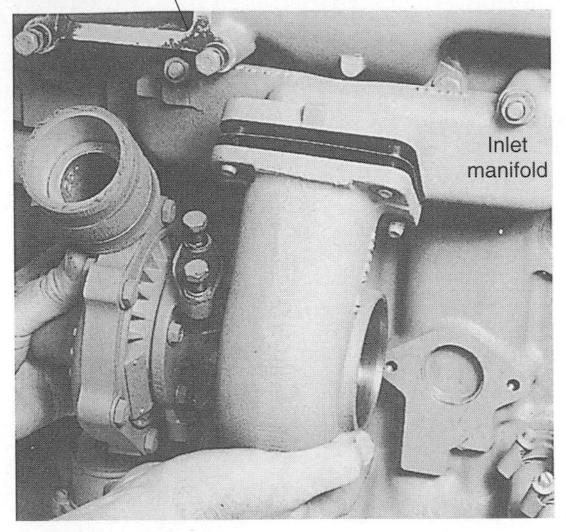

FIGURE 7-3. *Removing a turbocharger from the inlet and exhaust manifolds. (Courtesy Perkins Engines Ltd.)*

on the header tank to release the vacuum that forms when you drain the block.

4. Remove all equipment attached to the cylinder head. This includes inlet and exhaust manifolds, as well as the turbocharger and aftercooler (if present; see Figure 7-3).
5. Break loose all injection lines (delivery pipes) from their respective injectors (or the fitting on the outside of the cylinder head if they are internal) and from the injection pump (see

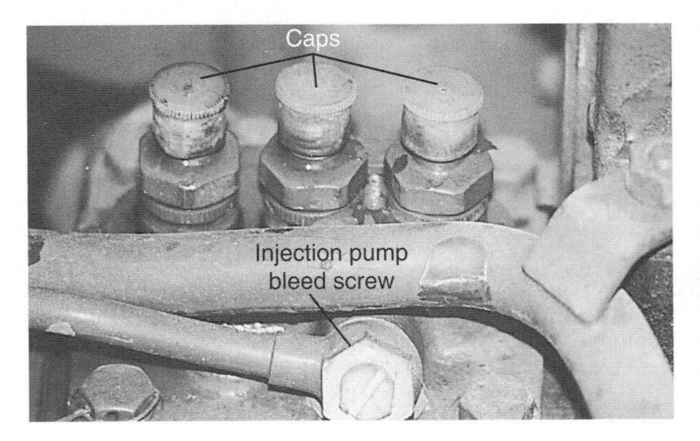

FIGURE 7-5. *Properly capped fuel lines on a Volvo MD 17C injection pump.*

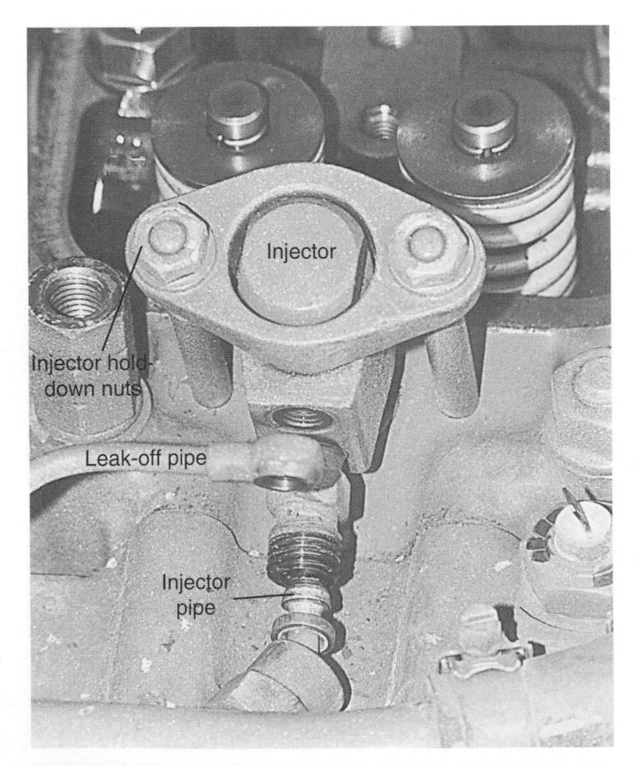

FIGURE 7-4. *A cylinder head with the injector lines broken loose.*

FIGURE 7-6. *An engine with the valve cover removed. This engine has individual cylinder blocks and heads for each cylinder.*

Figure 7-4). The minute you disconnect any fuel line, you must cap both the fuel line and the unit to which it is attached to prevent the entry of ANY dirt into the fuel system (see Figure 7-5). Number the fuel lines for ease of reassembly—a piece of masking tape and a felt-tip pen work fine. If the injectors are to be overhauled, taking them out now is easier than when the cylinder head is off (see the Injectors section, beginning on page 154).

6. Remove the valve cover. If it is stuck in place, don't try to pry it up. Instead, give it a good whack with a rubber mallet; this should jar it loose in a way that may let you reuse the gasket or O-ring.
7. Unbolt the rocker assembly (see Figures 7-6, 7-7, and 7-8). On some engines, it comes off in one piece; on others, each cylinder has a separate unit. Take out the push rods (if present) and lay them down in order (perhaps labeled, as with the injector lines).
8. On some engines with an overhead camshaft, gears drive the camshaft; on others it is powered by a belt or chain (see Figure 7-9). To remove the rockers or the cylinder head, you must first remove the belt or chain. Before you can do that, you must remove the crankshaft pulley and the timing case cover on the front of the engine. You will probably need a special puller to remove the pulley. If the pulley has any tapped (threaded) holes in its face, you can make a puller (see Figure 7-10). To remove the pulley:
   - Slack off the pulley's retaining nut. It may have a locking washer with a tab that you must first bend back out of the way. The nut may be difficult to break

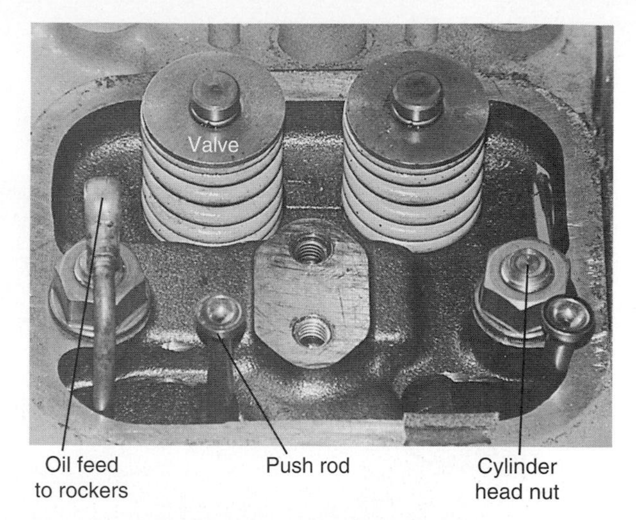

FIGURE 7-7. *A cylinder head with the rockers removed.*

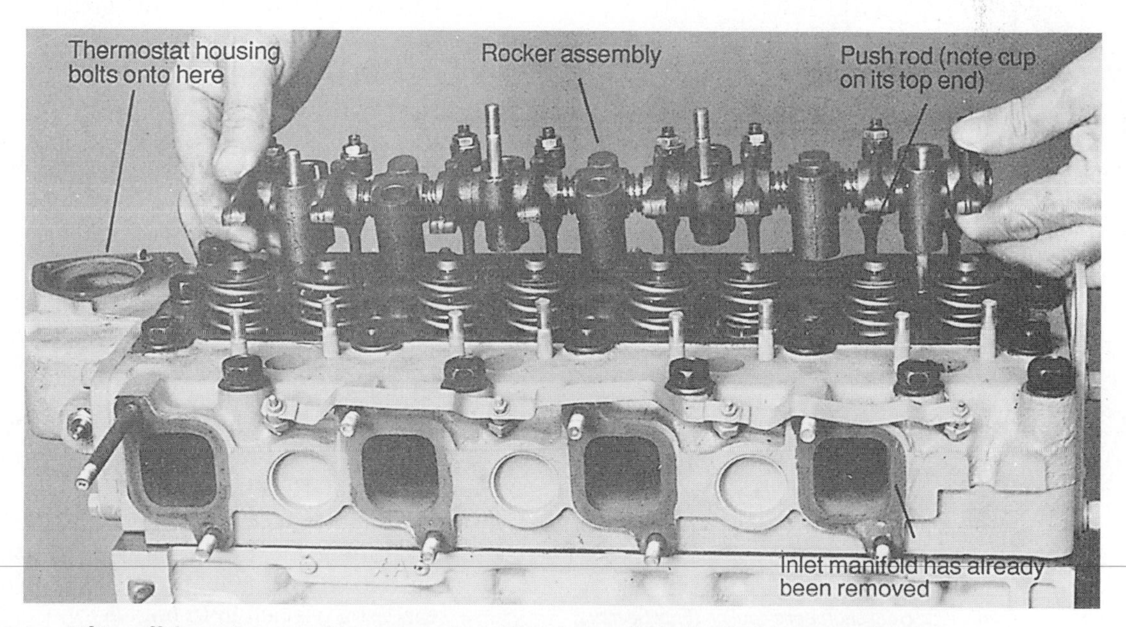

FIGURE 7-8. *Lifting off the rocker assembly on an engine with a single cylinder head. (Courtesy Perkins Engines Ltd.)*

FIGURE 7-9. *An overhead camshaft.*

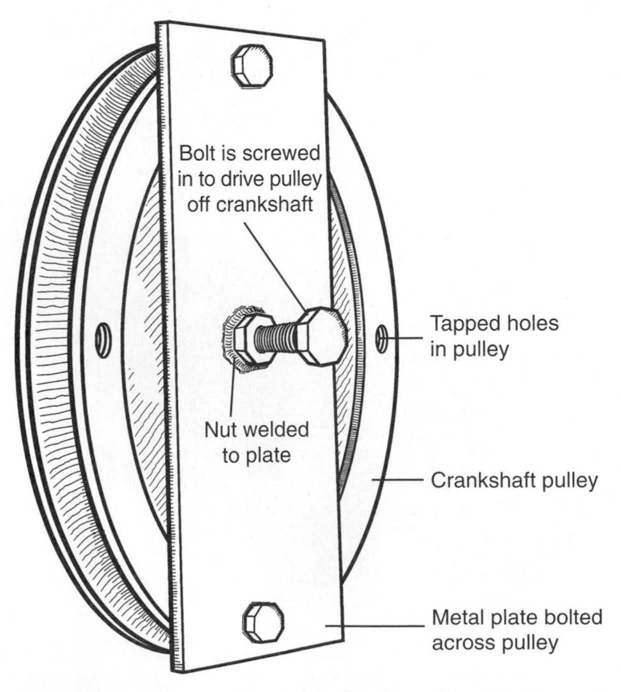

FIGURE 7-10. *A homemade crankshaft pulley puller. (Fritz Seegers)*

loose because the engine will turn over when you apply pressure to the wrench. If a smart blow on the wrench fails to jar loose the nut, grasp both sides of the water pump (or alternator) drive belt, hold it tightly, and try again. If the boat has a manual transmission, place the engine in gear and lock it by putting a pipe wrench on the propeller shaft, after first wrapping the shaft with a rag in order to avoid scoring the shaft.

- Once the nut is off, bolt a flat metal plate, with either a tapped hole drilled in its center or a nut welded over the hole, across the pulley, using the threaded holes in the face of the pulley.
- Screw a bolt down through the tapped hole or welded nut onto the end of the crankshaft. This will pull the pulley off the shaft. (Note: If the nut is placed on the other side of the plate, it doesn't need to be welded in place as long as you can get a wrench in to hold it.)

Anytime you remove a camshaft, both the fuel injection pump and valve timing will be disturbed and will need resetting. This timing is absolutely critical to engine operation. If you have any doubt about being able to reset either valve or injection pump timing, do not disturb the camshaft!

You are now ready to remove the cylinder head.

Cylinder Head Removal

A cylinder head is held down by numerous nuts or bolts spaced around each cylinder. In order to evenly relieve the pressure exerted on the cylinder head, loosen each nut (or bolt) a half turn or so in the sequence outlined in the engine manual. If no manual is available, you should loosen a nut (or bolt) at one end of the head a half turn, then loosen the one opposite it at the other end, and then another at the first end, and so on, working toward the center of the head. After you have released the initial tension, you can remove all the nuts (or bolts).

A cylinder head will frequently bond to the cylinder block, making it difficult to break loose. When this happens, the temptation to stick a screwdriver in the joint between the two and beat on it with a hammer is strong, but is also dangerous and must be resisted. The tremendous pressures concentrated on the tip of a screwdriver can result in a cracked head or block. Instead, if the injectors are still in place, try turning the engine over; the compression will often be enough to loosen the cylinder head. Failing this, firmly hold a solid block of wood to the head at various points and give it a moderate whack with a hammer or mallet. The shock should be enough to jar the head loose. Make sure that the wood contacts a large area of the head; any point loading can crack the head. Be sure to give the wood a smart blow. If the head still refuses to budge, check to see that all the fastenings have been removed—it is surprisingly easy to miss one, especially on dirty engines.

When the head starts to come up, you must lift it squarely to avoid bending the hold-down studs (see Figure 7-11). Do not drag it across the top of the studs because you may scratch the face of the head. Sometimes you can remove the head with the rocker assembly still in place and the push rods still in the engine. If you do this, the push rods may stick to the rockers when you lift off the head. On a few engines they may then drop into the oil sump, and you'll have to remove it to retrieve them! If any push rods stick, lift the head an inch or so, then carefully replace them on their cam followers before proceeding.

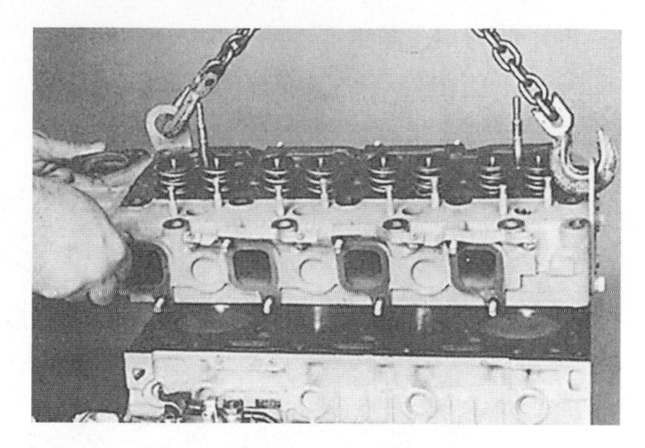

FIGURE 7-11. *Lifting off a cylinder head. (Courtesy Perkins Engines Ltd.)*

The injector tips of open combustion (direct injection) engines often protrude below the level of the cylinder head (see Figure 7-12). Do not rest the head on them, as this will damage the nozzles.

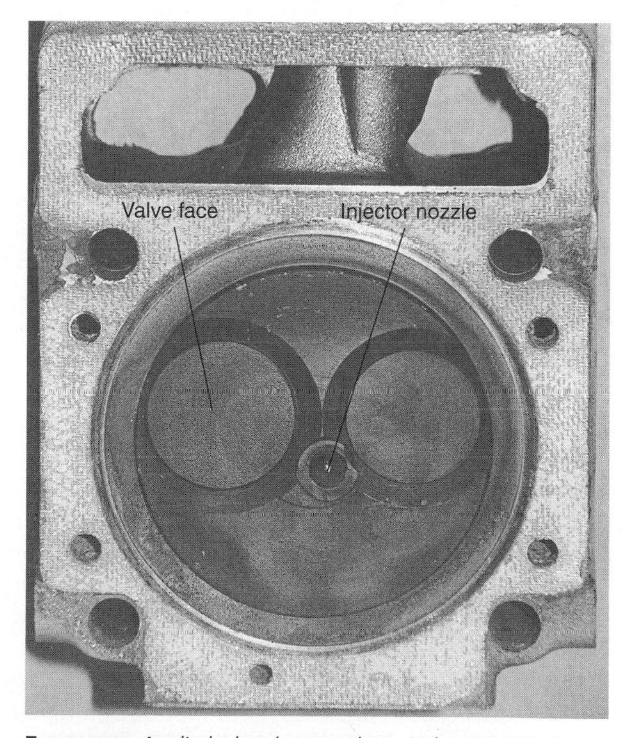

FIGURE 7-12. *A cylinder head removed on a Volvo MD 17C. Do not rest the head on the injector nozzle. (Note: This engine has a direct combustion chamber so it doesn't have a precombustion chamber.)*

FIGURE 7-13. *A Volvo MD 17C with the cylinder heads removed. Note the toroidal crown of the pistons, which is common to direct combustion engines.*

Anytime you remove a head, or any other piece of equipment, remember to block off all exposed passages and holes into the engine to prevent trash and engine parts from falling inside. It is unbelievably frustrating to drop a nut down an oil drain passage and into the sump, then have to remove the engine from the boat to drop the sump and recover the nut!

Once the head is off (see Figure 7-13), remove the old gasket material and other dirt from the face of the head and the block. Gaskets frequently become extremely well bonded to cylinder heads and blocks. You can buy a variety of scrapers from automotive stores to get them off, or you can use a paint scraper. An old chisel (about 1 inch wide) or any good-size pocket knife will do as well. The key, especially on aluminum, is to keep the blade at a shallow angle to the surface being cleaned, otherwise you risk scratching or gouging the metal. Particularly stubborn residues require patience. On cast iron, they can be softened with paint remover, but chemicals are not advised on aluminum.

With the head off and its face cleaned, now is probably a good time to check the head for warpage. Lay a straightedge (a steel ruler is excellent) across it at numerous points and attempt to slide a feeler gauge under it (see Figure 7-14). Feeler gauges are available from any automotive parts stores. They consist of a number of thin metal blades, precision ground to the specified thickness stamped on the blade face. A set from 0.001 to 0.025 inch (one thousandth to twenty-five thousandths of an inch), or the metric equivalent if you have a metric engine, is needed. (See Appendix D for conversion tables.)

Allowable warpage varies according to the size of the cylinder head. If the manufacturer's specifications

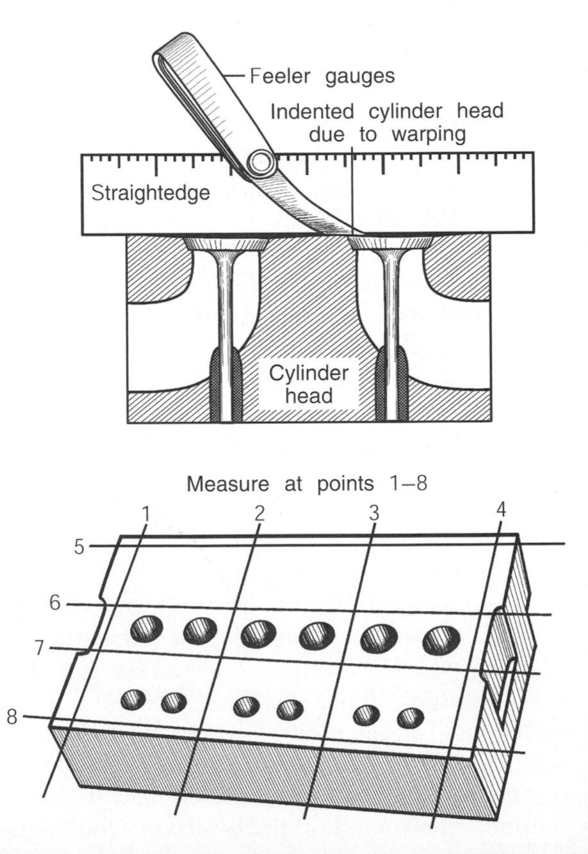

FIGURE 7-14. *Checking for cylinder-head warpage. (Fritz Seegers)*

are not available (as they almost certainly will not be), you can safely assume that on the engine sizes under consideration here, any warpage over 0.004 to 0.005 inch (four to five thousandths of an inch) is excessive.

VALVES

You can test valves for leaks by laying the head on its side and pouring kerosene or diesel into the valve ports. If a valve is bad, the liquid will dribble out where the valve rests on its seat. If no leak is present, you may be wise to leave the valve in place and merely clean the carbon off its face and out of its port.

VALVE REMOVAL

If a leak is present, the valve will have to be reground. Valves are usually held in place by two keepers (or collets), small semicircles of metal that lock in a slot cut into the valve stem (see Figure 7-15). A dished metal washer on top of the valve spring holds the keepers against the valve stem. To remove the keepers, you must compress the valve spring far enough to push the dished washer out of the way. Use a valve-spring compressor or clamp to do this, which is essentially a large C-clamp that fits over the cylinder head; one end rests on the face of the valve and the other slots over the valve stem and around the top of the spring (see Figure 7-16). Closing the clamp compresses the dished washer

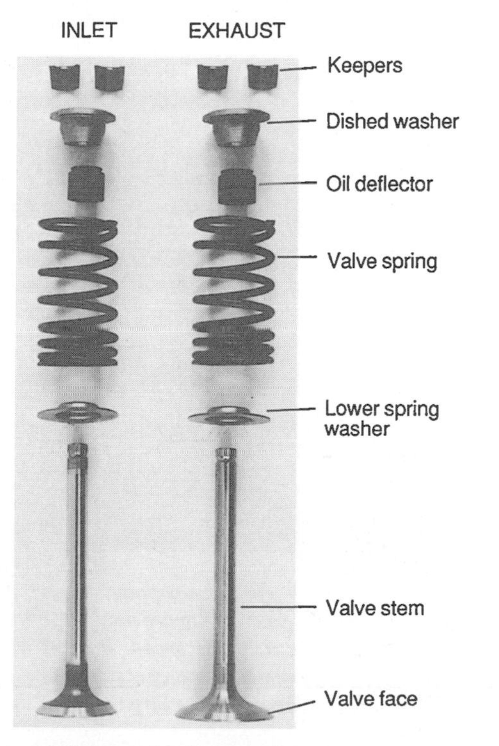

FIGURE 7-15. *Inlet and exhaust valve components. (Courtesy Perkins Engines Ltd.)*

FIGURE 7-16. ***Top:*** *Using a valve-spring compressor. (This is slightly different from the one described in the text, but it serves the same purpose.) (Courtesy Perkins Engines Ltd.)* ***Bottom:*** *A hydraulically operated valve-spring compressor in a large machine shop. (Courtesy Caterpillar)*

and spring down the valve stem, allowing the keepers to be picked off. Releasing the clamp permits the washer and spring to slide off the valve stem and the valve to be pushed out of the cylinder head.

Some automotive parts stores rent valve-spring clamps. Most adjust to fit a wide range of cylinder heads, but if at all possible, take the head to the store and check the clamps for the best fit.

In an emergency, it is possible (though not easy) to push down on the valve spring with a suitably sized box-end wrench, with a second person to remove the keepers. This is easier with older, slower-revving engines, which tend to have weaker valve springs. The trick is to pick off the keepers without allowing the wrench to slip off the spring, thus letting it shoot off the stem. The tiny keepers often get lost when this happens. Keepers are hard to buy and easy to lose so handle them with care.

Cleaning and Inspection

You can degrease a cast-iron cylinder head and all the steel parts (valves, springs, etc.) in a solution of 1 pound of caustic soda plus 8 ounces of detergent dissolved in 1 gallon of water. (Use noncaustic detergents to degrease aluminum.) If possible, put the solution in a tub large enough to hold the cylinder head, and heat the whole thing for an hour or so. Afterward, be sure to flush the head very thoroughly with copious amounts of fresh water.

The key area of a valve is the beveled region that sits on the valve seat in the cylinder head (refer back to Figure 7-15). Inspect valves for any of the problems illustrated in Figures 7-17, 7-18, and 7-19.

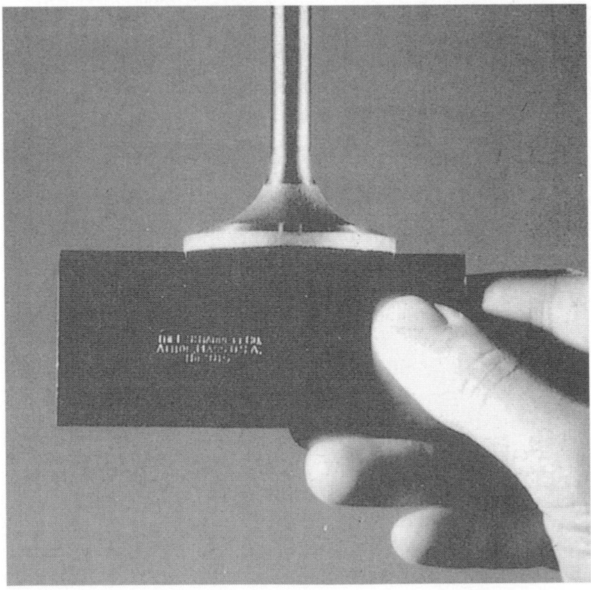

Figure 7-17. ***Bottom left:*** *Some cupping of valve faces is normal, but if it's excessive (more than 0.010 inch), as it is here, the valve must be discarded.* ***Bottom right:*** *Use a straightedge and feeler gauge to check for valve cupping.* ***Top right:*** *If the valve has large or deep marks near the edge, discard it. (Courtesy Caterpillar)*

FIGURE 7-18. *Valve failure.* ***Top left and right:*** *Corrosion from moisture and acids.* ***Bottom right:*** *Stress cracks caused by high temperature.* ***Bottom left:*** *Metal-to-metal transfer (galling) from a valve stem sticking in its guide. (Courtesy Caterpillar)*

If the valve or seat is pitted in the area of contact, the head will have to go to a machine shop for regrinding of the seat and refacing of the valve. Exhaust valves need checking more closely than inlet valves because they are subject to much higher temperatures and the exhaust gases tend to burn them. Note that on a marine engine, the exhaust valve closest to the exhaust manifold exit pipe is sometimes corroded as a result of water vapor from the water-cooled exhaust coming back up the exhaust pipe.

If the seat and valve face are reasonably smooth, you can frequently lap in the valve by hand, as discussed in the next section, although doing this to

FIGURE 7-19. *Valve failure.* ***Top left:*** *A valve badly burned by escaping gases.* ***Top right:*** *A beaten and battered valve stem end from excessive tappet clearance.* ***Bottom left:*** *Damage caused by a foreign object bouncing around in the combustion chamber.* ***Bottom right:*** *A slightly bent valve. Notice the uneven grind marks on the face. (Courtesy Caterpillar)*

the specially hardened valves and seats on modern engines is less feasible than it used to be. Lapping hardened valves and seats by hand may alleviate minor problems, but problems that cannot be solved in this fashion will likely require new valves and probably new seats, since machining either in any way will cut through the hardened surfaces.

LAPPING

To lap in a valve, apply a thin band of medium grinding paste (available from any automotive parts store) around the seating surface. Drop the valve into the cylinder head, and place a lapping tool (also available from any automotive parts store) on the face of the valve. (Some valves have a screwdriver slot in them, in which case the lapping tool is unnecessary.) Spin the handle backward and forward between your palms while you maintain a gentle downward pressure to hold the valve against its seat (see Figure 7-20). Every so often, lift the valve off its seat, rotate the valve a quarter of a turn or so, drop it back down, and work backward and forward some more. This

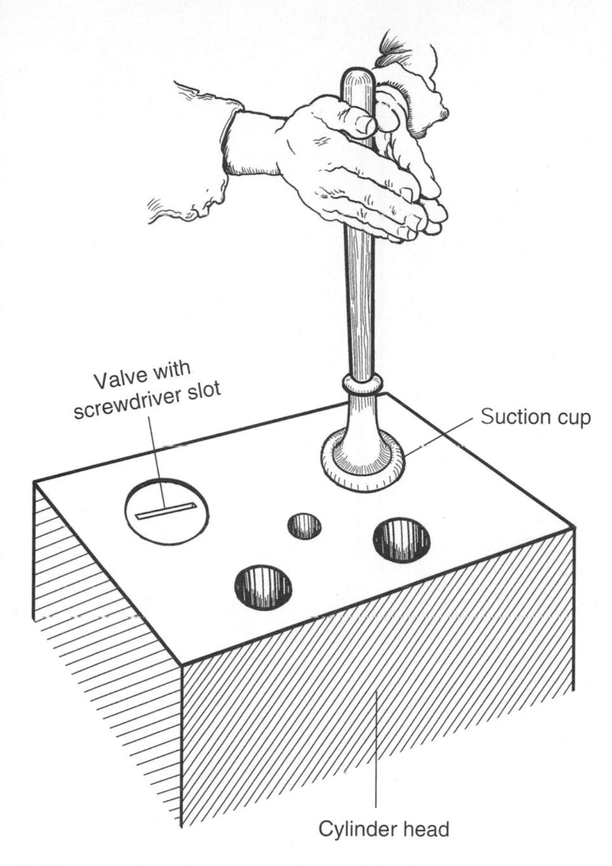

FIGURE 7-20. *Hand-lapping a valve. (Fritz Seegers)*

ensures an even grinding of the valve and its seat, regardless of the position of the valve.

Continue this procedure until a line of clean metal is visible all the way around the valve and its seat. (Add more grinding paste as needed.) As soon as this line appears, polish the surfaces in the same manner, using a little fine grinding paste.

You can check the fit of a valve in its seat by making a series of pencil marks across the face of the valve about ⅛ inch apart, then dropping the valve onto its seat. All the pencil marks should be cut by the seat (see Figure 7-21).

Do not overgrind the valves. The objective is a thin line of continuous contact between the valve and its seat, not a perfect fit. Overgrinding lowers the valve in the head, which increases the size of the combustion chamber and leads to a loss of compression. Once the valves and seats have been ground beyond a certain point, the loss of compression becomes unacceptable and requires a major cylinder head overhaul.

The valve seats of some engines are separate inserts pressed into the cylinder head, which can be removed and replaced (this is machine shop work). Combined with new valves and valve guides (see below), this produces, for all intents and purposes, a new cylinder head.

VALVE GUIDES

A valve guide holds the valve in alignment in its cylinder head. A buildup of carbon around the valve stem and in the guide sometimes causes a valve to stick in its guide. These areas must be carefully cleaned during decarbonizing.

Excessive wear in a guide allows lubricating oil from the rocker arm to run down the valve stem into the valve port, where it is sucked into the engine and burned (in the case of an inlet valve), or burned by the hot exhaust gases (in the case of an exhaust valve). Check wear on the valve guides and stems by attempting to rock the valve from side to side in its guide. No lateral movement is permissible—the wear limit is normally around 0.005 inch but requires a dial indicator to measure. Most engines have replaceable valve guides, which are pressed into the cylinder head, but this is a job for the machine shop.

VALVE REPLACEMENT

Before you replace any valves, visually check the springs for cracks, nicks, and corrosion (see Figure 7-22). Check the length of each spring against the manufacturer's specifications, if possible, or compare it to a new spring. Replace the spring if it is damaged or short.

Valve replacement is the reverse of removal. First, wash away all traces of grinding paste by thoroughly flushing the cylinder head and components with diesel or kerosene. After you've refit the valves, perform the kerosene test to check their seating.

At this point, you should probably test older engines for overgrinding of the valves. To do this, lay a straightedge across the face of the cylinder head, over the top of the valve, and use a feeler gauge to measure how far the valve is recessed into the head (see Figures 7-23 and 7-24). Check the degree of valve indentation against the manufacturer's specified limits to determine if the head needs new valves and seats.

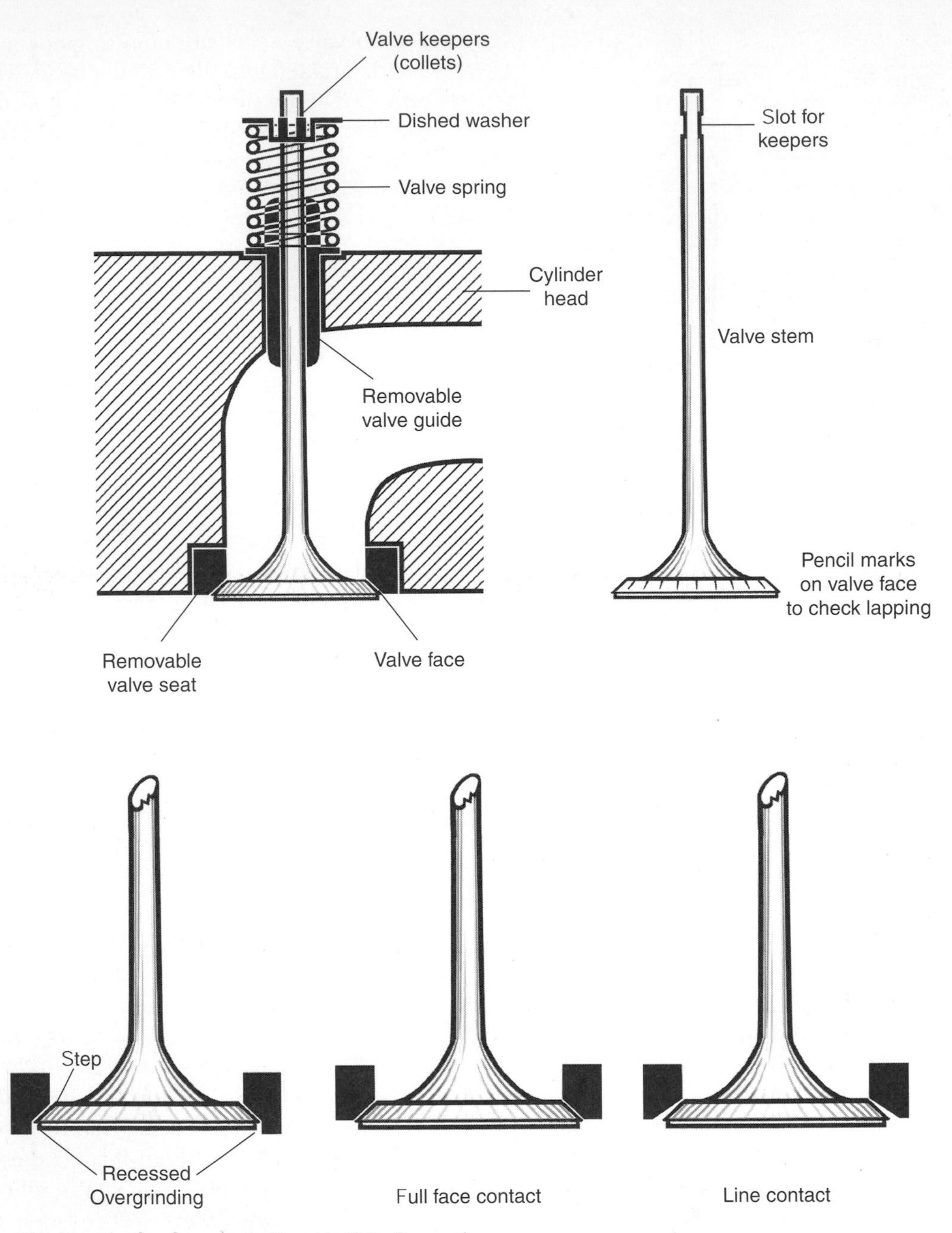

Figure 7-21. *Checking the fit of a valve in its seat. (Fritz Seegers)*

Cylinders

Cylinder Inspection

When a piston is at the top of its stroke, its topmost ring is still a little way down the cylinder. Because a cylinder wears only where it is in contact with the rings, this top part of the cylinder will be unworn. A significantly worn cylinder bore will have a step right below this. In order to check for wear, rotate the crankshaft until the piston is at or near the bottom of its stroke. Clean away the carbon that has collected at the top of the bore with a suitable

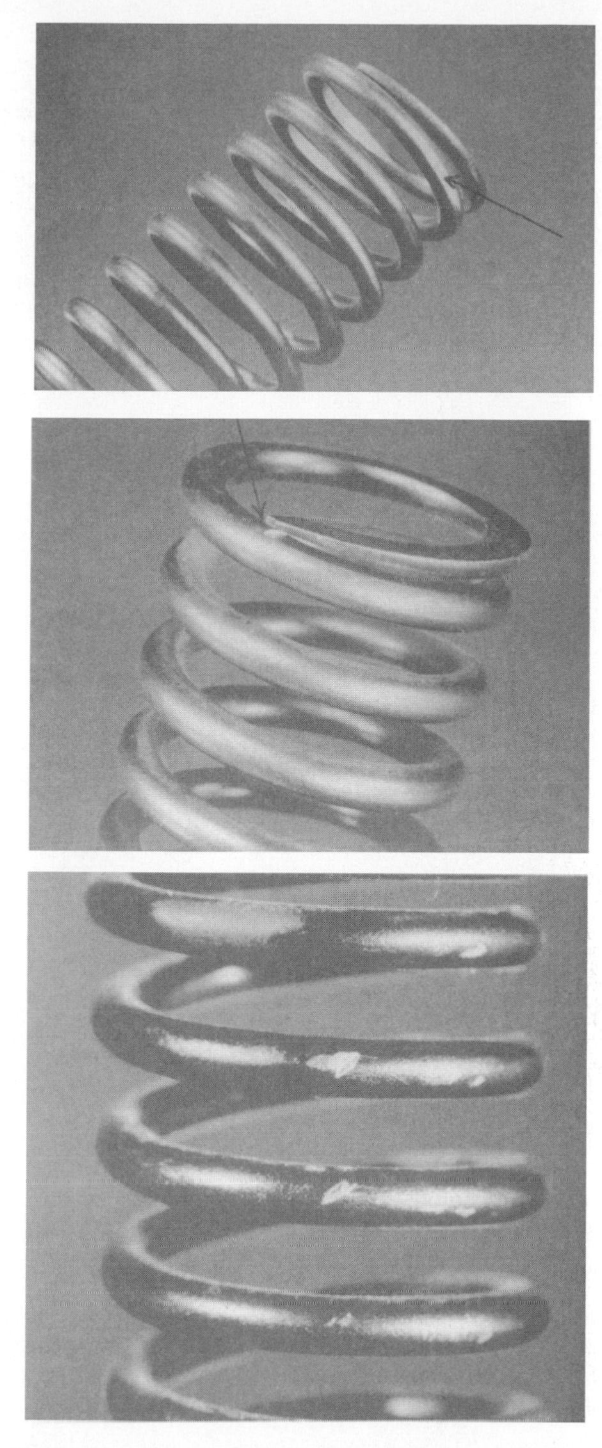

FIGURE 7-22. *Wear on the sides of a valve spring (top) is not acceptable; discard it. Notches at the end of a valve spring (middle) are not acceptable; discard it. If a valve spring has deep nicks or notches (bottom), it must be discarded. (Courtesy Caterpillars)*

scraper or knife, finishing off with a piece of wet-or-dry sandpaper dipped in diesel. Wipe the cylinder wall clean and run your fingernail up and down the top part of the bore. If the step approaches anywhere near the thickness of a fingernail, you should have the bores professionally measured to determine whether the time has come for a cylinder renewal, which also includes new pistons and rings. (See pages 187, 188 for a simple technique to approximately measure cylinder wear.)

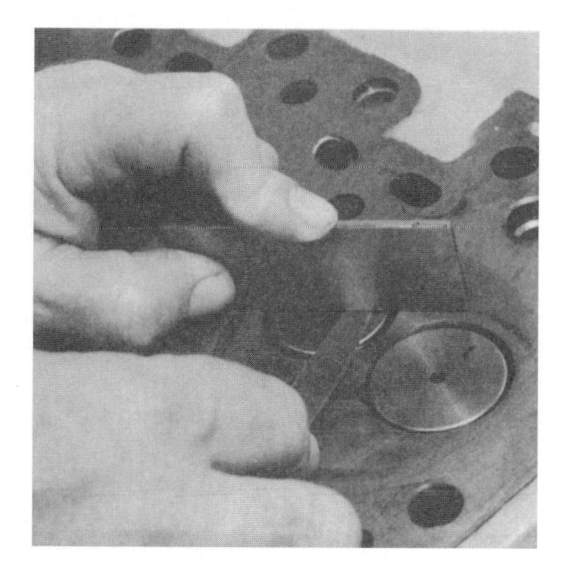

FIGURE 7-23. *Checking for overgrinding of the valves. (Courtesy Perkins Engines Ltd.)*

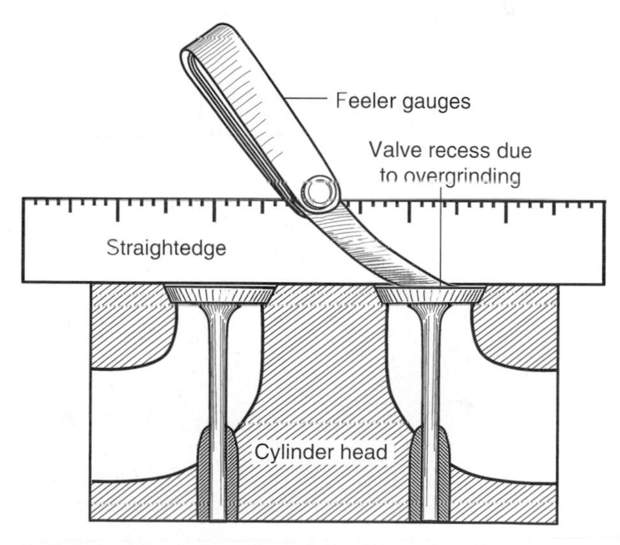

FIGURE 7-24. *A cutaway view of checking valves for overgrinding. (Fritz Seegers)*

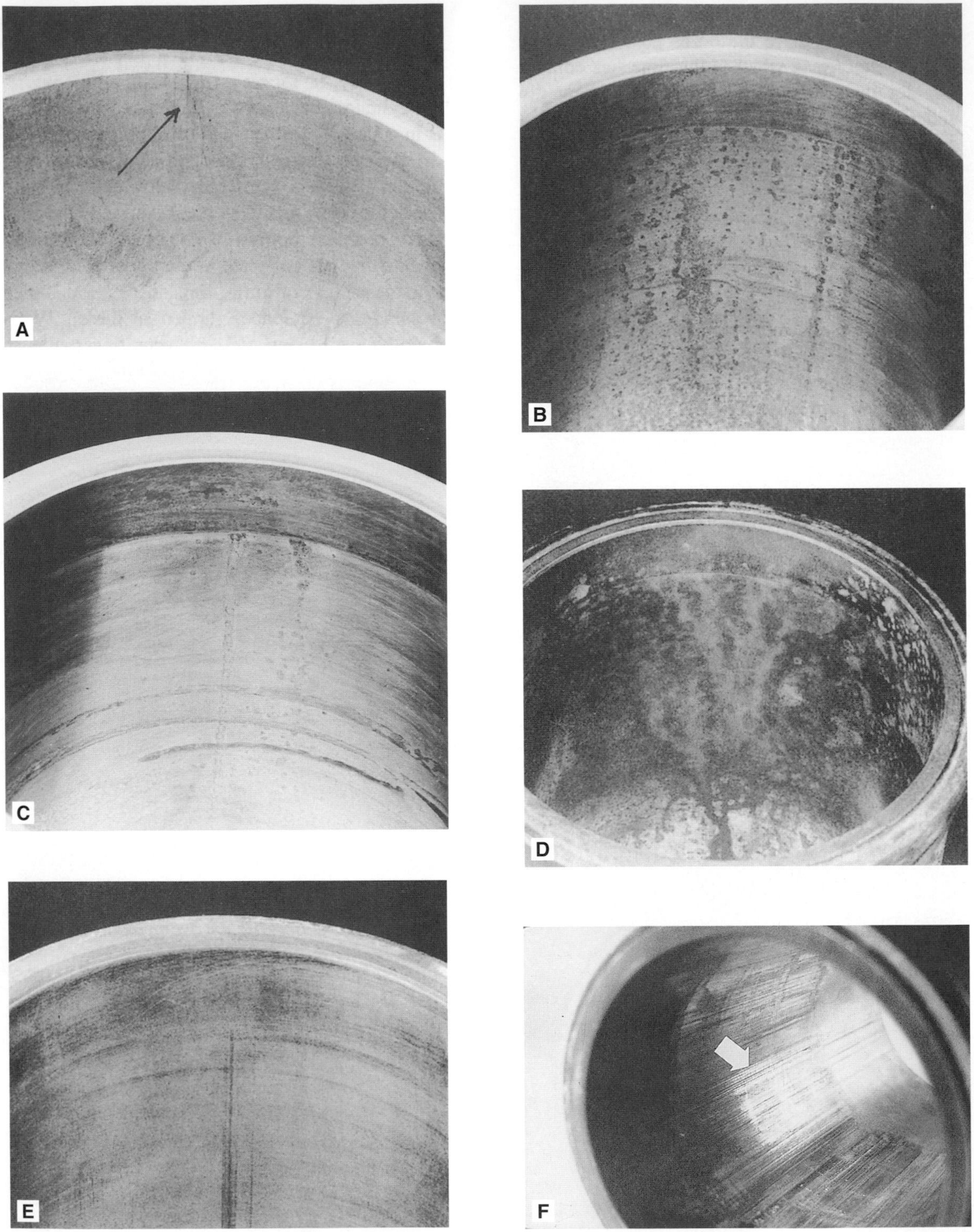

FIGURES 7-25A to 7-25F. *A cracked cylinder (A)—discard. Pits on the inside surface (B)—discard. Water spots on inside surface (C)—can be used again. Heavy rust (D)—discard. Deep grooves on the inside surface (E)—discard. A scuffed liner from piston seizure (F)—discard. (Courtesy Caterpillar)*

Other indications that the engine needs a new cylinder are any cracks, however small, or evidence of holes in the cylinder wall (such as erosion on the top flange of a wet cylinder liner). Air filter failure will cause vertical scratches in the cylinder wall. Improper injection, resulting in fuel hitting the cylinder, will burn away the affected area. If the engine has experienced a piston seizure, the softer aluminum of the piston (most pistons are aluminum) will frequently peel off and stick to the cylinder wall. All these problems will probably also necessitate a new cylinder (see Figures 7-25A to 7-25F).

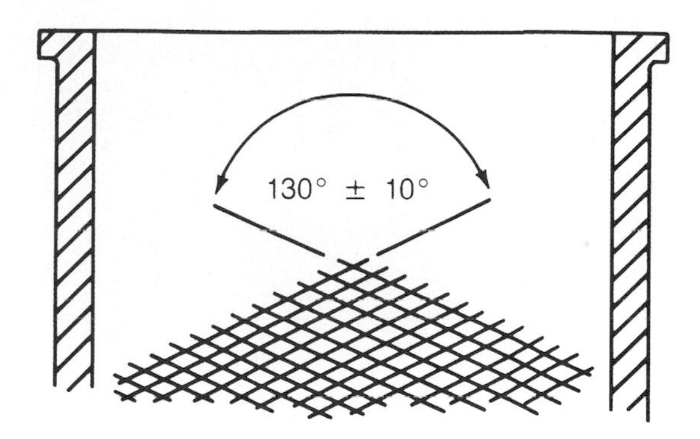

FIGURE 7-26. *Correct crosshatch pattern produced by honing—no pits, rust, scratches, or shiny areas. (Courtesy Caterpillar)*

CYLINDER HONING

Long hours of low-load running often result in a cylinder wall becoming glazed—very smooth and shiny. If there is very little other wear, glazing can be broken up by a process called honing, using a flexible nylon brush with an abrasive material on the tips. This is known as a flex hone. Wet liners can be pulled out of the block and taken to a machine shop; dry liners can be honed with the block and cylinder still in the boat, but first the pistons must be removed (see below). The hone will generate a good bit of abrasive dust, so cover the crankshaft below the cylinder with a large rag.

You used to be able to rent an appropriately sized hone from automotive parts store, but they are pretty hard to find these days, so you will probably have to call a machine shop.

If you can rent a hone, put it into a slow-speed electric drill (350 to 500 rpm), lubricate it with kerosene, and run it up and down inside the cylinder at the rate of approximately 30 strokes a minute (1 second down, 1 second up). On no account hold the hone stationary—it will score the cylinder wall. The idea is to produce a crosshatch pattern of very light scratches angled at around 130 degrees to one another (see Figure 7-26), which cover the entire piston-ring contact area. You can adjust the angle of the crosshatch by changing the speed of the drill or the rate at which you move the hone up and down the cylinder. If the slope of the crosshatch is less than 130 degrees, increase the speed of the drill or decrease the rate at which you move it up and down. If the angle is more than 130 degrees, slow the drill or increase the rate of movement. Use the flex hone for approximately 30 seconds only, certainly no more than 60 seconds; the time will depend on how badly glazed the cylinder is and how worn the hone is.

After you hone the cylinder, clean it thoroughly. Swab it down with kerosene and wipe it with paper towels until the towels remain spotless. Immediately after cleaning, wipe the cylinder with a towel soaked in oil—rust can begin to form in minutes on a clean, dry cylinder wall. After you remove the rag from the crankcase, thoroughly flush the area to remove all traces of the abrasive material shed by the hone.

If you removed a wet liner (see Figure 7-27) from the engine, fit new O-ring seals before replacing it. Lubricate the seals with liquid soap (dishwashing liquid works well). If the liner has a light scale or corrosion on just one side, rotate it 90 degrees from its previous position when you reinstall it.

PISTONS AND CONNECTING RODS

Pistons are sealed in their cylinders by piston rings. Significant wear occurs in the cylinder wall (see above), the outer surface of the piston ring, and the width of the piston-ring groove. This groove widens over time as a result of the ring working up and down as the piston moves in the cylinder. Piston rings can only be checked after a piston has been removed from its cylinder.

PISTON REMOVAL

To remove a piston from its cylinder, you must detach its connecting rod from the crankshaft, then push the piston/connecting rod assembly out through the top

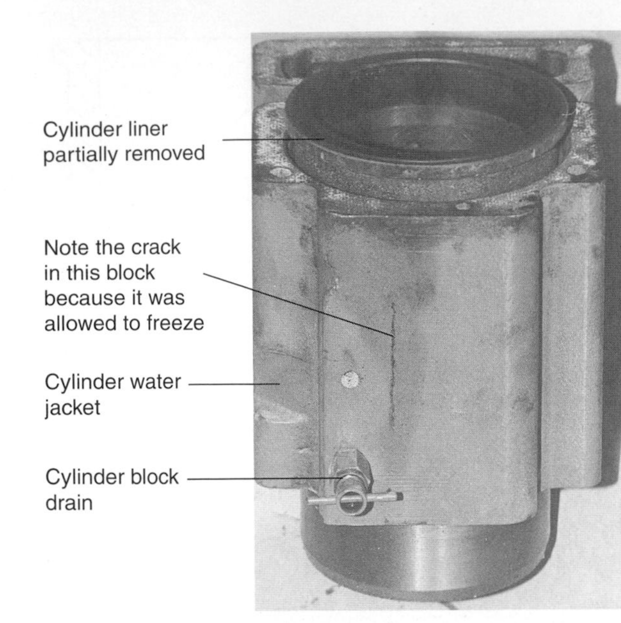

FIGURE 7-27. *A wet cylinder liner.*

of the cylinder. A cap, fastened with two bolts, holds connecting rods to the crankshaft (see Figure 7-28). In the majority of engines, you must remove the engine's oil pan to gain access to these bolts. In most boats, this requires lifting the engine off its bed to gain access to the pan. A few smaller marine diesels, and many larger ones, provide access to the connecting rod caps through hatches in the side of the crankcase, allowing you to remove the pistons without disturbing the engine.

Lifting an engine from its bed is a major undertaking because it involves breaking loose fuel lines, electrical connections, the exhaust system, the propeller coupling, and other equipment. You'll also need some form of overhead crane or hoist. The main boom of a sailboat, if it's long enough and adequately supported, can often be used with an appropriate block and tackle. The boom of a trawler yacht's riding sail is another possibility.

Before you take a piston from its cylinder, you have to remove the ridge of carbon at the top of the cylinder. Also, before undoing the connecting rod cap, get hold of the connecting rod where it clamps around the crankshaft *journal* (the machined section of the crankshaft around which the bearing is clamped) and work the piston up and down and backward and forward. You will notice some lateral movement along the crankshaft journal, but otherwise this bearing should have no appreciable play, which is usually accompanied by knocking, and means the bearing needs replacing.

After the connecting rod bolts are removed, sometimes separating connecting rod caps from their rods is difficult. Never pry them apart. Tap

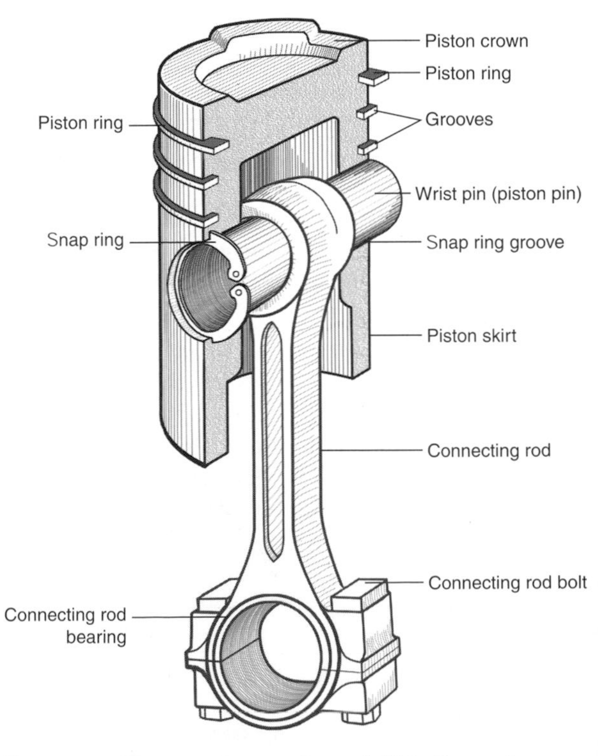

FIGURE 7-28. *Piston and connecting rod. (Fritz Seegers)*

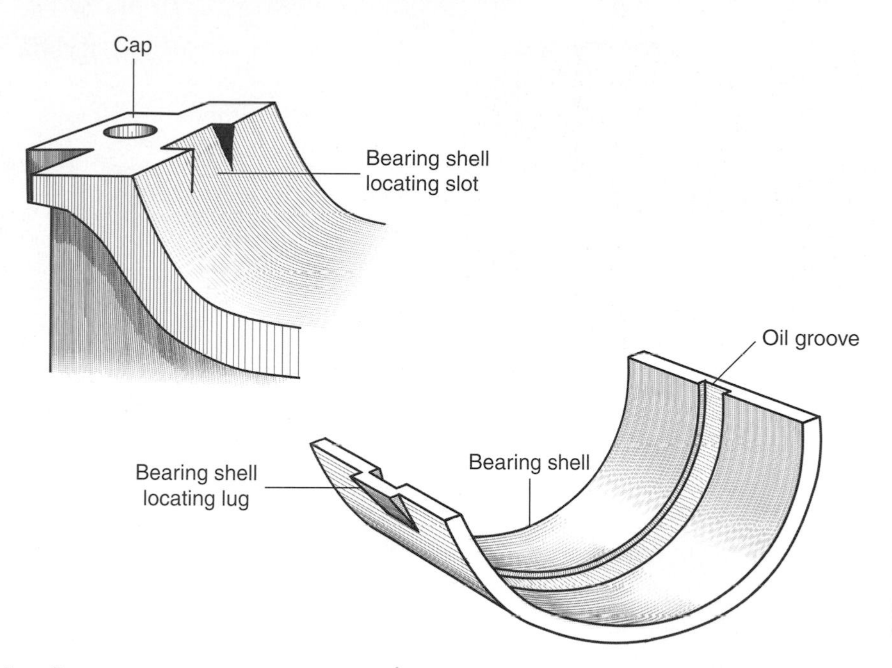

FIGURE 7-28. *(continued)*

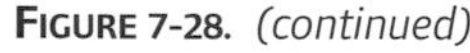

the cap gently with a soft hammer or a block of wood while you pull down on it; this will invariably break it loose.

Pistons must go back in the cylinders they came from, and they must not scratch the cylinder liners. The piston must face in the same direction; the connecting rod cap must go on the same way and the connecting rod bolts must go into the same holes. Some engines require that you fit new bolts every time you remove the caps. This is a good practice for any engine. The piston crown (top) may already be marked with its cylinder number and forward face (see Figure 7-29), and the connecting rod and cap may also be numbered and marked. If not, you have to make some kind of identification.

CLEANING AND INSPECTION

Pistons are cleaned commercially by either using various solvents or blasting with glass beads. If you have to clean them yourself, give them a good soaking in diesel to help loosen carbon and other deposits, which you can then remove with 400- to 600-grit wet-or-dry sandpaper, keeping it constantly wetted out with diesel. Do not scratch the pistons, most of which nowadays are made of aluminum. If you have cast-iron pistons, you can soak them in the caustic soda solution previously recommended for cast-iron cylinder heads.

After you've cleaned the pistons, check them for excessive wear or damage. The following problems,

FIGURE 7-29. *The markings on a piston crown. (Courtesy Perkins Engines Ltd.)*

Figures 7-30A to 7-30G. *Generalized overheating of the crown. The top of the piston side has started to peel away and stick to the cylinder wall (A); the piston cannot be used again. Cracking of the piston crown through overheating, in this case concentrated where the gases blow down out of the precombustion chamber (B); the piston cannot be used again. Severe erosion of the piston crown due to erratic combustion caused by a plugged air filter (C); this piston cannot be used again. The stainless steel plug in the crown, found only in high-performance engines, helps dissipate heat from the precombustion chamber. Carbon cutting around the piston top from the ring of carbon at the top of the cylinder (D); although the piston is scratched, it isn't breaking up and can be used after it has been cleaned. The rings on this piston are stuck in their grooves, leading to overheating, blowby, and serious erosion of the side of the piston (E); it cannot be used again. The skirt (the area below the piston pin) of this piston has been scuffing (rubbing without lubrication) on the cylinder wall because of the engine's overheating (F); the skirt has begun to break up, so the piston cannot be used again. This piston has been seizing from top to bottom through serious overheating or lack of lubrication (G); it cannot be used again. (Courtesy Caterpillar)*

with their likely causes, indicate that new pistons are called for (see Figures 7-30A to 7-30G):

1. A severely battered piston crown. This is generally caused by a broken or sticking valve, or a broken glow plug or injector tip.
2. Excessive cracking of the piston crown. The crown takes the full force of combustion, and some hairline cracking is common on modern high-speed diesels. On engines with precombustion chambers, this generally is concentrated at the point on the piston crown where the combustion gases drive out of the precombustion chamber and hit the piston. However, extensive crazing or deep cracks mean that the piston top has overheated, and the piston must be replaced. The most likely cause is faulty fuel injection.
3. Parts of a piston crown may be eaten away, also as a result of faulty injection. Injector dribble causes late combustion and detonation, which leads to burned exhaust valves and erosion of pistons, especially in the area of the crown closest to the exhaust valve. The latter is a result of the continuing combustion during the exhaust cycle. A plugged air filter or defective turbocharger also causes improper combustion and can lead to more widespread burning of the piston crown.
4. A piston may become severely worn all around its sides from the crown down to the top ring. This indicates that the above problems have resulted in generalized overheating of the piston crown, causing it to expand and rub against the cylinder wall. Some scratching of this portion of the piston is normal because it rubs against the carbon ridge at the top of the cylinder. Excessive wear requires a new piston. If the piston rings are also damaged or stuck in their grooves, blowby of hot gases is likely to spread this wear (scuffing) down the sides of the piston.
5. The same scuffing on the base (skirt) of a piston indicates widespread overheating, most likely due to a failure of the cooling system or a lack of lubrication. Left unattended, this will probably lead to a piston seizure, causing the surface of the piston to break up and stick to the cylinder wall.

Piston Pins and Bearings

Check for wear in the piston-pin bearing. The piston will be free to move from side to side on the pin, but if you detect any up-and-down movement in the bearing, you will have to replace the piston-pin bushing in the connecting rod.

Piston Pin Removal

The piston pin is normally held in place by a snap ring at each end. These are spring-tensioned rings that expand into a groove machined into the piston. To remove these rings, you need snap-ring pliers, which have hardened steel pins set in the end of each jaw. At each end of a snap ring is a small hole. Insert the tips of the pliers into the snap-ring holes and squeeze (see Figure 7-31).

If you don't have snap-ring pliers, you may be able to accomplish the job by grinding the jaws of a pair of needle-nose pliers to the proper size to fit the snap-ring holes (see Figure 7-32). Pinch the snap rings only enough to remove them. Excessive squeezing and compression of snap rings is a major cause of failure after an engine is put back in service.

You only need to remove the snap ring from one end of the piston. If the piston pin does not slide

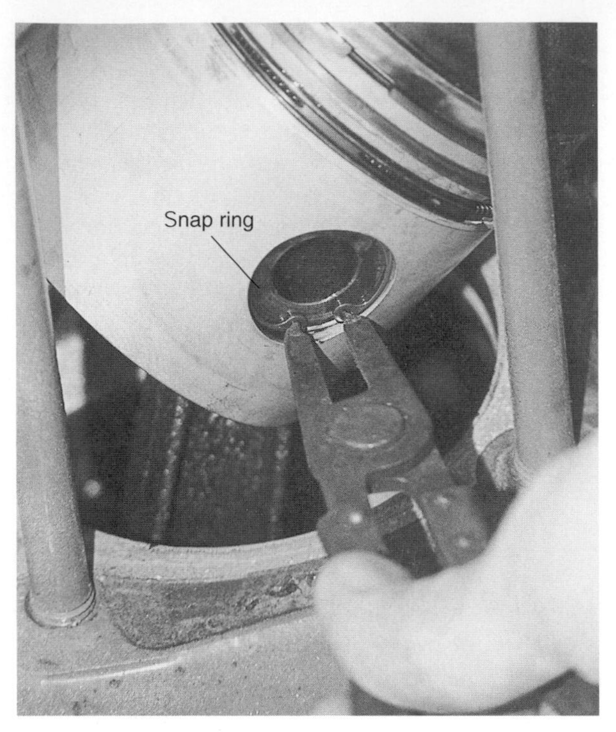

FIGURE 7-31. *Snap-ring pliers in action.*

out easily (some are an *interference* fit), dip the piston into near-boiling water to expand it before you tap out the pin. Any attempt to force the pin out of a cold piston is likely to distort the piston permanently. When the piston pin is out, check it for ridging or burning due to oil starvation (burning will turn the

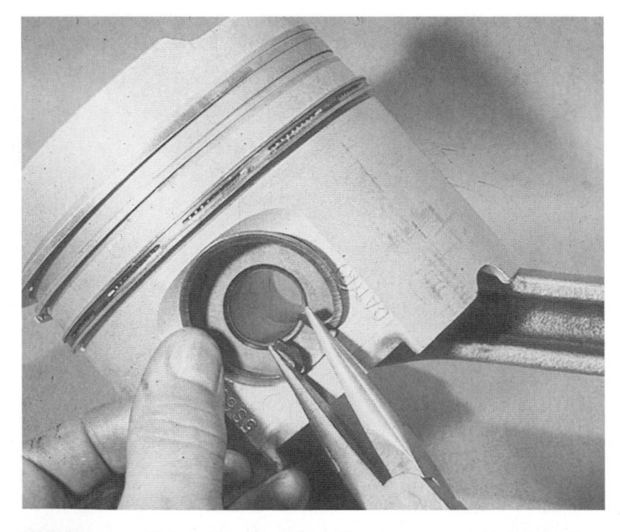

FIGURE 7-32. *Using needle-nose pliers to remove a snap ring. (Courtesy Caterpillar)*

piston pin a blue color). If either is present to any marked degree, replace the pin.

Bushing Replacement

In order to push the old piston-pin bushing out of the connecting rod, you may also have to heat the rod in near-boiling water. Firmly support the connecting rod on blocks of wood, place the new bushing against the old one, hold a block of wood to the new bushing, and gently tap the block with a hammer. The new bushing will drive out the old one. The procedure will go more easily if you cool the new bushing in a freezer for a few minutes before starting. *Note that most bushings have an oil hole that must line up with an oilway in the connecting rod.*

Now check the fit of the piston pin in the new bushing. Some bushings require reaming before the pin fits. This is really work for a machine shop, but in the absence of specialized tools, you can do it by wrapping a piece of 400- to 600-grit wet-or-dry sandpaper around a suitably sized piece of doweling, lubricating with diesel, and working back and forth. Be sure you hold the dowel square to the bushing and work around the whole bushing evenly. Do this until the piston pin will slide smoothly into place; when the fit is correct, a lubricated piston pin will slide slowly through the bushing under its own weight.

Reassembling a piston to its connecting rod is the reverse of disassembly, and it may require the piston (but not the piston pin) to be heated. Again, put the pin into a freezer or pack it in ice, and it will slide right into place.

CONNECTING ROD BEARINGS

Now is also the time to replace the connecting rod bearings, if this is necessary. These bearings consist of a precision-made steel shell lined with a special metal alloy (babbitt or lead bronze). You can remove the shells from the connecting rod and its cap by pushing on one end of each shell—they should slide around inside the rod or cap and slip out the other end. A locating lug on the back of each shell at one end ensures that they can only be pushed out, and new ones inserted, in one direction (see Figures 7-28 and 7-33).

If this is the engine's first major overhaul, it will almost certainly have standard-sized bearings. But an older engine may have had its crankshaft reground, in which case the crankshaft journals will be smaller than standard and the bearing shells will

FIGURE 7-33. *Aligning a bearing shell locking tab with its housing. (Courtesy Caterpillar)*

be correspondingly thicker. The back of the shells will be stamped STD, 0.010, 0.020, or 0.030, indicating the size of the new shells required.

If the engine has been knocking badly or the old bearings are seriously worn or scored (see Figures 7-34A to 7-34F), the crankshaft journal may be damaged or worn into an ellipse (see Figures 7-35A to 7-35E). The latter can only be measured with the appropriate micrometers, which will require a specialist's help.

If the crankshaft is damaged or excessively worn, fitting new bearings is pointless because they will only last a short while. The crankshaft will have to be removed and reconditioned, which is beyond the scope of this book. If the old bearings failed due to oil starvation, the crankshaft may have got excessively hot and will show heat discolorations. If heat has destroyed the special hardening process applied to crankshaft bearing journals, a new crankshaft is needed.

When fitting new bearing shells to a connecting rod and its cap, the backs of the shells and the seating surfaces on the rod and cap must be spotlessly clean. You can push the new shells directly onto their seats or slide each around inside the housing until the lug seats in the slot. The shells must seat squarely, and the lugs must be correctly positioned. *Make sure that any bearing shell that has an oil hole is fitted to the appropriate housing and lined up with its oilway* (see Figure 7-36).

PISTON RINGS

Piston ring failure is generally caused by poor installation practices (see below), excessively worn piston-ring grooves, the ring hitting the ridge of carbon at the top of the cylinder, water in the cylinders, or detonation as a result of improper fuel injection or excessive use of starting fluid. In all instances, the top ring is the most vulnerable.

You must replace broken rings. Check unbroken rings for wear, and check the fit of new and old rings in their grooves. The rings may be stuck in the grooves with carbon and other gummy deposits, so the first task is to clean them.

REMOVING PISTON RINGS

Piston rings are extremely brittle and easily broken. Loosen them in their grooves by carefully cleaning off excess carbon, using plenty of penetrating fluids. After you have freed the rings, expand the ends (pry them apart) to enlarge the diameter sufficiently to lift them off the piston. Piston-ring expanders are available for this and should be used if at all possible (see Figure 7-37). If you cannot lay your hands on a piston-ring expander (available from automotive parts stores), a few strips of thin metal slipped under the ring as you expand it out of its groove should enable you to slide it off the piston (see Figure 7-38). Old hacksaw blades, carefully ground down to remove all sharp edges, work well enough. You must slide off the ring evenly; if you cock it, it will probably break.

While this is a simple procedure, you must take great care when easing the rings out of their grooves. Expand them just enough to slide them off the piston. The two ends of a piston ring often have sharp points; do not let these scratch the piston as you remove and reinstall the rings.

Incorrect removal and installation procedures are a major cause of piston ring failure. In general, rings should be removed only if strictly necessary and should then be replaced with new ones to be on the safe side.

CLEANING AND INSPECTION

Once you have removed the rings, clean them and the ring grooves in the piston. The grooves must not be scratched or widened because this will allow gases to blow past the rings. Although it is frequently done (and I have done it many times), I

FIGURES 7-34A to 7-34F. *Extensive scratching from dirt in the oil (A); these shells cannot be used again. This shell got so hot that it began to melt (B). These shells are breaking up as a result of oil starvation (C). This shell had a paint chip behind it due to improper cleaning at the time of installation, which resulted in severe localized overheating (D); it cannot be used again. Another case of oil starvation (E). This crankshaft will have to be removed from the engine and reground. A set of bearing shells in good condition (F); light scratching is normal, as long as the bearing journal on the crankshaft is shiny and smooth. (Courtesy Caterpillar)*

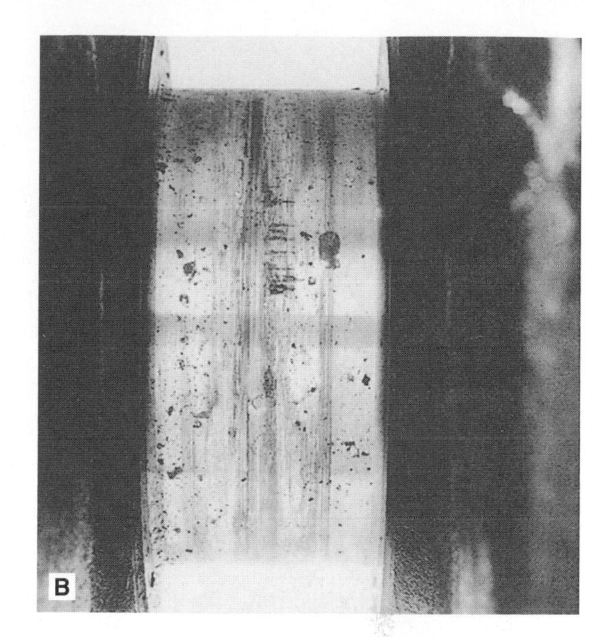

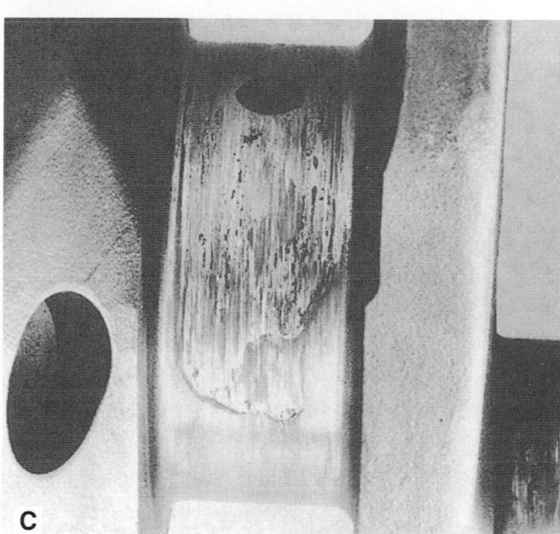

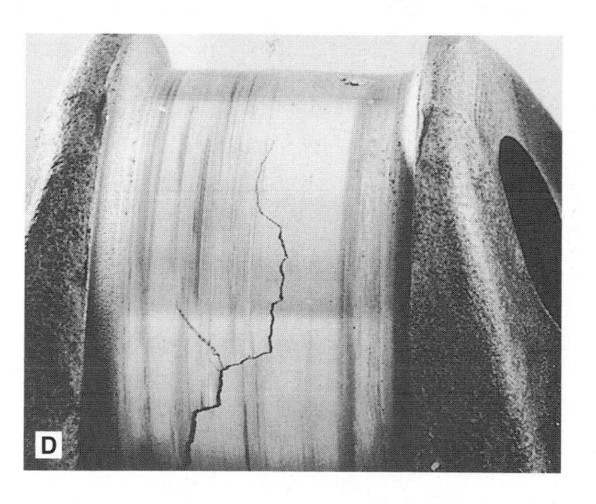

FIGURES 7-35A to 7-35E. *Heat distortion (A)—discard. Corrosion of crankshaft journal (B)—discard. Heat distortion, plus material from the bearing shell is adhering to the journal (C)—discard. Cracked crankshaft (D)—discard. Smearing of bearing shell around the crankshaft (E)—discard. (Courtesy Caterpillar)*

FIGURE 7-36. *Aligning the oil hole in the bearing shell with its oilway. (Courtesy Caterpillar)*

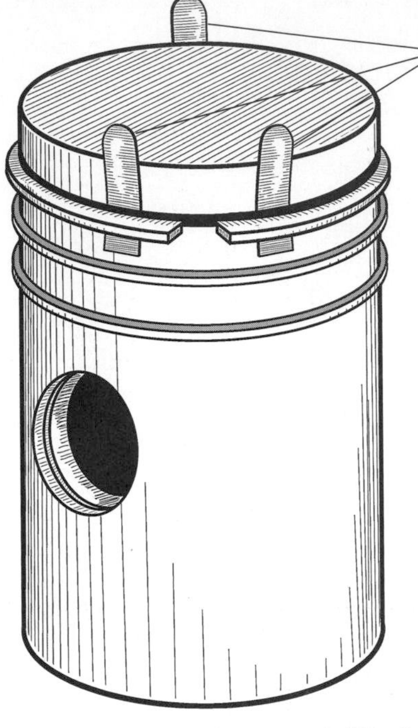

FIGURE 7-38. *Piston ring removal. (Fritz Seegers)*

FIGURE 7-37. *A piston-ring expander. (Courtesy Detroit Diesel)*

don't recommend using a piece of an old ring for cleaning out the ring grooves. Instead, make a scraper out of a piece of hardwood fitted to the grooves (see Figure 7-39).

Piston rings are made of cast iron, generally with a facing of chrome where they contact the cylinder

FIGURE 7-39. *Cleaning piston-ring grooves with a piece of hardwood. (Courtesy Caterpillar)*

wall. Anytime this chrome is worn through, the ring should be replaced. Because the action of a ring rubbing on the cylinder wall polishes its face, it is often hard to tell whether or not the chrome is gone. You can get an indication of ring wear by looking at the ring's side profile. Rings are either flat-faced, tapered, rounded (barrel-faced), or double-faced (e.g., oil scraper rings, always the bottom ring on a piston). Tapered and barrel-faced rings contact a cylinder only at the top of the taper or barrel. As wear increases, the point of contact grows wider. If these rings are worn flat, with the whole ring face in contact with the cylinder wall, it is time to replace them. Double-faced rings can be compared with new ones to gauge the extent of wear.

Measuring Wear

To check piston ring wear more accurately, insert the ring into a cylinder and push it down to the bottom (no wear takes place at the bottom of the cylinder). Use an upside-down piston as a plunger; it will keep the ring square, which is important for accurate measurements. Now, measure the gap between the ends of the ring (end gap), using a feeler gauge (see Figure 7-40). As a ring wears and pushes out on a cylinder wall, this gap increases. Compare the size of the gap to the manufacturer's specifications to see if a new ring is called for. As a general rule of thumb, the gap should be between 0.003 and 0.006 inch per inch of cylinder diameter.

Note that it is also possible to get some idea of the extent of cylinder wear by comparing the end gap just measured with the gap on the same ring when measured just below the lip at the top of the cylinder. The difference between these two measurements is the cylinder's increase in circumference due to wear. To convert the circumference increase to a diameter increase, divide this number by 3.142.

Fit piston rings to the pistons by reversing the removal procedure. Most rings have a top and a bottom—the upper face should be marked T; sometimes the bottom face is marked BTM. In any event, when you take them off, note which side is the top and install them the same way up and in the same grooves.

Measure the wear of the ring groove by sliding the appropriate feeler gauge into the groove between the piston and the ring (see Figure 7-41). This clearance should be about 0.003 to 0.004 inch.

Figure 7-40. *Checking piston-ring gap. Note: The ring is shown at the top of the cylinder for clarity. You should measure the ring gap with the ring at the bottom of the cylinder unless the cylinder is brand-new. (Courtesy Perkins Engines Ltd.)*

Compare your measurements with the maker's specifications. Excessive clearance means that you will need a new piston.

Pistons are sometimes supplied in sets, complete with connecting rods. This is to keep the engine in balance and cut down on vibration. Each piston and rod assembly is machined to the same weight as all the others in the set, so if one piston needs replacing, you may have to change all of them.

Because of the extremely close tolerances between the piston crown and the cylinder head, new pistons for some older engines are made oversize and later machined in a lathe for an exact fit in the cylinder. This is known as topping. If you have this type of piston, you will have to have this work done by a specialist.

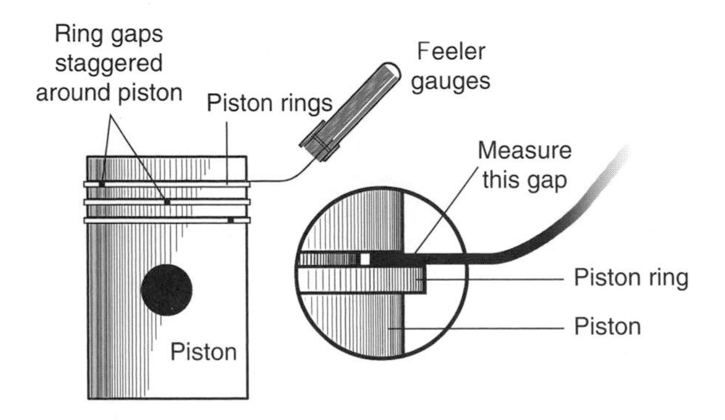

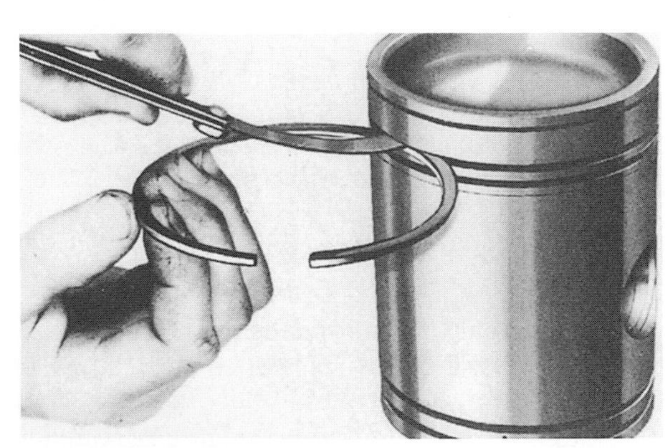

FIGURE 7-41. *Measuring the vertical clearance between the ring and the groove. (Courtesy Caterpillar, top; Fritz Seegers, middle; Detroit Diesel, bottom)*

FIGURE 7-42. *Lowering a piston into the cylinder. (Courtesy Perkins Engines Ltd.)*

REPLACING PISTONS AND CONNECTING RODS

When you place a piston into the cylinder from which it came, the crank for that cylinder should be at bottom dead center (BDC). Coat the cylinder, piston, and rings with oil, then gently lower the piston into the bore until the bottom ring (the oil scraper ring) rests on the top of the cylinder (see Figure 7-42). At this point, arrange all the rings so that their end gaps are staggered around the piston (see Figure 7-43). This prevents blowby through lined-up gaps.

Ring grooves on Detroit Diesel two-cycles have a pin in them at one point. This is placed in the middle of the end gap and prevents the rings from turning on the piston, which might result in the ends of the rings lining up with the air-intake ports at the bottom of the cylinder. If this were to happen, the rings would try to spring out into the ports and would break.

INSTALLING A PISTON

You can rent a piston-ring clamp from an automotive parts store, if you know the piston's diameter. This tool holds the rings tightly in their grooves while you slide the piston into its cylinder (see

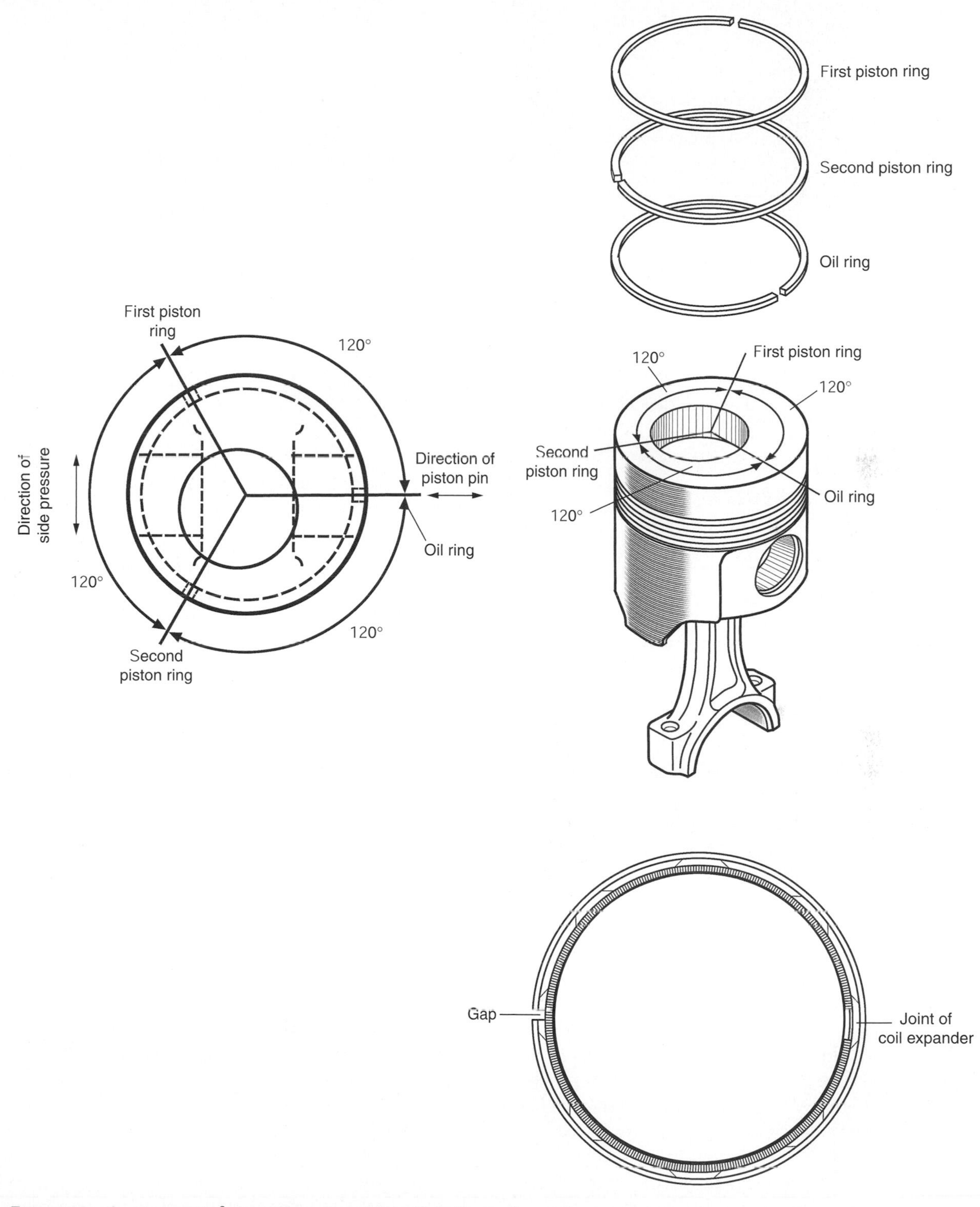

FIGURE 7-43. *Arrangement of piston rings on a piston. (Fritz Seegers)*

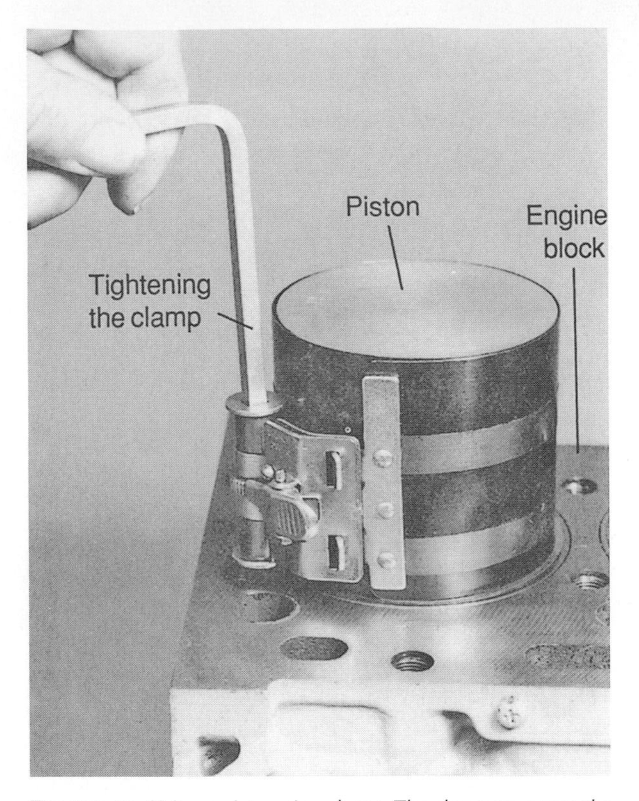

FIGURE 7-44. *Using a piston-ring clamp. The clamp squeezes the piston rings into the grooves. (Courtesy Perkins Engines Ltd.)*

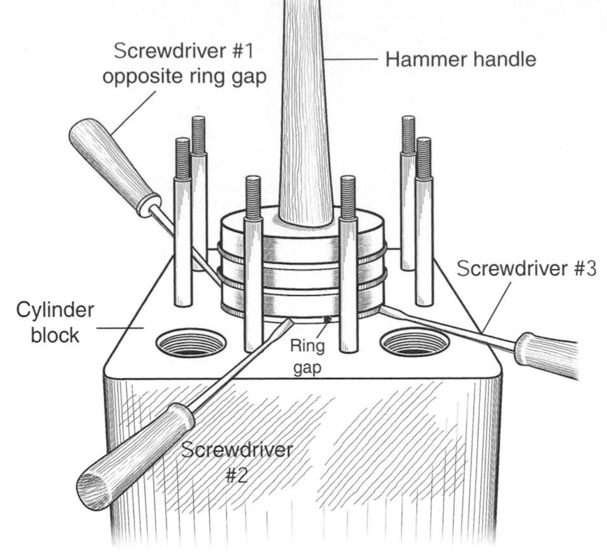

FIGURE 7-45. *Replacing a piston without a ring clamp. (Fritz Seegers)*

Figure 7-44). You can dispense with the clamp, however, if you have to. You will need a helper and three screwdrivers, or similar blunt instruments (see Figure 7-45).

Using one screwdriver, push the ring on which the piston is sitting into its groove at its center point (the point opposite the ring gap). Your helper then works around the ring in one direction, easing the ring into its groove until almost at one end. He or she holds the ring at this point with the second screwdriver. Then he or she works around the ring in the other direction, easing it into its groove until almost at the other end, holding it at this point with the third screwdriver. The ring should now be all the way into its groove, and you still have one hand free to tap the piston gently down into the cylinder, using the handle of a hammer or similar piece of wood (see Figure 7-45).

The piston will slide down until the next ring sits on top of the cylinder; repeat the procedure. Never use force—it will result in broken rings. Once all the rings are in the cylinder, push the piston down from above and guide the rod onto the crankshaft from below. Make sure the connecting rod is lined up squarely with the crankshaft journal. The crankshaft journal (bearing surface) must be spotlessly clean (use lint-free rags) and well oiled. Replace the connecting rod cap.

It is a good practice to fit new cap bolts, whether they are called for or not. These bolts are subjected to very high loads, and if one fails, a tremendous amount of damage will result.

TORQUING PROCEDURES

When the cap nuts and bolts are replaced, they must be tightened to a very specific torque. This will be given in the manufacturer's specifications (if not available, see Tables 7-1 and 7-2 for general guidelines). You will need a torque wrench. These wrenches indicate exactly how much pressure is being applied to the nut or bolt.

Torque wrenches come in two basic types. You adjust the torque setting of the more expensive wrenches by screwing the end of the handle in and out, lining up a pointer with a scale on the body of the wrench. The scale indicates the torque pressure at which the wrench is now set. As you tighten the bolt or nut, the wrench will click when you reach this torque setting. You can hear and feel the click.

TABLE 7-1. Cap Bolt and Nut Specifications

Grade Identification Marking on Bolt Head	SAE Grade Designation	Nominal Size Diameter (in.)	Minimum Tensile Strength (psi)
None	2	No. 6 through $\frac{3}{4}$	74,000
		Over $\frac{3}{4}$ to $1\frac{1}{2}$	60,000
	5	No. 6 through 1	120,000
		Over 1 to $1\frac{1}{2}$	105,000
	7	$\frac{1}{4}$ through $1\frac{1}{2}$	133,000
	8	$\frac{1}{4}$ through $1\frac{1}{2}$	150,000

Cheaper wrenches have a flexible handle with a pointer attached to it, the end of which moves over a scale set across the wrench. As you apply pressure to a nut or bolt, the handle of the wrench flexes and the pointer moves across the scale. It is hard to use these wrenches with any degree of precision.

When you torque nuts or bolts, first be sure the threads are clean, free-running, and generally oiled (sometimes a manual will specify a dry torque setting, without oil). Friction in the threads and sudden jerks on the wrench will give a false torque reading. You must pull the nuts or bolts down using an even, steady pressure. On the most important nuts and bolts, overtighten them by just a few pounds then back off and retorque to the specified point. This ensures that everything is correctly pulled down. There should be some means of locking the nuts, such as a locking washer that bends over onto one of the flats on the nut, or a nylon insert on the nut.

At all critical bolt-tightening points in engine work, bolts must be tightened evenly: partially tighten the first one, then do the same to the opposite bolt, then apply a few more pounds of pressure to each one. Continue in this fashion until the final torque setting is reached. Do this in a minimum of three stages.

When you replace bearing caps, turn the shaft they enclose one full revolution by hand after each increase in the tightening pressure until the full torque is reached. This ensures that there are no tight spots or binding. Tight spots must be removed; this procedure is especially important when fitting new bearing shells.

TABLE 7-2. Standard Cap Bolt and Nut Torque Specifications

Thread Size	Grade 2		Grade 5	
	ft. lb.	Nm	ft. lb.	Nm
$\frac{1}{4}$-20	5–7	7–9	7–9	10–12
$\frac{1}{4}$-28	6–8	8–11	8–10	11–14
$\frac{5}{16}$-18	10–13	14–18	13–17	18–23
$\frac{5}{16}$-24	11–14	15–19	15–19	20–26
$\frac{3}{8}$-16	23–26	31–35	30–35	41–47
$\frac{3}{8}$-24	26–29	35–40	35–39	47–53
$\frac{7}{16}$-14	35–38	47–51	46–50	62–68
$\frac{7}{16}$-20	43–46	58–62	57–61	77–83
$\frac{1}{2}$-13	53–56	72–76	71–75	96–102
$\frac{1}{2}$-20	62–70	84–95	83–93	113–126
$\frac{9}{16}$-12	68–75	92–102	90–100	122–136
$\frac{9}{16}$-18	80–88	109–119	107–117	146–159
$\frac{5}{8}$-11	103–110	140–149	137–147	186–200
$\frac{5}{8}$-18	126–134	171–181	168–178	228–242
$\frac{3}{4}$-10	180–188	244–254	240–250	325–339
$\frac{3}{4}$-16	218–225	295–305	290–300	393–407
$\frac{7}{8}$-9	308–315	417–427	410–420	556–569
$\frac{7}{8}$-14	356–364	483–494	475–485	644–657
1-8	435–443	590–600	580–590	786–800
1-14	514–521	697–705	685–695	928–942

Adapted from tables supplied by Detroit Diesel Corp.

REPLACING CYLINDER HEADS

The cylinder head and block must be spotlessly clean before you replace the head (see Figure 7-46). First be sure to remove any pieces of rag and so on used to block off the oil, water, and other passages in the block and head. Set a new gasket on the block and lower the head onto it. Some gaskets have a top and will be appropriately labeled.

Although most manufacturers do not recommend it, metal gaskets often benefit from a little

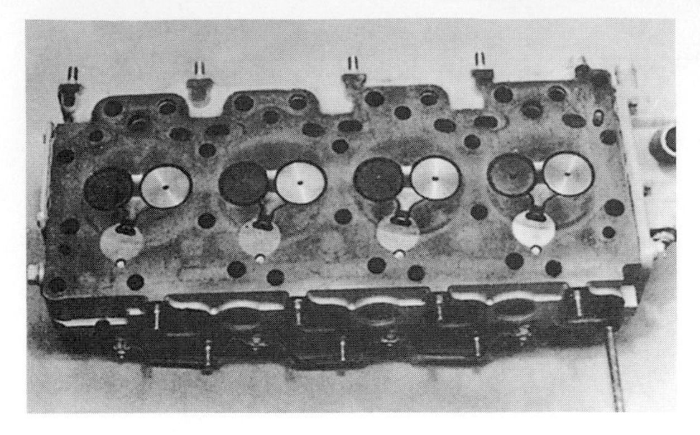

FIGURE 7-46. *A well-cleaned cylinder-head face. (Courtesy Perkins Engines Ltd.)*

jointing paste (compound) smeared on them, taking care not to get any down the passages. Jointing paste is available at automotive parts stores. Be sure it is made for high temperatures, is resistant to water and oil, and will withstand high pressure. Most fiber gaskets (increasingly the norm) are fitted without any paste. Always fit a new gasket whenever possible, even if the old one looks fine. It is tremendously aggravating to reassemble an engine with an old gasket only to find that the gasket leaks.

The head nuts (bolts) must be tightened evenly, a bit at a time as outlined above, until the manufacturer's specified torque setting is reached (see your owner's manual and Figure 7-47). Proper torquing procedures are more important here than on gasoline engines, owing to the much higher cylinder pressures generated by diesel engines.

If you don't follow the correct bolt-tightening sequence, uneven pressure may develop and lead to a blown head gasket or a warped cylinder head. If the manufacturer's recommended torque sequence is not available, you can safely assume that the center nuts are pulled down first. From then on, work out to the ends of the cylinder head, tightening a nut on one side of the center to a particular torque setting,

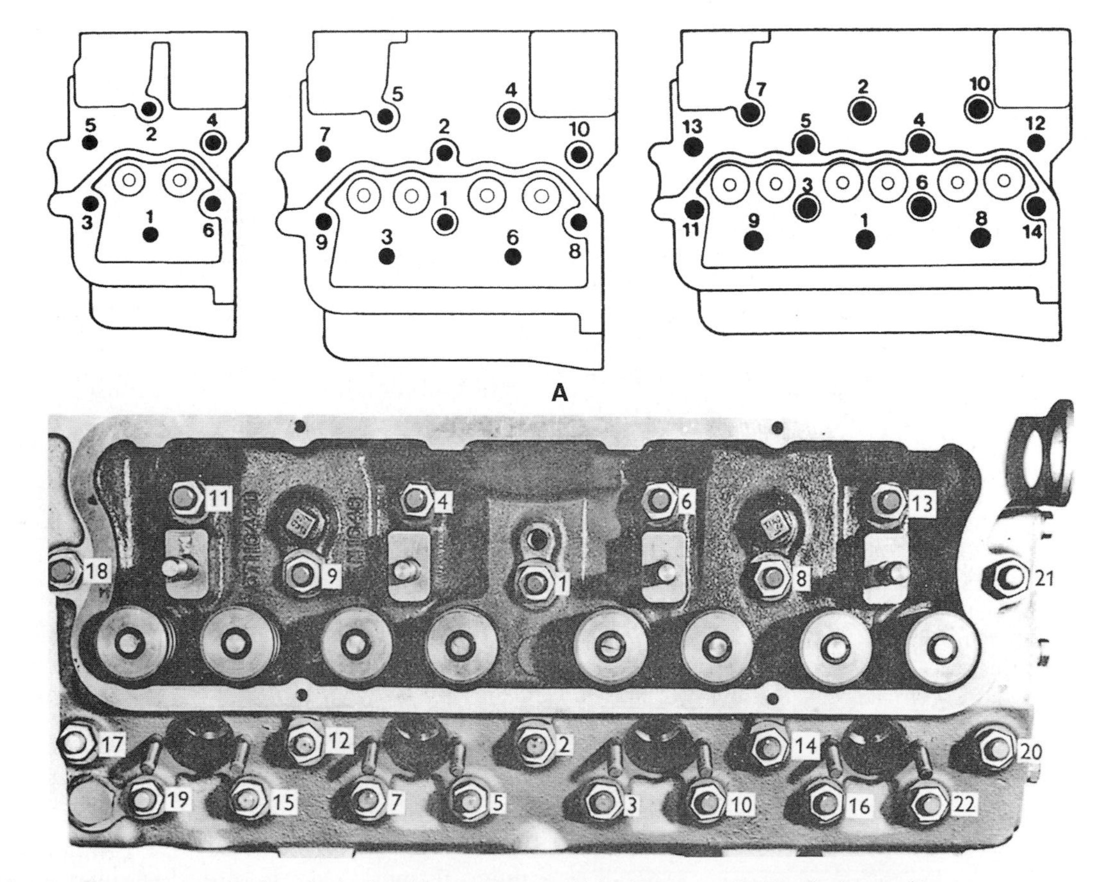

FIGURE 7-47. *Torquing sequence for cylinder-head bolts. (Courtesy Volvo Penta and Perkins Engines Ltd.)*

then one on the other side, and so on until all are done. Next, increase the torque setting and do this again. Once again, torquing should be done in a minimum of three stages.

The correct torque setting is especially critical on engines with dissimilar metals (e.g., a cast-iron block and an aluminum cylinder head), because the metals have differing rates of expansion and contraction. The correct torque setting is more important on engines with a greater number of cylinders (e.g., six as opposed to four). Always recheck the torque settings after an engine has been reassembled and run for a while. A head gasket, especially a metal one, will occasionally settle, loosening the head nuts and creating the potential for a blown gasket.

Replacing Push Rods and Rockers

Before you replace the push rods, roll them on a flat surface to make sure they are straight. A chart table or galley countertop is level enough; the bed of a table saw or drill press is even better. Any bend will be immediately apparent.

Place the push rods in their respective holes in the cylinder head and block (round ends down, cupped ends up) and be sure each is properly seated in a hollow (called the cam follower; see Figure 7-48).

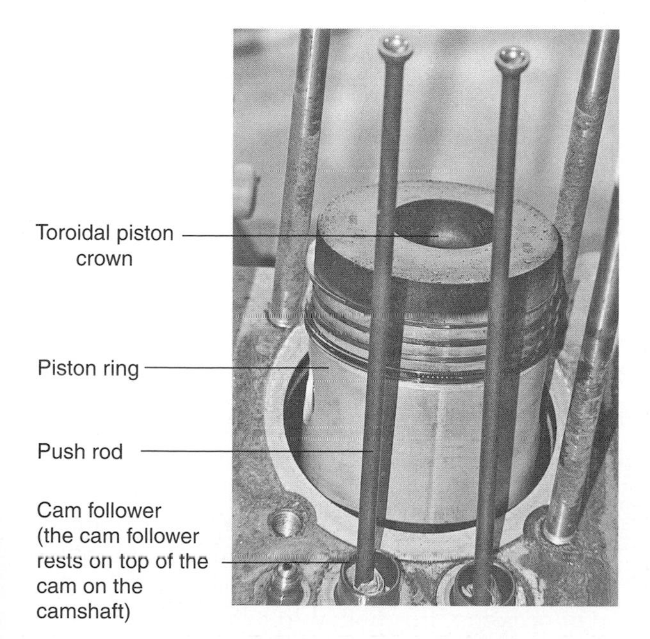

Figure 7-48. *Push rods and cam followers. (The cylinder has been removed and the push rods have been wedged in place with paper to illustrate their locations.)*

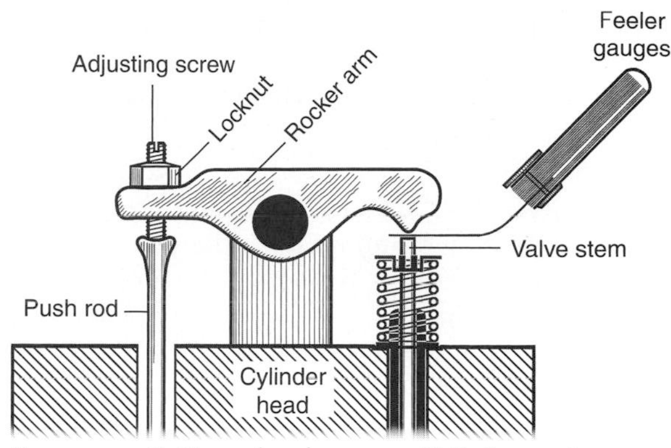

Figure 7-49. *Setting valve clearances. (Fritz Seegers)*

You can't see this seat, but if you've missed it, the push rod will be cockeyed and will probably be resting on the rim of the cam follower. In some engines, the push rods share a common space, which makes it possible to miss the cam follower altogether or even to hit the wrong one. In most engines this cannot be done. If the push rod is centered in its hole in the cylinder head and feels firmly cupped at its lower end, it is seated correctly. Note that some push rods will be sticking up more than others.

The rockers go on next, but before fitting them, loosen the locknut at the end of each rocker arm and undo the screw it locks a couple of turns (see Figure 7-49). This is a safety precaution in case the valve timing has been upset or valve clearances radically changed. It prevents any risk of forcing a valve down onto the piston crown and bending the valve stem when the rocker bolts are tightened up. Torque down the rocker assembly to the manufacturer's setting.

Retiming an Engine

Decarbonizing will not disturb the timing of an engine with push rods (in which case this section can be skipped), but anytime an overhead camshaft is removed, valve timing is upset and will need to be reset. Since the fuel injection pump is tied in with the valve timing on all engines, the following procedure for retiming an overhead camshaft coincidentally describes how to retime a fuel injection pump on any engine.

General Principles

Engine timing involves the timing drive gear, which is keyed to the end of the crankshaft; the camshaft drive gear, which operates the valve timing; and the fuel injection pump drive gear. On two-cycle engines, these gears are all the same size because the camshaft and fuel injection pump rotate at the same speed as the engine. On four-cycle engines, the timing gear is half the size of the other two because the crankshaft must rotate twice for every complete engine cycle.

Engine timing is set by getting these three gears in exactly the correct relationship to one another. Each of the gears involved in engine timing has a punch mark or line somewhere on its face. When engine timing is belt driven or chain driven, these marks line up with corresponding marks on the timing-gear housing. When engine timing is transmitted through intermediate gears, the intermediate gears have punch marks that line up with the marks on the timing gears (see Figure 7-50). This alignment should be exact; if it is not, something is wrong.

Specific Procedures

Timing is always done at top dead center (TDC) on the compression stroke of the #1 cylinder (the one at the front end, or timing-gear end, of the engine). When the engine is at TDC on the #1 cylinder, the keyway in the end of the crankshaft, which positions the timing drive gear, will also be at TDC.

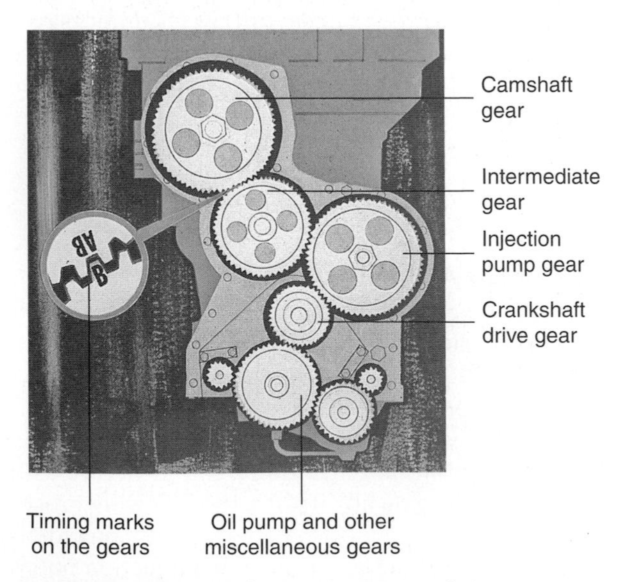

Figure 7-50. *Timing marks. (Courtesy Caterpillar)*

Two-cycle engines have only one TDC because the engine fires on every revolution of the crankshaft. Four-cycle engines, however, have two TDCs: one on the compression stroke, and one on the exhaust stroke. You time the engine at TDC on the compression stroke. Normally you can determine which stroke it is by looking at the position of the valves, but with the camshaft off this is not possible.

If the fuel injection pump timing has not been disturbed, the mark on its drive gear will line up with a corresponding mark on the timing gear housing, or on an intermediate gear, when the engine is at TDC on the compression stroke of the #1 cylinder. This will tell you that you have the correct TDC.

If the fuel pump timing has been disturbed, then it does not matter at which TDC on the #1 cylinder you set the timing, because the camshaft and fuel pump will be timed together. You must, however, time the fuel injection pump and the camshaft at the same TDC—which one is immaterial. Otherwise it would be possible to have the injection pump injecting the cylinders when the pistons were at the top of their exhaust stroke, and the engine would never run. This is known as the timing being out 180 degrees; although the engine is a full revolution out (360 degrees), the camshaft and injection pump turn at half engine speed on four-cycle engines, so one of them would be out 180 degrees.

With the engine at TDC on the #1 cylinder and the crankshaft keyway also at TDC, line up all the gear marks with their corresponding marks on the gear housing or intermediate gears, then install the belt, chain, or intermediate gears (see Figure 7-51). Always double-check that all the marks still line up after you tension the belt or chain; sometimes one of the gears will move around by one tooth, in which case timing will have to be repeated. If the timing is belt driven, I recommend fitting a new belt. If chain driven, compare the chain's length, stretched, to the manufacturer's specified length to see if it needs replacing.

This completes the basic timing. All that remains to do is to fine-tune the fuel injection pump. In almost all instances, the injection pump will be bolted to the other side of the timing gear housing. (On occasion, it is bolted to a little platform of its own.) The flange on the pump that bolts up to the gear housing has machined slots for its bolts, which means that even after the gear timing has been set,

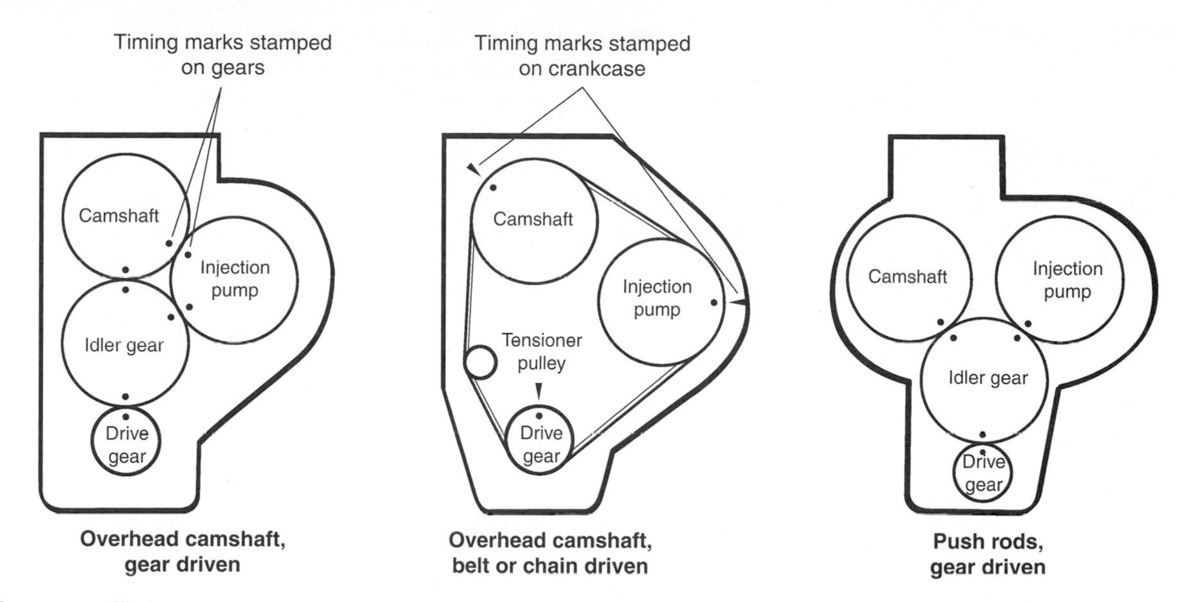

FIGURE 7-51. *Timing arrangements. (Fritz Seegers)*

the pump can still be rotated to the extent allowed by the slots. This rotation does not move the timing gear but turns the pump around its driveshaft.

The line scribed on the pump flange and the one on the timing gear housing must be exactly lined up before the pump flange is tightened (see Figure 7-52). This completes the injection pump timing.

Detroit Diesel two-cycle injection timing requires a special timing gauge and specific instructions (see Figure 7-53). It is straightforward enough, but you will need the relevant shop manual and gauge.

VALVE CLEARANCES FOR FOUR-CYCLE ENGINES

All valves have a small clearance between the valve stem and rocker arm when they're fully closed. It is important to maintain the manufacturer's specified clearance. Too little clearance causes a valve to stay slightly open at all times as the engine heats up and the metal parts expand. This results in lost compression and a burned seat and valve. If the clearance is too great, the valve will open slightly late, won't open far enough, and will close a little too soon.

General Principles

The inlet valve of a four-cycle engine opens on the downward stroke of its piston. Both valves are closed during the next upward (compression) stroke and for

FIGURE 7-52. *A vertically mounted fuel injection pump (CAV type DPA). Timing marks scribed on the pump's mounting flange and engine's timing cover (1). Idle-speed adjusting screw (2). Maximum-speed screw (3). (Do not tamper with the maximum-speed screw. Although you cannot see it in this photo, the screw has a seal on it, and breaking it automatically voids the engine's warranty.) (Courtesy Perkins Engines Ltd.)*

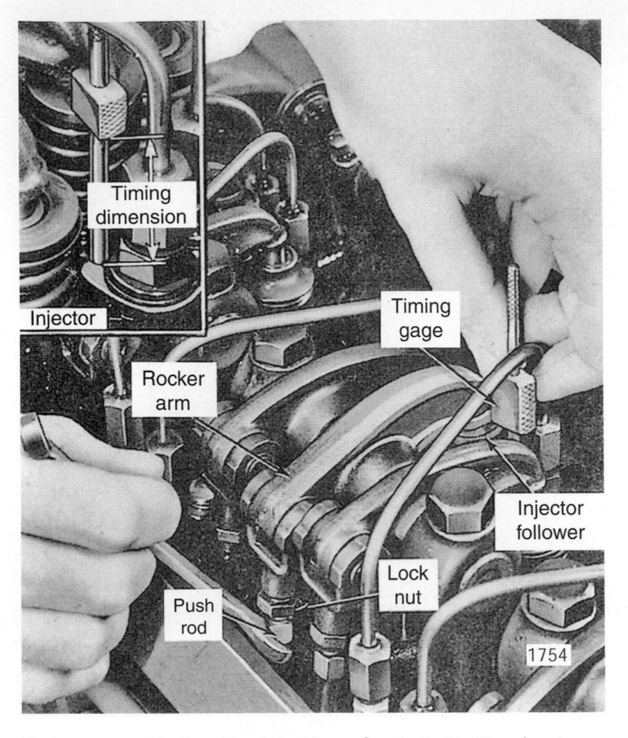

FIGURE 7-53. *Timing the injection of a Detroit Diesel using a special gauge. (Courtesy Detroit Diesel)*

most of the following (power) stroke. The exhaust valve then opens and remains open on the next upward (exhaust) stroke. At the top of this stroke, the exhaust valve is closing at the same time that the inlet valve is opening. This is known as valve overlap; the valves are said to be rocking. By watching the movement of the rocker arms while you slowly rotate the engine, you can determine where each piston is in the cycle and therefore the position of the cam that operates each push rod.

Set the valve clearance when the valve is fully closed at TDC on the compression stroke. On engines with overhead camshafts, you can see the cams; set the valve clearances when a cam is 180 degrees away from the rocker it operates. On engines with push rods, where you cannot see the operation of the camshaft, use the following method to establish the correct point for setting valve clearances.

Specific Procedures

To find TDC for any cylinder, slowly rotate the crankshaft in its normal direction of rotation. Watch the inlet valve's push rod as it moves up and down. When it is almost all the way down, the piston is at the bottom of its inlet stroke. Mark the crankshaft pulley, and turn the engine another half a revolution. Now the piston will be close to TDC on its compression stroke, and you can set the valve clearances on this cylinder. On some engines the crankshaft pulley is marked for TDC on the #1 cylinder, but on older engines, you should not rely on any mark—someone may have changed things around at some time. Many other engines have timing marks on the flywheel (see Figure 7-54).

The manufacturer's specifications will indicate valve clearances in thousandths of an inch or millimeters, and whether the valves should be adjusted hot or cold. If they are to be set when the engine is hot, make an initial adjustment when the engine is cold then check again after running the engine. Place the appropriate feeler gauge between the top of the valve stem and the rocker arm (see Figure 7-55). Tighten the adjusting screw until the arm just begins to pinch the feeler gauge. Now, tighten the locknut on the adjusting screw and check the clearance again in case something slipped. This valve is set; adjust the other one on the same cylinder. (Note: Although most valves are adjusted this way, some engines have threaded push rods, which are screwed in and out to adjust valve clearances, whereas other engines have an adjusting nut in the center of the rocker assembly.)

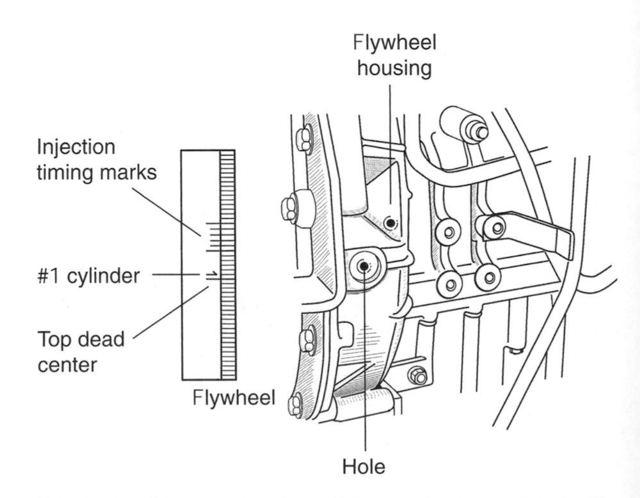

FIGURE 7-54. *Many engines have timing marks stamped on the flywheel. You can see them through a hole in the flywheel housing. (Fritz Seegers)*

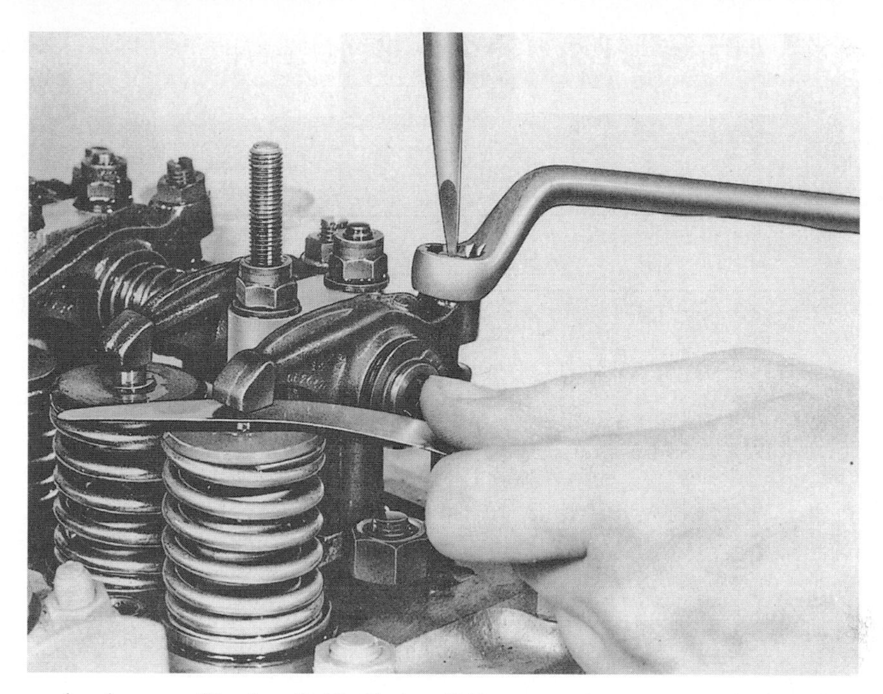

FIGURE 7-55. *Setting a valve clearance. (Courtesy Perkins Engines Ltd.)*

On a two-cylinder engine, when one piston is at TDC, the other is at BDC. After setting the valves on the first cylinder, do a half turn to bring the piston on the other cylinder to TDC. Take a quick glance at the valves to see if this is TDC on the exhaust stroke (the valves will be rocking) or the compression stroke. If it is the former, rotate the engine another full turn. You can now set the valve clearances on the second cylinder.

On three-cylinder and six-cylinder in-line engines, one-third of a revolution will always bring another piston to TDC. The pistons of six-cylinder in-line engines move in pairs—normally #1 and #6 together, #2 and #5, and #3 and #4 (see Figure 7-56). When one piston of a pair is at TDC on its exhaust stroke (valves rocking), the other piston in the pair will be at TDC on its compression stroke, and you can set its valves.

The pistons of four-cylinder, in-line engines also move in pairs—#1 and #4 together and #2 and #3—with a half turn separating TDC between the pairs. When the valves on either one of a pair of pistons are rocking, the other piston in the pair is at TDC on its compression stroke and its valves can be set.

If you are at all unsure of how this works, write down all the cylinders on a piece of paper, showing which ones operate together. Determine the inlet and exhaust valves from their respective manifolds and turn the engine over slowly a few times to familiarize yourself with the valve opening and closing sequence. After some careful attention to the logic of the situation, you will soon see where the pistons

# OF CYLINDERS	VALVES "ROCKING" ON:	SET VALVE CLEARANCE ON:
4	4 2 1 3	1 3 4 2
6	6 2 4 1 5 3	1 5 3 6 2 4

FIGURE 7-56. *Sequence for setting valve clearances on four-cycle (not two-cycle) engines.*

are and when to set valve clearances. Be sure to turn the engine over in its normal direction of rotation.

Valve Clearances for Two-Cycle Engines

Detroit Diesel two-cycle engines have no inlet valves; instead they have two or more exhaust valves per cylinder. Remember that these valves open when the piston is near the bottom of the power stroke and close when the piston is partway up the compression stroke. From the point of closure, another one-third of a turn brings the piston more or less to TDC. At this time, the rocker arm operating the injector for this cylinder will be fully depressed (i.e., on its injection stroke), and you can set the valve clearances.

Accessory Equipment

The final step in decarbonizing is to refit all the fuel lines, manifolds, valve cover, turbocharger, and anything else that was removed. If no other part of the fuel system has been broken loose, a few turns of the engine should push diesel up to the injectors. Otherwise the fuel system will need bleeding as explained on pages 96–101.

CHAPTER 8

MARINE TRANSMISSIONS

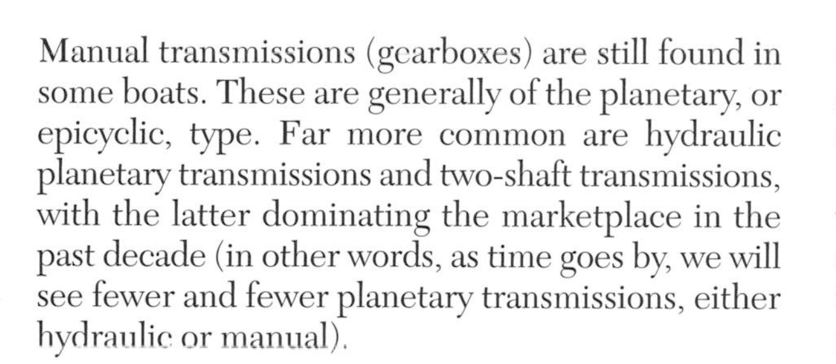

Manual transmissions (gearboxes) are still found in some boats. These are generally of the planetary, or epicyclic, type. Far more common are hydraulic planetary transmissions and two-shaft transmissions, with the latter dominating the marketplace in the past decade (in other words, as time goes by, we will see fewer and fewer planetary transmissions, either hydraulic or manual).

Planetary Transmissions

With a planetary transmission (manual or hydraulic), the engine turns a geared driveshaft (the drive gear) that rotates constantly in the same direction as the engine. Deployed around and meshed with a gear on this shaft are two or three gears (the first intermediate gears) on a carrier assembly. These mesh with more gears (the second intermediate gears), also mounted on the carrier assembly. The second intermediate gears engage a large geared outer hub. The carrier assembly, with its collection of first and second intermediate gears, is keyed to the output shaft of the transmission (see Figure 8-1).

The forward clutch is on one end of the driveshaft. Engaging forward locks the entire driveshaft and carrier assembly together—the whole unit rotates as one, imparting engine rotation to the output shaft of the transmission via the carrier assembly (see Figure 8-2). This is very efficient because no gears are involved in imparting drive to the output shaft.

Reverse is a little more complicated. The forward clutch is released and a second clutch is engaged. This locks the outer geared hub in a stationary position. Meanwhile the drive gear is still rotating the intermediate gears. Unable to spin the outer hub, the intermediate gears rotate around the inside of the hub in the opposite direction to the drive gear. The carrier assembly imparts this reverse motion to the output shaft (see Figure 8-3). There are some energy losses in the gearing.

Manual and hydraulic versions of a planetary box are very similar. The principal difference is that the reverse clutch of a manual box consists of a brake band that is clamped around the hub (see Figure 8-4), whereas in a hydraulic box, a second clutch, similar to the forward clutch, is used (see Figure 8-5).

A manual box uses pressure from the gearshift lever to engage and disengage the clutches. A hydraulic box incorporates an oil pump; the gearshift lever merely directs oil flow to one or the other clutch, with the oil pressure doing the actual work. While quite a bit of pressure is needed to operate a manual transmission, shifting gear with a hydraulic transmission is a fingertip affair.

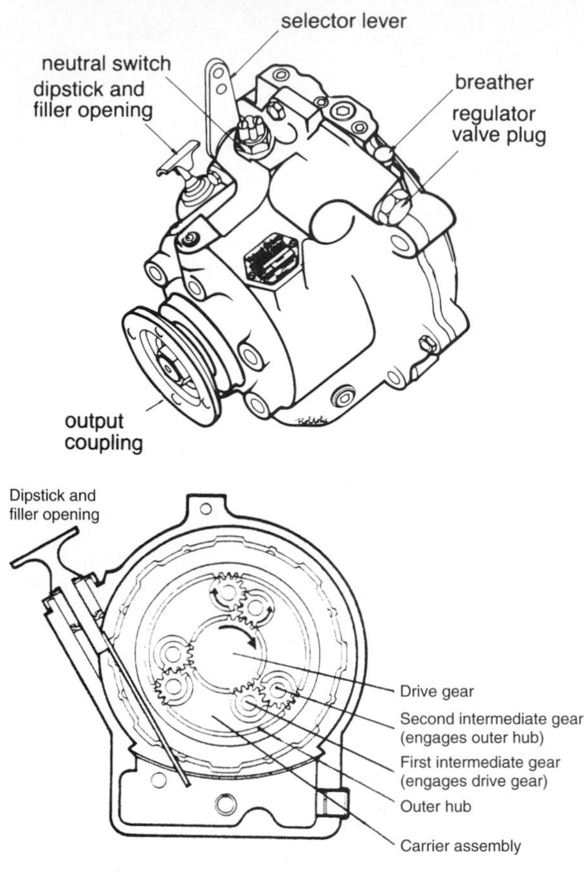

FIGURE 8-1. *A typical planetary transmission (top). The internal view (bottom) shows the gears as having teeth only partway around, to illustrate how they mesh, but the gears actually have teeth all the way around. (Courtesy BorgWarner)*

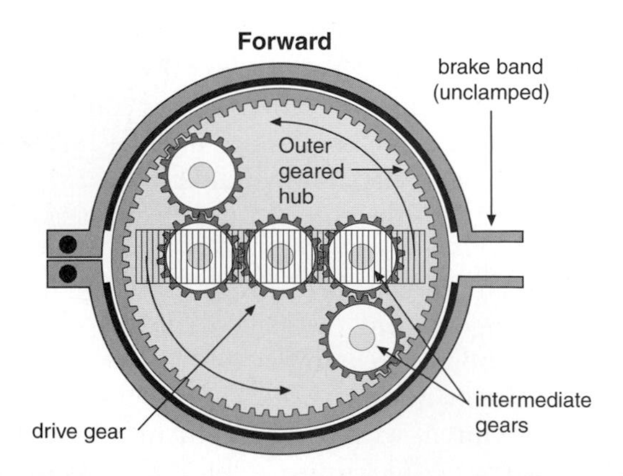

FIGURE 8-2. *In forward gear in a planetary transmission, the brake band is unclamped while the forward clutch locks the drive gear, carrier assembly, and geared hub together so that they rotate as one (schematically represented by the band across the center of the illustration). Engine rotation is imparted to the output shaft. (Michael D. Ryus)*

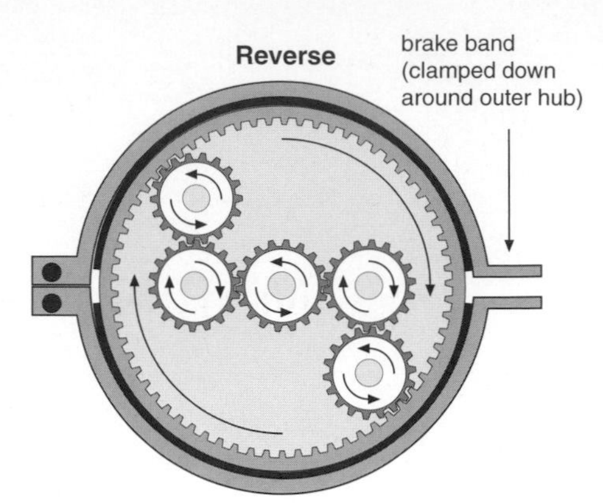

FIGURE 8-3. *When in reverse in a planetary transmission, the forward clutch is released, leaving the carrier assembly free to rotate around the drive gear, while the brake band is clamped down, locking the geared hub. The carrier assembly is driven around the hub in the opposite direction to the drive gear, imparting reverse rotation to the output shaft. (Michael D. Ryus)*

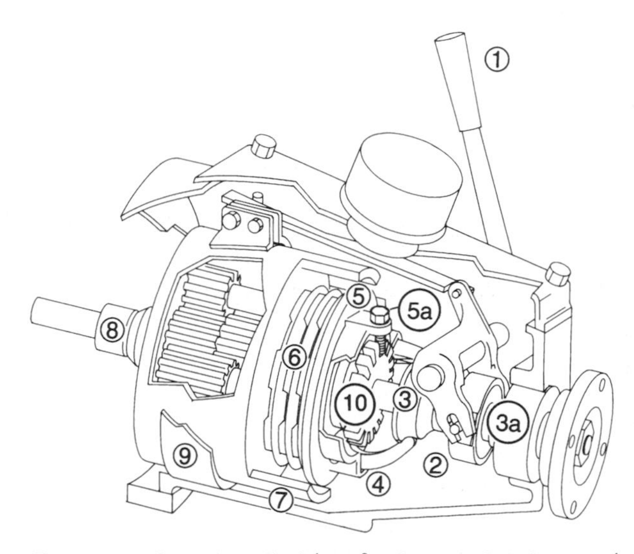

FIGURE 8-4. *Operating principles of a Paragon S-A-O manual planetary transmission. For forward, the gear lever (1) is pushed toward the engine; the shift yoke (2) moves back the opposite way (toward the coupling), sliding the shift cone (3) along the output shaft (3a), thus moving out the cam levers (4). The cam levers bear on the pressure plate (5), compressing the friction discs (6) in the clutch. The friction discs press against the gear carrier (7), causing the motion of the input shaft (8) to be transmitted to the output shaft (3a). The pressure plate is adjusted by backing out the lock bolt (5a), screwing up the castellated nut (10), and retightening the lock bolt. For reverse, the gear lever (1) is moved the opposite way, compressing the reverse band (9), which locks around the gear carrier like a huge hose clamp. (Jim Sollers)*

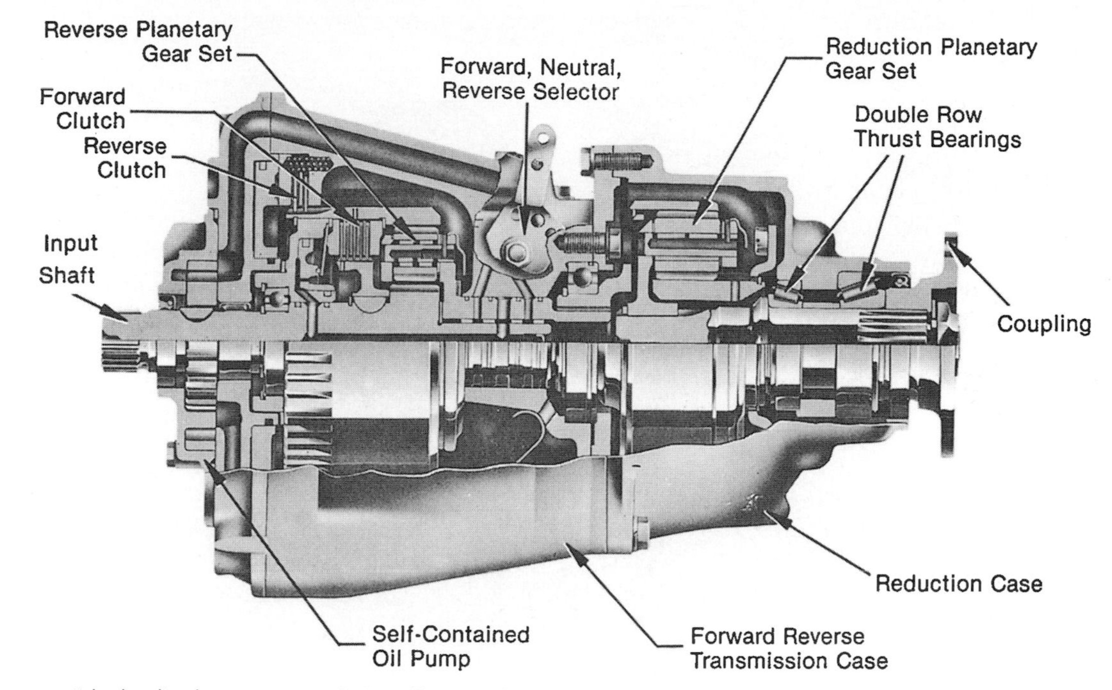

FIGURE 8-5. *A hydraulic planetary transmission. (Courtesy BorgWarner)*

TWO-SHAFT TRANSMISSIONS

The simplest form of a two-shaft transmission (see Figure 8-6) has the engine coupled to the transmission's input shaft, to which two gears are keyed, one at either end. A second shaft, the output shaft, has two more gears riding on it; one of these engages one of the input gears directly, and the other engages the second input gear via an intermediate gear (see Figure 8-7). These two output gears are mounted on bearings and freely rotate around the output shaft.

The drive gears impart continuous forward and reverse rotation to the output gears. Each output gear has its own clutch, and between the two clutches is an engaging mechanism. Moving the engaging mechanism one way locks one gear to the output shaft, giving forward rotation; moving the mechanism the other way locks the other gear to the shaft, giving reverse rotation (see Figures 8-7 and 8-8).

Larger two-shaft transmissions may have more shafts and gears in order to achieve gear reductions that lower the speed of the output shaft and perhaps change the angle of the output shaft. However, if you mentally strip away the extra gears, the underlying operating principle is as described above. On smaller two-shaft transmissions, the case is in two halves that come apart to reveal all the contents; on larger ones, the end housings are generally unbolted (see Figure 8-9).

In a manual two-shaft transmission, when the clutch-engaging mechanism is first moved to either forward or reverse, it gently presses on the relevant clutch. This initial friction spins a *disc carrier*, which holds some steel balls in tapered grooves. The rotation drives the balls up the grooves. Because of the taper in the grooves, the balls exert an increasing pressure on the clutch, completing the engagement (see Figure 8-10). Only minimal pressure is needed to set things in motion, and thereafter, a clever design supplies the requisite pressure to make the clutch work, but without the necessity for oil pumps or oil circuits. Shifting gear is once again a fingertip affair.

In a hydraulic two-shaft transmission, an oil pump provides the pressure for operating the clutches, just as in a hydraulic planetary transmission.

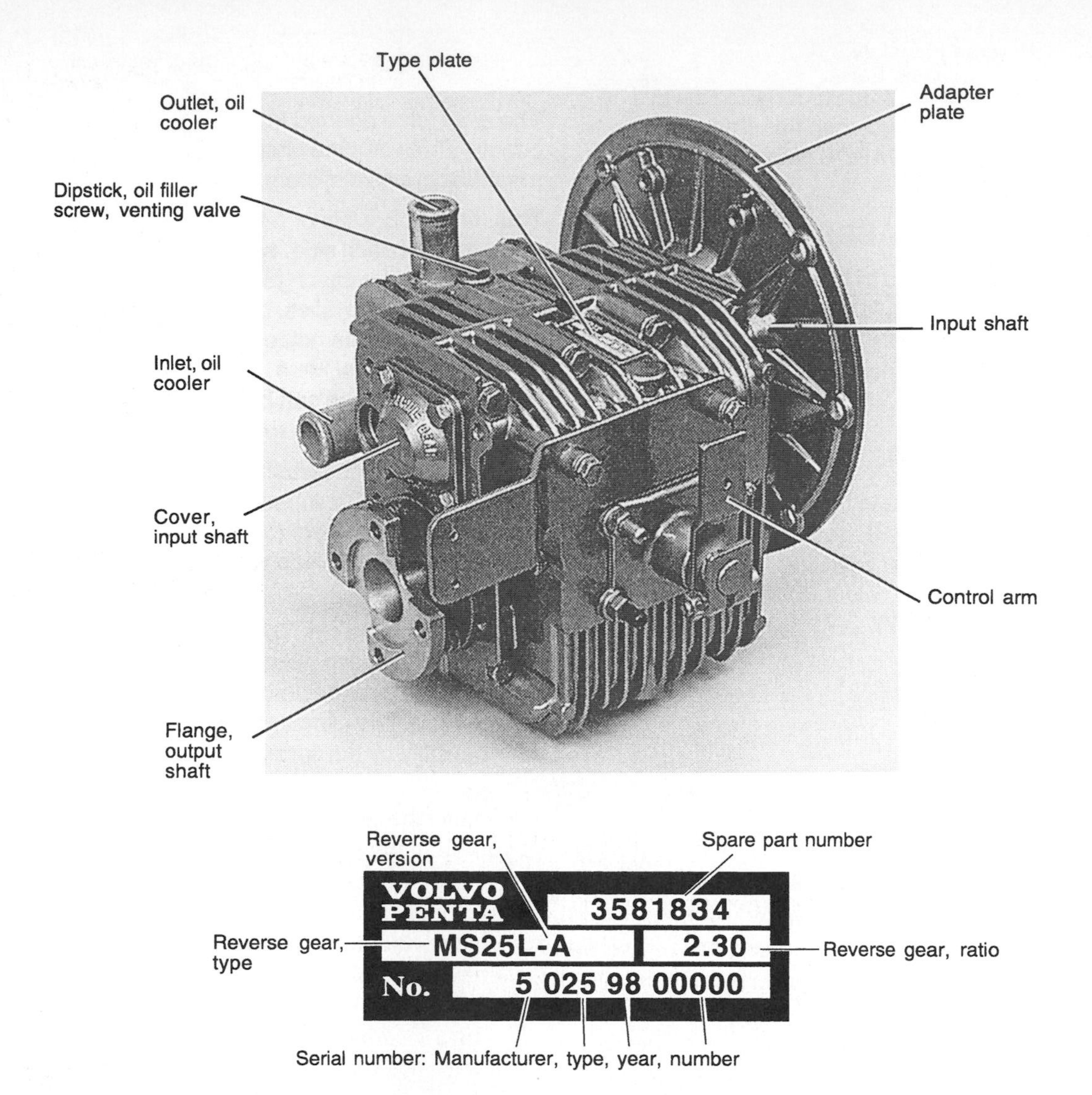

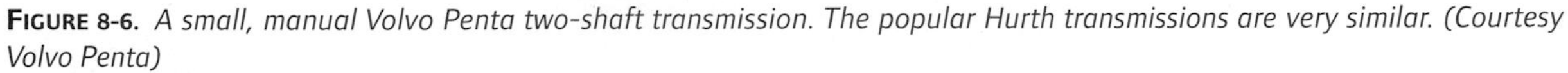

FIGURE 8-6. *A small, manual Volvo Penta two-shaft transmission. The popular Hurth transmissions are very similar. (Courtesy Volvo Penta)*

VARIATIONS ON A THEME

V-DRIVES

A V-drive is an arrangement of gears that allows the engine to be installed "backward," placing it directly over the output shaft (see Figure 8-11). This enables more compact engine installations to be made, and in particular, enables sportfishing boats and other planing-type hulls to have the engine installed in the stern of the boat, which is the best spot from the point of view of weight distribution. All transmission types can be used with V-drives.

INBOARD/OUTBOARDS

Inboard/outboards (see Figures 8-12 and 8-13) are exactly what their name implies: an inboard engine coupled to an outboard motor–type drive assembly and propeller arrangement with a universal joint

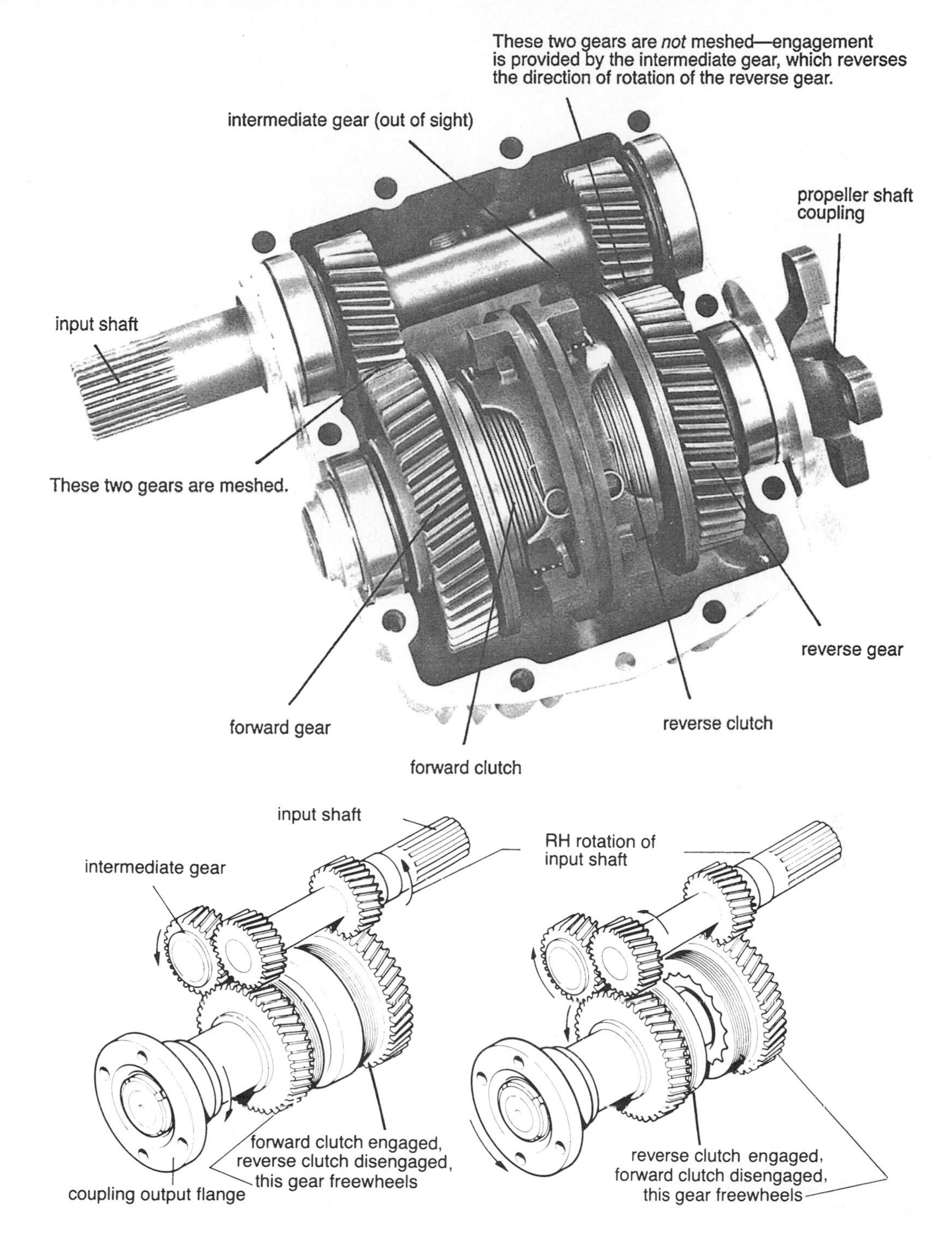

FIGURE 8-7. *A two-shaft Hurth transmission with the cover removed (top). With the forward clutch engaged (bottom left), the output shaft rotates in the opposite direction to the input shaft. With the reverse clutch engaged (bottom right), the output shaft rotates in the same direction as the input shaft via the intermediate gear. (Courtesy Hurth)*

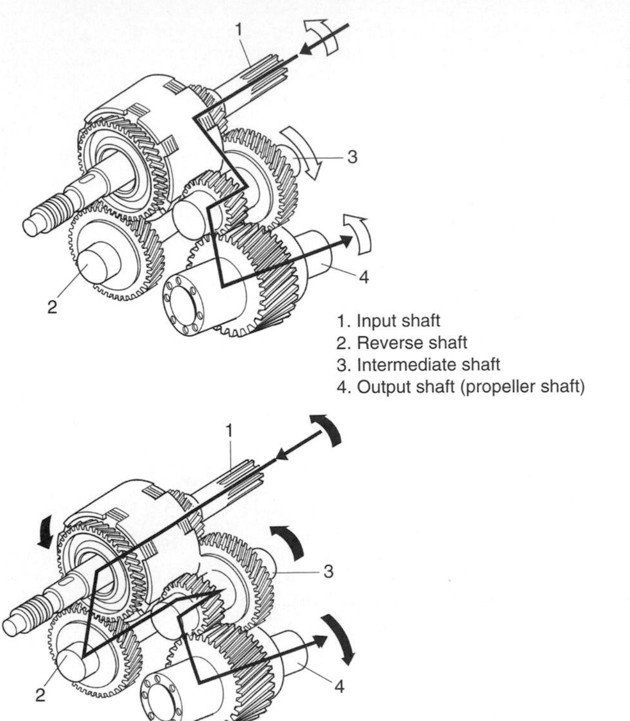

FIGURE 8-8. *Forward (top) and reverse (bottom) in a two-shaft transmission with additional reduction and shaft-angle-changing gearing. (Courtesy Volvo Penta)*

FIGURE 8-9. *Removing the end cover from a larger two-shaft transmission. It would have been a good idea to clean the exterior before working on it! Note the extensive rusting of the internal parts due to saltwater intrusion (the boat in which the transmission was installed sank).*

(U-joint) that allows the drive unit to be raised and lowered and turned from side to side. These units have definite advantages in planing craft, notably:

- Inboard/outboards allow an engine to be mounted in the stern of the boat, which is often the best place in terms of weight distribution on many planing hulls.
- The outboard unit can be pivoted up and down hydraulically. This enables these boats to take full advantage of their shallow draft to run up onto beaches. It also permits infinite propeller-drive-angle to hull attitude adjustments for changes in boat trim and makes trailering easy.
- The whole outboard unit turns for steering, greatly increasing maneuverability and removing the need for a separate rudder.
- There is no propeller shaft, stern tube, or stuffing box to leak into the boat.

Naturally, there are drawbacks:

- The extra gearing and sharp changes in drive angle absorb more power than a conventional transmission.
- The U-joints in the transmission tend to have a relatively high failure rate if driven hard.
- Above all, most outboard units are built of relatively corrosion-susceptible materials. If the boat is dry-docked between use, or the outboard unit is raised out of the water, this is not a great problem; but if the unit is left in the water, it can be (see the Aluminum Castings and Corrosion sidebar).

Observing all maintenance schedules is important, particularly oil changes, greasing of U-joints, and replacing those all-important zincs. The U-joint is protected by a rubber boot or *bellows* that needs to be checked for signs of cracking or other defects. If you find any problems, immediately replace the boot to avoid expensive damage to the U-joint.

Note: When changing the oil in a sterndrive unit, check the manual for the procedure. Most have to be tilted up (e.g., Volvo Pentas, to 35 degrees). If the old oil is discolored, there is a problem.

Aluminum Castings and Corrosion

There are two common methods for making the aluminum cases for inboard/outboard drive legs (and also saildrive legs): pressure die-casting, and using permanent molds. The nature of the two processes requires the use of different alloys, with the permanent-mold approach using more corrosion-resistant alloys. This is important to the boatowner!

You cannot tell by looking at any particular unit what alloy has been used. As far as I know, all MerCruiser units use pressure die-casting, as do Volvo Penta units on gasoline-powered inboard/outboards, whereas the Volvo Pentas coupled to diesel engines use the more corrosion-resistant permanent molding.

Following manufacture, all units are given special paint jobs that are designed to seal the surface of the aluminum and prevent corrosion. They are then protected with sacrificial zinc anodes (and sometimes with impressed-current cathodic protection, which is beyond the scope of this book). It is absolutely critical to maintain the zincs, replacing them when they are no more than 50% depleted. For boats operating in fresh water, Volvo Penta recommends exchanging the zinc anodes for magnesium anodes (available from Volvo Penta).

If the paint barrier on an outboard drive (or saildrive) gets damaged, you will need to repair it. This requires that you clean down to bare metal, using aluminum oxide sandpaper to avoid leaving contaminants that would promote corrosion.

The next step is to acid-etch and seal the aluminum before repainting; it's best to get advice from the inboard/outboard (or saildrive) manufacturer on what to use and how to apply it. Also, check with the manufacturer for compatible bottom (antifouling) paints. Copper-based paints are likely to be destructive to the aluminum.

Any boat that has aluminum below the waterline and an onboard AC electrical system is asking for trouble whenever the boat is plugged into shore power if the shore-power inlet on the boat does not have a capacitor-type galvanic isolator, or better yet an isolation transformer. (See my *Boatowner's Mechanical and Electrical Manual* for an extensive discussion of this.)

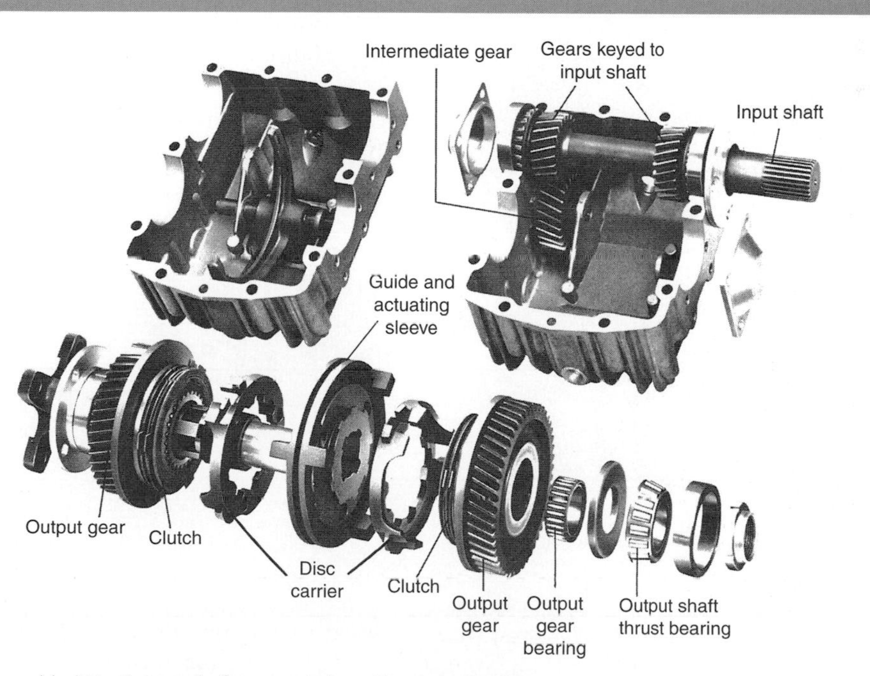

FIGURE 8-10. *A disassembled Hurth two-shaft transmission. (Courtesy Hurth)*

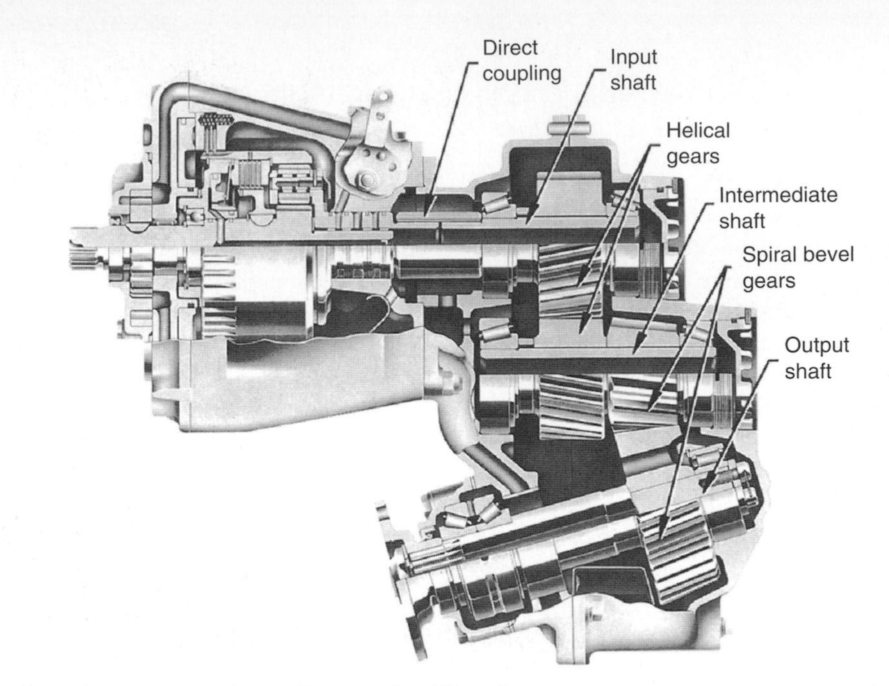

FIGURE 8-11. *A V-drive planetary transmission. (Courtesy BorgWarner)*

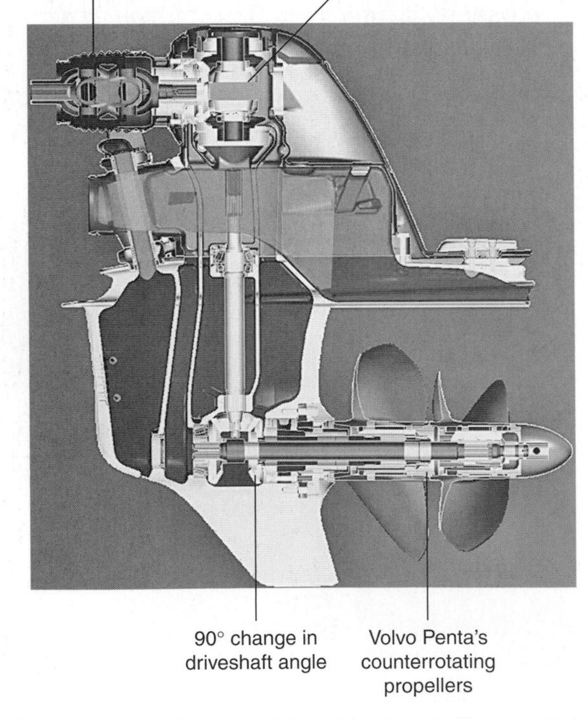

FIGURE 8-12. *A Volvo Penta inboard/outboard. (Courtesy Volvo Penta)*

SAILDRIVES

Saildrives are an adaptation of an inboard/outboard in which the drive leg is permanently immersed and cannot be tilted up and down (see Figure 8-14). At the top of the drive leg, a gear and clutch arrangement similar to that in a two-shaft transmission enables the unit to be put into forward and reverse (see Figure 8-15).

Many boatbuilders, especially in Europe, like saildrives because they are much easier to install than a traditional engine and take up less room.

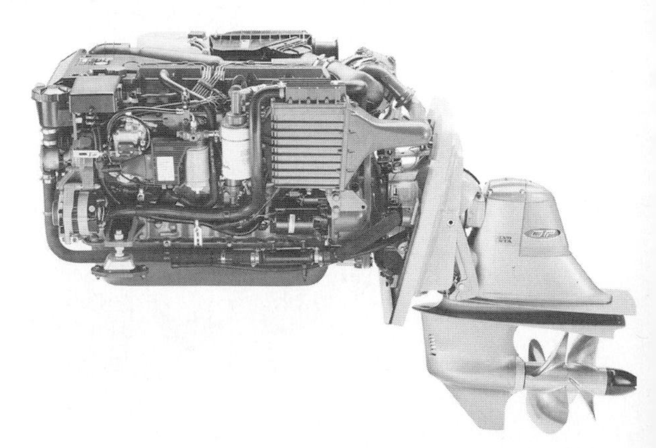

FIGURE 8-13. *An inboard/outboard unit coupled to an inboard engine. (Courtesy Volvo Penta)*

Essentially, you cut a big hole in the bottom of the boat, then bond in an adapter plate. The engine and its leg bolt to this adapter plate, which is sealed into the bottom of the boat by a large rubber diaphragm. There is no stern tube, shaft seal, or stern (Cutless) bearing; no engine alignment is required. The total cost of an installation (engine, transmission, and installation) is significantly less than that of a traditional installation.

From the boatowner's perspective, a saildrive eliminates the hassles associated with the shaft seal found in a conventional propeller shaft installation. Saildrives are also soft mounted, eliminating much vibration and making for quieter operation. They are reputed to offer less drag under sail (I can't confirm this, except to note that many racing boats use them).

Saildrives may have higher mechanical losses than a transmission and shaft seal because of the two right-angle changes of drive direction. But because the propeller is put in the water in the horizontal plane (as opposed to the propeller shaft being angled downward, as it is in most conventional installations, to give the propeller sufficient

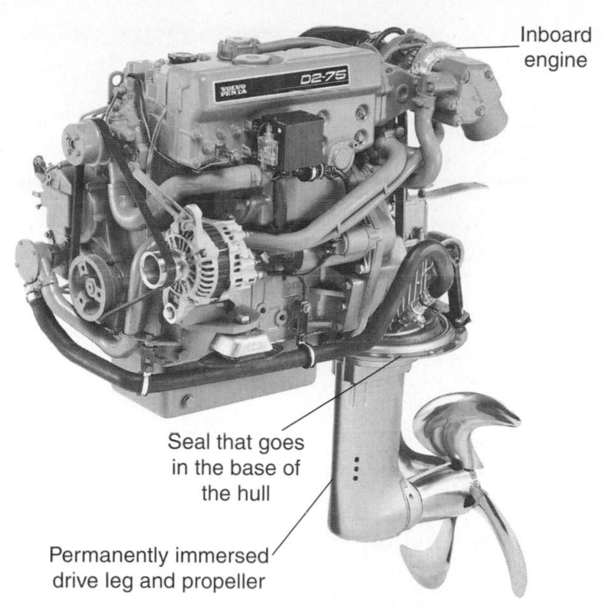

FIGURE 8-14. *A Volvo saildrive unit. (Courtesy Volvo Penta)*

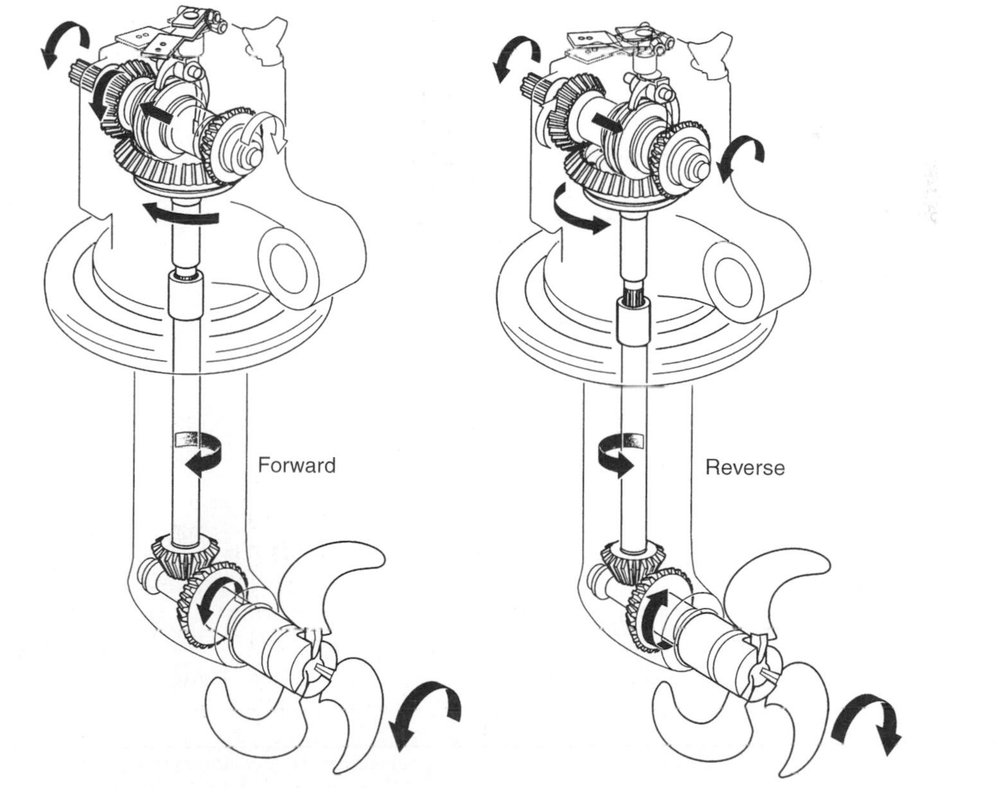

FIGURE 8-15. *The operating principles of a saildrive. (Courtesy Volvo Penta)*

clearance beneath the hull), the propeller is more efficient at moving the boat (saildrive manufacturers claim 10% to 12% improved efficiency).

The single biggest drawback of saildrives is corrosion (see the Aluminum Castings and Corrosion sidebar). The drive leg is aluminum and, if not properly installed, it can rapidly become a giant galvanic anode with respect to the rest of the boat's underwater metal (and that of other boats when plugged into shore power without some form of galvanic isolation). Thus the drive leg should be electrically isolated and adequately protected with sacrificial zincs.

Saildrives come with attached zincs, which need regular attention and must be replaced before 50% of the anode has been consumed; if neglected, loss of the zinc can result in expensive damage. In terms of other maintenance, Volvo Penta recommends replacing the diaphragm that seals the saildrive in the bottom of a boat every seven years, but few owners do this. Many saildrives over twenty years old are still operating with the original diaphragms.

Other than corrosion, the most likely source of saildrive problems is getting monofilament fishing line up the back of the propeller and winding this into the shaft seal, damaging it. The boat will have to be hauled to replace the seal.

Note: There have been a number of reports regarding Volvo Penta saildrives and total zinc loss as a result of galvanic corrosion between the zinc and its fasteners (causing the zincs to fall off). A much better way to manufacture zincs and prevent their loss is to cast them around a steel plate that the fasteners go through. Keep an eye on those zincs!

Volvo Penta Inboard Performance System (IPS)

The Volvo Penta Inboard Performance System (IPS) was introduced in 2005. It packages an electronically controlled common rail diesel engine with a specially constructed saildrive unit that has twin counterrotating, forward-facing propellers (see Figures 8-16 and 8-17). It is only available for twin-engine installations, and includes electronic (fly-by-wire) throttle, gear shifting, and steering.

The net result is a simple installation that has significant efficiency and performance benefits over a traditional engine and propeller shaft installation, with lowered emissions and greater maneuverability at both low and high speeds. Volvo Penta's trademark counterrotating propellers (invented by Volvo and pioneered on their inboard/outboard units) cancel out any tendency for prop walk. The two propellers result in a greater blade area than on a single propeller installation. This makes the propellers less sensitive to loading (i.e., less cavitation and lower tip losses—see Chapter 9). The forward-facing propellers operate in relatively undisturbed water, which further improves efficiency. Of course,

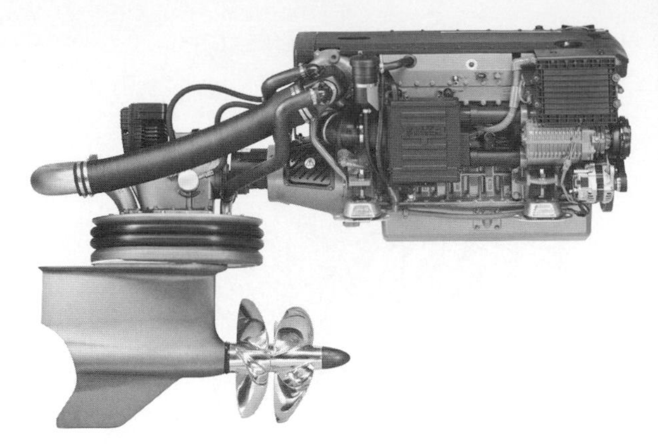

Figure 8-16. *The Volvo Penta Inboard Performance System. (Courtesy Volvo Penta)*

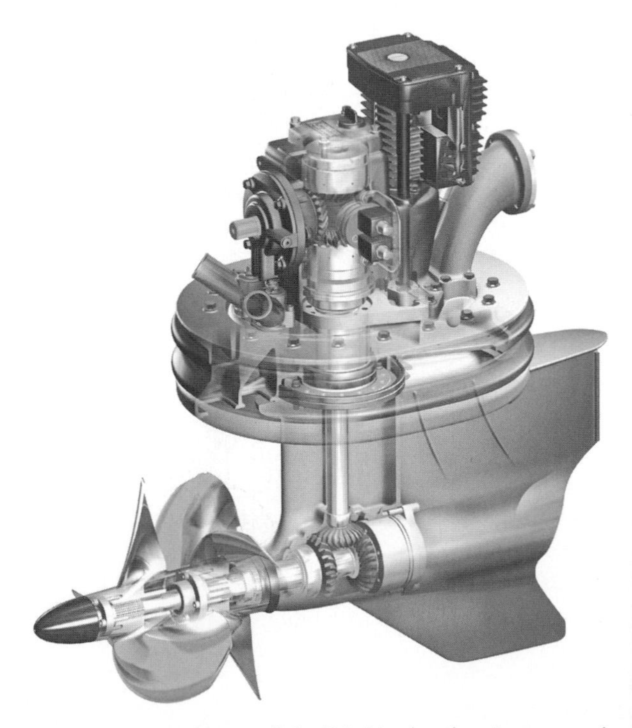

Figure 8-17. *A cutaway of the IPS drive leg that, in essence, is an adaptation of a saildrive leg. (Courtesy Volvo Penta)*

if you don't have a keel, they are also the first thing to hit the bottom if you run out of water.

Other than the counterrotating propellers, the various technical innovations—common rail injection, electronic engine controls, and a saildrive—have been covered elsewhere (see Chapter 2 and above). The case of the drive leg is manufactured from nibral (a nickel-bronze-aluminum alloy), which is rugged, highly corrosion resistant, and naturally resistant to biofouling (it will need scrubbing once in a while).

SHAFT BRAKES

When a boat is being towed or is under sail with the motor shut down, the flow of water over a fixed-blade propeller (for more on propellers, see Chapter 9) will spin the propeller unless the propeller shaft is locked in some way. This is of little concern with manual and two-shaft transmissions, except that it creates unnecessary wear on bearings, oil seals, and the shaft seal. But on some hydraulic transmissions, it will lead to a complete failure of the transmission since no oil is pumped to the bearings when the engine isn't running. Some output-shaft oil seals (particularly rawhide seals found on some really old transmissions) also will fail. Freewheeling propellers can also make quite a racket.

You can lock the propeller shaft on a manual transmission simply by putting the transmission in gear. The same holds for two-shaft transmissions. Note that you must always lock the transmission in the opposite direction to that in which the boat is moving (e.g., if sailing or being towed forward, put the transmission in reverse). If you don't, the clutch may slip and burn up (the critical speed at which this happens on Hurth transmissions and Hurth look-alikes seems to be around 6 knots).

A hydraulic transmission cannot be locked by putting it in gear. Without the engine running, the oil pump will not be working, and there will be no oil pressure to operate the clutches. You need a shaft brake (except with folding and feathering propellers), and you have essentially two choices.

HYDRAULIC UNITS

Hydraulically operated shaft brakes all have a spring-loaded piston riding in a cylinder that has an oil line plumbed to the transmission oil circuit. When the engine is at rest, the spring forces the piston down the cylinder to operate the brake. When the engine is started, oil pressure from the transmission oil pump forces the piston back up the cylinder against the spring pressure, releasing the brake. These units are thus automatic in operation, overcoming the main objection to older manual units, which was the probability of leaving the brake on, putting the transmission in gear, and burning up the brake.

Four variations on the hydraulic theme are widely available:

1. *Caliper disc.* These operate as on a car. A disc is bolted between the two propeller shaft coupling halves. A hydraulically operated caliper grips the disc to stop rotation (see Figure 8-18).
2. *Cam disc.* Similar to the above except that the disc has several cams. A hydraulically operated arm locks into the cams.

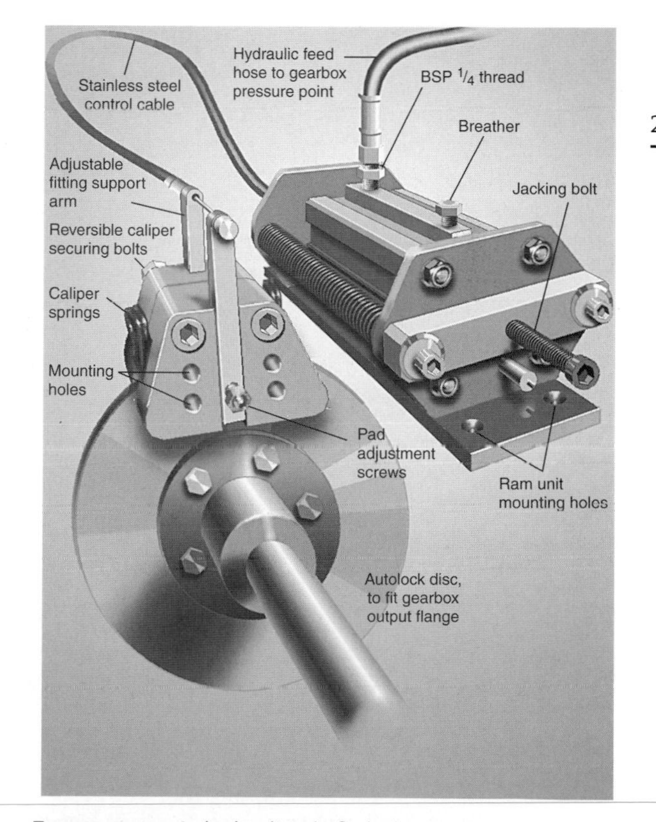

FIGURE 8-18. *A hydraulic shaft lock. (Courtesy Brunton's Propellers)*

3. *Brake band.* A hydraulically operated brake band clamps around the propeller shaft coupling.
4. *Plunger type.* A slotted sleeve is clamped around the shaft, and a hydraulically operated plunger locks into one of the slots.

MANUAL UNITS

On older manual units, a brake pad clamps around a block on the shaft or coupling. There is the obvious inherent risk that an owner will forget to disengage the brake when he or she next starts the engine and will burn up the brake. Newer manual units, however, have a notched disc clamped to the propeller shaft, with a plunger engaging the notch. If the brake is left engaged when the engine is started and put in gear, the plunger is simply forced out of the notch and held back by another spring-loaded pin until it is set manually once again (see Figures 8-19 and 8-20).

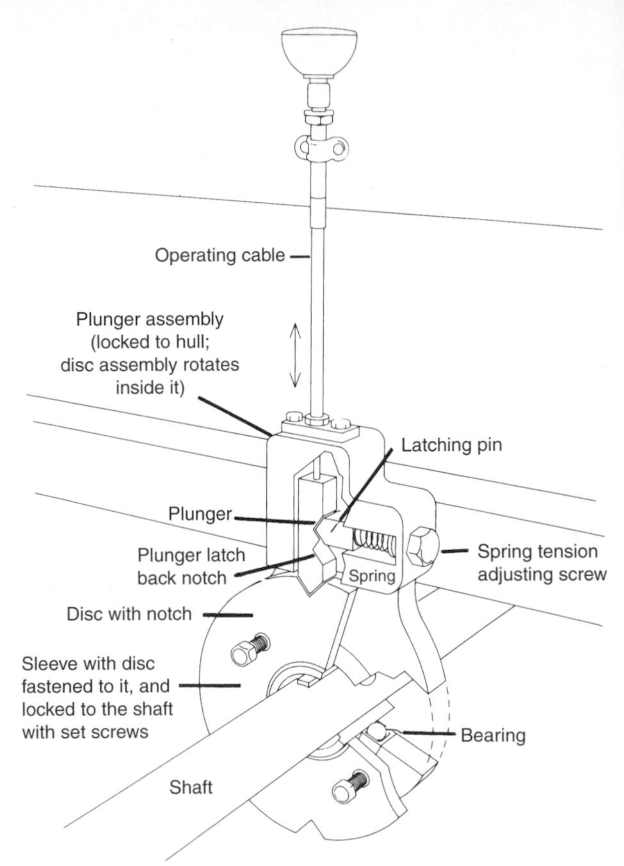

FIGURE 8-20. *Anatomy of a manual shaft lock. (Jim Sollers)*

FIGURE 8-19. *A manual propeller shaft lock. (Courtesy Shaft Lok)*

PROBLEM AREAS

- On hydraulic units, any oil leaks through faulty piston seals, connections, or hoses will cause the loss of the transmission oil and ultimately transmission failure. Adding a hydraulic shaft lock may, in fact, void a transmission warranty.
- The manual plunger types have a tendency to jump out during hard sailing. Spring tension on the latching pin can be increased to hold the plunger in place, but then it becomes more difficult to disengage the plunger.
- The cam disc, hydraulic plunger, and manual notched-disc plunger-style units can be set only at low propeller speeds.

Ignoring this is likely to cause damage to the first unit; with the second and third units, the plunger will jump out. It may be necessary to slow the boat and reduce propeller freewheeling before engaging the device; in other words, these devices are shaft locks rather than shaft brakes.

MAINTENANCE

Hydraulic Units

Pay close attention to the hydraulic lines and inspect them regularly for any signs of leaks around the piston seals. Check the brake linings on caliper and brake band units for wear. Don't allow them to slip—a lot of heat will be generated. Check any sleeve mounting bolts from time to time.

Manual Units

The control cable is a Morse cable. Figure 8-27 on page 217 shows a number of points to watch for with Morse cables. The plunger unit is mounted on a bearing within which the propeller shaft rotates. The bearing is sealed for life. Check for undue play once in a while, and while you're at it, also check the setscrews that lock the central sleeve to the propeller shaft. If water is being thrown around this area (generally from a leaking shaft seal), the bearing will rust out over time.

TRANSMISSION MAINTENANCE

Transmission maintenance is minimal (see Figure 8-21). It generally boils down to the following:

- Keep the exterior of the transmission clean, which is important for detecting oil leaks.

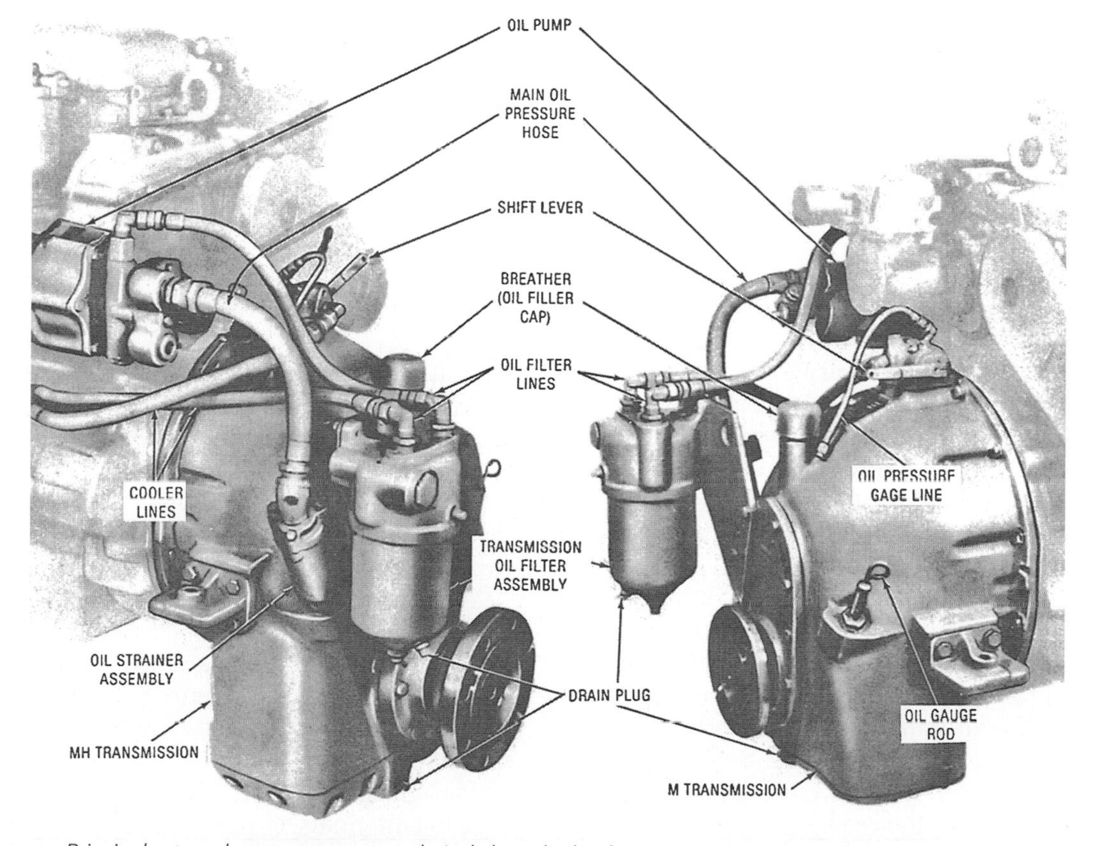

FIGURE 8-21. *Principal external components on a relatively large hydraulic transmission. (Courtesy Detroit Diesel)*

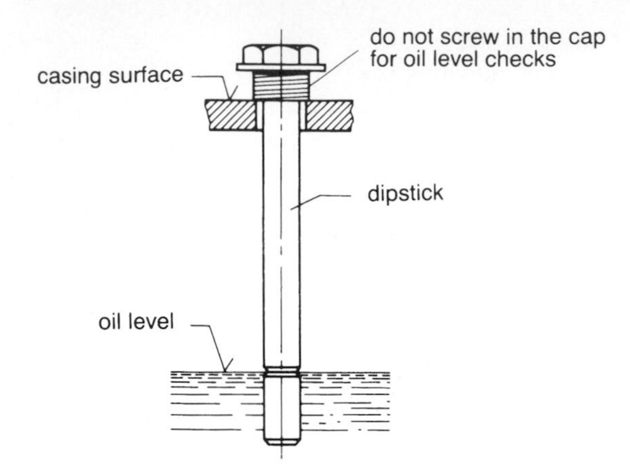

FIGURE 8-22. *Checking the oil level on a Hurth transmission. (Courtesy Hurth)*

- Periodically check the oil level (see Figure 8-22). Unless there is a leak, it should never need topping off.
- Check for signs of water contamination. Water emulsified in engine oil gives the oil a creamy texture and color. In automatic transmission fluid, it looks more like a strawberry frappe!
- Change the oil annually. Check the color—a slipping clutch will turn the oil black.

Most transmissions operate on 30-weight engine oil or automatic transmission fluid (ATF type A—normally Dextron II or Dextron III; the latter is an improved formulation of the former and is interchangeable with it). Note that mixing some different types of transmission fluids (e.g., A and F types) can destroy transmission seals, so be sure to stick with whatever the manufacturer specifies.

If there is an oil screen or a magnetic plug (or both) in the base of the transmission, inspect them for any signs of metal particles or other internal damage when changing the oil. Some transmissions have an oil filter beneath the oil filler cap that will need replacing. Before checking the new oil level, run a hydraulic transmission for a couple of minutes, then shut it down to make the check. On some transmissions, the dipstick must be screwed completely in to check the level; on others the threaded portion is kept on top of the housing—check the manual.

If the transmission has a raw-water-cooled oil cooler with a sacrificial zinc anode, you must check the zinc anode regularly and change it when only partly eaten away.

TROUBLESHOOTING AND REPAIRS

The majority of transmission difficulties arise as a result of improper clutch adjustments (manual transmissions) or problems with the control cables, rather than from problems with the transmission itself (see below; manual two-shaft transmission clutches, in particular, are very sensitive to improper cable adjustments). However, before discussing these issues, there are a couple of problems peculiar to hydraulic boxes that I want to note.

A buzzing sound indicates air in the hydraulic circuit, generally as a result of a low oil level. This will lead to a loss of pressure and slipping clutches. Most hydraulic transmissions also have an oil pressure regulating valve that passes oil back to the suction side of the oil pump if excess pressure develops. If the valve sticks in the open position, the clutches may slip or not engage at all. On the other hand, if the valve sticks closed, the clutches will engage roughly (as they also will do if gears are shifted at too high an engine speed). The valve is generally a spring-loaded ball or piston screwed into the side of the transmission—removing it for inspection and cleaning is easy (see Figure 8-23).

MANUAL PLANETARY TRANSMISSION CLUTCH ADJUSTMENTS

The top of most manual transmissions unbolts and lifts off. Inside are adjustments for forward and reverse gears. On those models with a brake band for reverse (the majority in small boats), reverse is easier.

To adjust the reverse clutch, move the gear lever into and out of reverse—the brake band will be clearly visible as it clamps down on the hub and unclamps (refer back to Figure 8-4). On one side of the band will be an adjusting bolt. If the transmission is slipping in reverse, tighten the bolt a little at a time, engaging reverse between each adjustment. Adjustment is correct when the gear lever requires firm pressure to go into gear, and clicks in with a nice, clean feel.

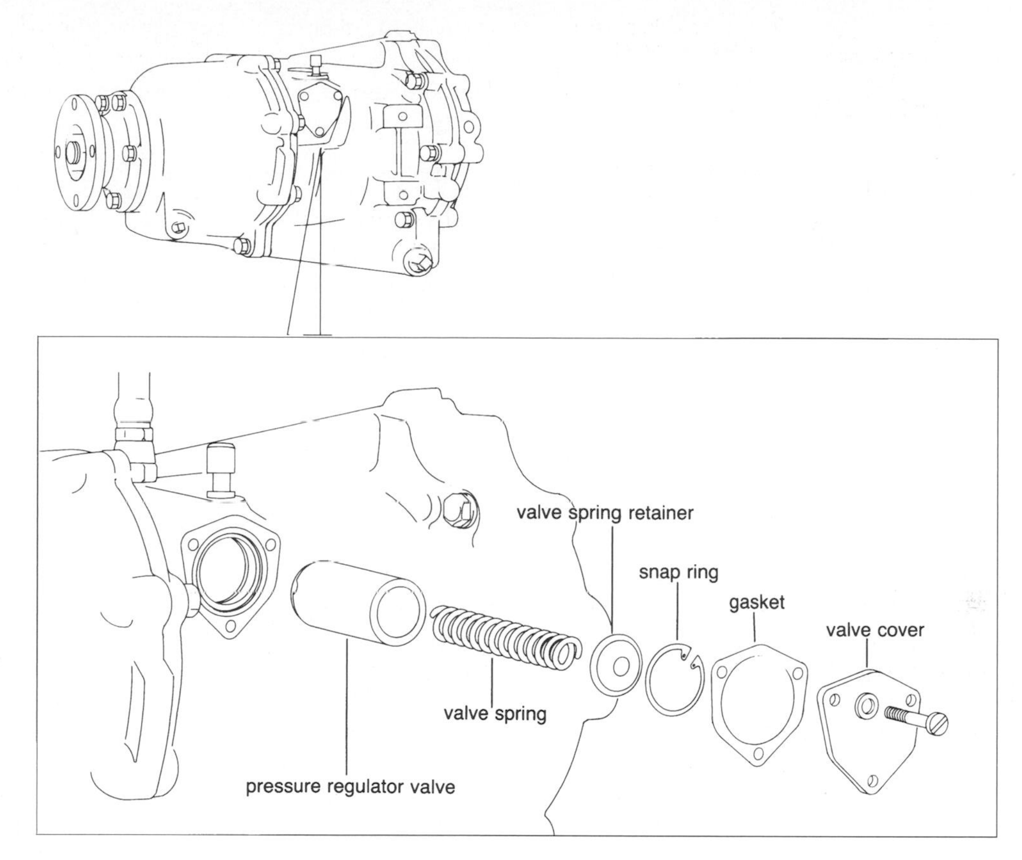

FIGURE 8-23. *An oil pressure regulating valve. (Jim Sollers)*

It is important not to overdo things. Put the box in neutral and spin the propeller shaft by hand. If the brake band is dragging on the hub, it is too tight—the box is going to heat up and wear will be seriously accelerated. If no amount of adjustment produces a clean, crisp engagement, the brake band is worn out and needs replacing—or at least relining.

To adjust the forward clutch, first put the transmission into and out of gear a few times to see what is going on. The main plate on the back of the clutch unit (it pushes everything together) will have either one central adjusting nut or between three and six adjusting nuts spaced around it. Tighten the central nut by one flat. For multiple adjusting nuts, put the box into neutral and turn it over by hand. As each adjustment nut becomes accessible, tighten it by one-sixth of a turn. After going all the way around, try engaging the gear again. Repeat until the gear lever goes in firmly and cleanly. Lock the adjusting nuts. Do not tighten to the point at which the clutch drags in neutral (the propeller shaft will turn slowly when the engine is running); if overtightening seems necessary, the friction pads on the clutch plates are probably worn out and need replacing.

HYDRAULIC AND TWO-SHAFT TRANSMISSION CONTROL CABLES

With the exception of manual boxes that have a gearshift handle, and modern electronically controlled engines that employ fly-by-wire gear shifting, most two-shaft transmissions use a push-pull cable to move the shift lever on the box (see Figure 8-24). A push-pull cable pushes the lever in one direction and pulls it in the other. More transmission problems are caused by cable malfunctions than anything else. Faced with difficulties, always suspect the cable before blaming the box.

If the transmission operates stiffly, fails to go into either or both gears, stays in one gear, or slowly turns the propeller when in neutral (clutch drag),

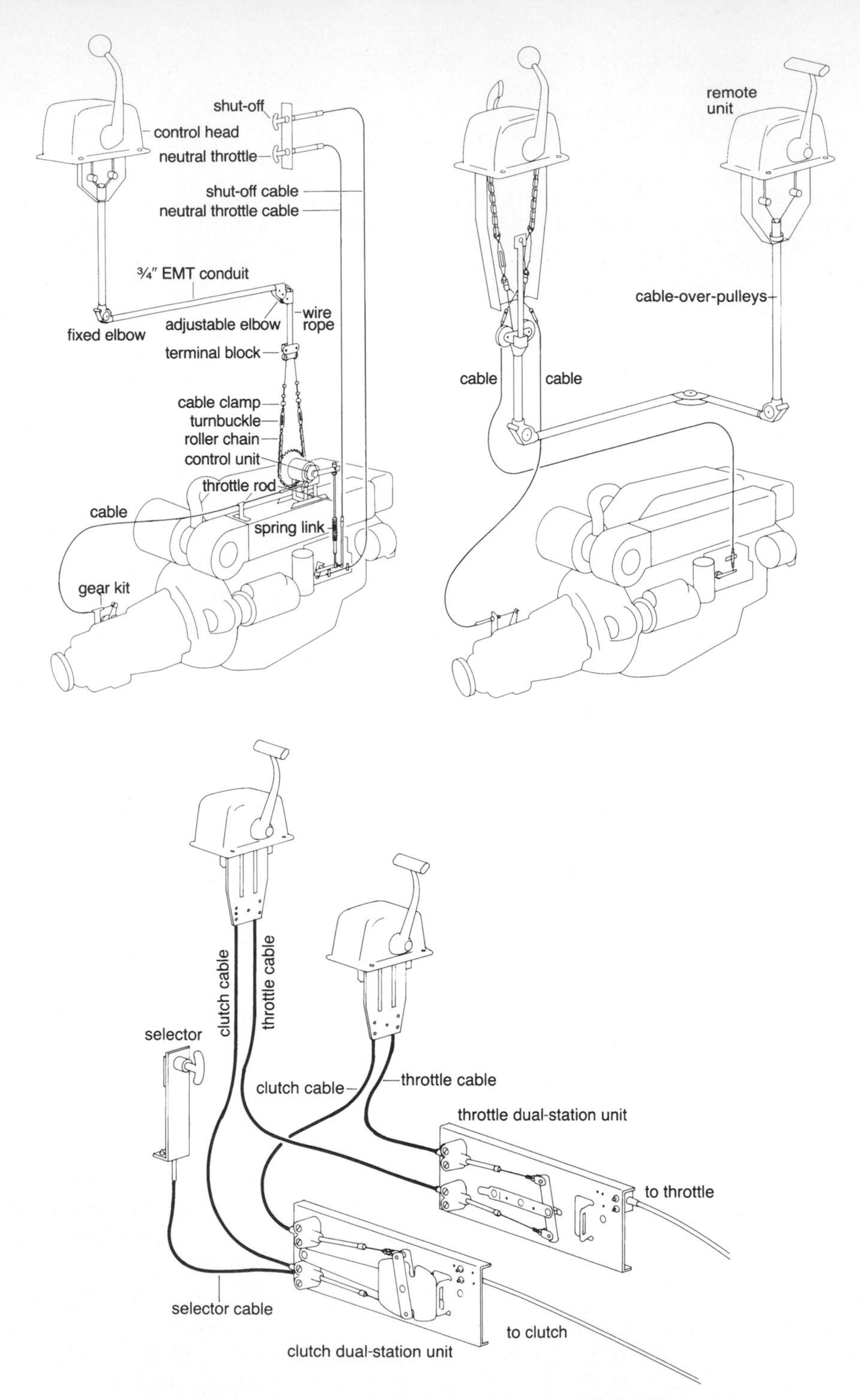

FIGURE 8-24. *Typical remote engine and transmission controls comprising three sections: the cockpit or pilothouse control, the cable system, and the engine control unit.* ***Top left:*** *An enclosed cable-over-pulley system. Input motion is transmitted from the pilothouse control via the cables to the actuating mechanisms on the engine and transmission.* ***Top right:*** *A dual-station installation, with the main station using push-pull cables, and the remote station using a cable-over-pulley system.* ***Bottom:*** *A dual-station installation using remote bellcrank units and single-lever controls. On many modern boats, the cockpit control unit incorporates an electronic device that sends a signal to an operating unit at the engine and transmission (fly-by-wire). The only physical connection between the cockpit and engine is the data cable. (Courtesy Morse Controls, adapted by Jim Sollers)*

Troubleshooting Chart 8-1. Transmission Problems **Symptoms:** Failure to engage forward or reverse; clutch drag in neutral; or tendency to stick in one gear.	
Move the remote control lever through its full range a couple of times. Does it move the operating lever on the transmission itself through its full range? YES	NO Check for a broken, disconnected, slipping, or kinked cable.
Is the remote control lever free-moving? YES	NO Break the cable loose at the transmission and try again. If still stiff, remove the cable from its conduit, and clean, grease, and replace. If the cable moves freely when disconnected from the transmission, move the transmission lever itself through its full range. If binding, the transmission needs professional attention.
When the remote control is placed in neutral, is the transmission lever in neutral? YES	NO Adjust the cable length.
Is the transmission oil level correct? (Most transmissions have a dipstick; hydraulic transmissions frequently make a buzzing noise when low on oil.) YES	NO Add oil and run the engine in neutral to clear out any air.
Does the transmission output coupling turn when the transmission is placed in gear? YES	NO The transmission needs professional attention.
Does the propeller shaft turn when the transmission coupling turns? YES	NO The coupling bolts are sheared or the coupling is slipping on the propeller shaft. Tighten or replace setscrews, keys, pins, and coupling bolts as necessary.
There must be a fault with the propeller: 1. It may be missing or damaged. 2. A folding propeller may be jammed shut. 3. A controllable-pitch propeller may be in the "no pitch" position. 4. If this is the first trial of the propeller, it may simply be too small and/or have insufficient pitch.	

perform the following checks (Troubleshooting Chart 8-1):

- See that the transmission actuating lever (on the side of the transmission) is in the neutral position when the remote control lever (in the cockpit or wheelhouse) is in neutral (see Figures 8-25 and 8-26).
- Ensure that the actuating lever is moving fully forward and backward when the remote control is put into forward and reverse. This is particularly important on two-shaft transmission boxes—the travel should be an equal distance in both directions, without the operating lever snagging the transmission case at any point. It may be necessary to undo the clamp that holds the lever to its shaft and slide the lever out a little bit.
- Disconnect the cable at the transmission and double-check that the actuating lever on the box is clicking into forward, neutral, and reverse. Note: Two-shaft transmission boxes do not have a distinct click when a clutch is engaged. However, the lever must move through a *minimum* arc of 30 degrees in either direction. Less movement will cause the clutches to slip. More is OK. As the clutches wear, the lever must be free to travel farther (up to 45 degrees). If the transmission actuating lever is stiff, or not traveling far enough in either direction, make sure that it is not rubbing on the transmission housing or snagging any bolt heads (see above).
- While the cable is loose, operate the remote control to see if the cable is stiff. If so, replace the cable. Note that forcing cables through too tight a radius when routing them from the control console to the engine is a frequent cause of problems.
- Inspect the whole cable annually, checking for the following (see Figure 8-27):
  a. seizure of the swivel at the transmission end of the cable conduit
  b. bending of actuating rods
  c. corrosion of the end fittings at either end
  d. cracks or cuts in the conduit jacket
  e. burned or melted spots
  f. excessively tight curves or kinks (the minimum radius of any bend should be 8 inches)

FIGURE 8-26. *Clutch adjustment for many Volvo Penta transmissions. The travel of the control arm from 0 to A and from 0 to B must be equal. When the control cable has traveled no more than 1.2 inches (30 mm) in either direction, the relevant gear (forward or reverse) should be engaged. If the travel from 0 to A and 0 to B is not equal, adjust the position of the cover until it is as equal as possible. The travel increases in the direction in which the cover is moved. (Courtesy Volvo Penta)*

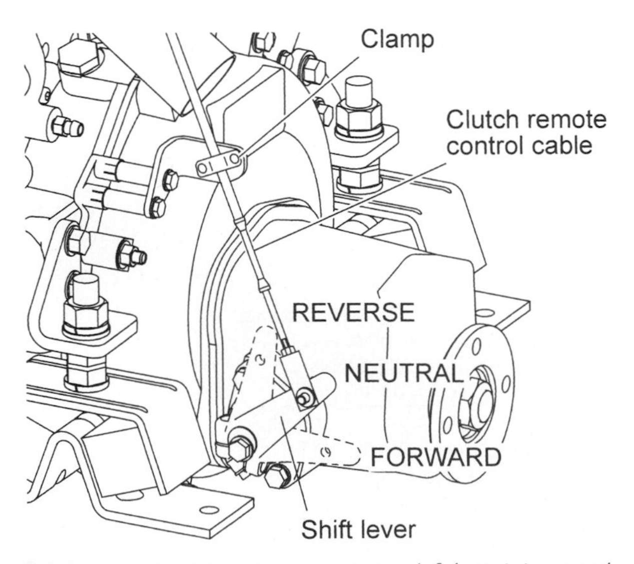

FIGURE 8-25. *Check that the transmission shift lever is in neutral when the remote control is in neutral, and that the lever can move freely through a minimum of 30 degrees in either direction. (Courtesy Yanmar)*

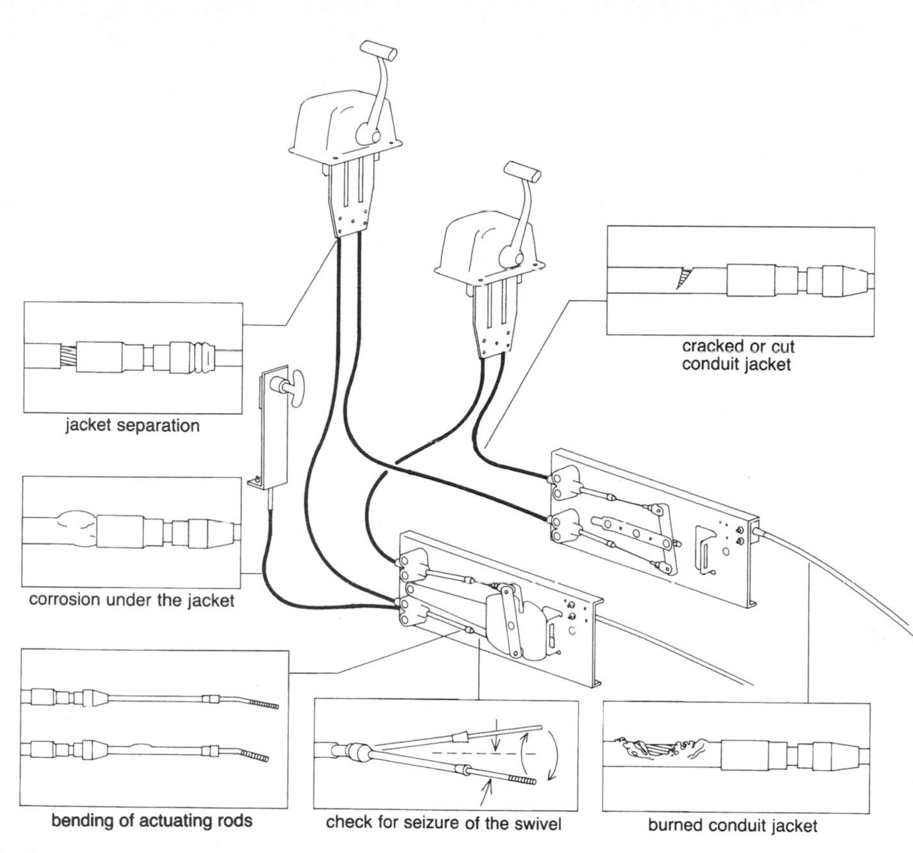

FIGURE 8-27. *Checking transmission control cables. (Courtesy Morse Controls, adapted by Jim Sollers)*

g. separation of the conduit jacket from its end fittings
h. corrosion under the jacket (it will swell up; see Figure 8-28)

FIGURE 8-28. *A corroded engine control jacket. The inner cable is stainless steel and is just fine, but the spiral-wound outer case is steel that has rusted right through.*

If at all possible, remove the inner cable and grease it with a Teflon-based waterproof grease before replacing. Replace cables at least every five years and keep an old one as a spare.

MISCELLANEOUS OPERATING PROBLEMS

If the oil is kept clean and topped off and the clutch or cables are properly adjusted, problems tend to be few and far between.

Overheating

Heavily loaded transmissions, especially hydraulic transmissions, tend to get hot (too hot to touch). In fact, many that do not have an oil cooler would benefit from the addition of one. Excessively high temperatures, however, are likely to arise only if the oil level is low (a smaller quantity of oil has to dissipate the heat generated); the clutches are slipping (creating excessive friction); or an oil cooler is not operating properly.

A slipping clutch should be evident from a loss of performance. The intense heat generated will soon warp clutch plates and burn out clutch discs. The oil in the transmission will take on a characteristic black hue and may well smell burned.

Oil cooler problems may arise on the water and oil sides. Transmission oil generally remains pretty clean, but a slipping clutch and other problems occasionally create a sludge that can plug up the oil side of a cooler. And if the clutch is slipping, the oil will be darkened in color. If the cooler is raw-water cooled, a more likely scenario is that silt, corrosion, and scale are interfering with the heat transference on the water side. Note that some entry-level transmission coolers are nothing more than a box bolted to the outside of the transmission case with raw water circulating through the box—i.e., there are no heat exchanger tubes inside the cooler.

Water in the Transmission

If the transmission has an oil cooler, the cooler is the most likely source of water, especially if it is a raw-water type. Pinholes form in cooler tubes just as in engine oil coolers. Regularly inspecting and changing sacrificial zinc anodes is essential. The only other likely source of water ingress is through the transmission output seal. For this to happen, the seal must be seriously defective and the bilges must have large amounts of water slopping around, both of which were far more common years ago when leather seals and wooden boats were the norm.

If you discover the water in a reasonable time and remove it, eliminate the source, and change the transmission oil a couple of times to flush the transmission, you stand a good chance of preventing any lasting damage.

Loss of the Transmission Oil

The rupture of an external oil line will produce a sudden, major, and catastrophic loss of oil, which will immediately be obvious. Less obvious will be the loss of oil through a corroded oil cooler. If it is raw-water cooled, the oil will go overboard to form a slick; if freshwater cooled, it will rise to the top of the header tank (see page 142).

Although the seal around the clutch actuating lever or the seal on a hydraulically operated shaft brake occasionally leaks small quantities of oil, the most likely candidate for this kind of leak is the output-shaft oil seal. This is particularly true if the engine and propeller shaft are poorly aligned (which leads to excessive vibration), or if the propeller shaft has been allowed to freewheel when the boat is under sail. Alignment checks are covered in Chapter 9; seal replacement is dealt with below.

On rare occasions, the oil gets pumped through some ruptured seal into the flywheel housing.

Slow Engagement

With a two-shaft transmission, a delay in engagement when shifting into gear normally indicates wear in the thrust washers. The transmission should be professionally serviced before too long.

Replacing an Output Shaft Seal

1. Unbolt and separate the two halves of the propeller coupling (see Figure 8-29). Mark both halves so you can bolt them back together in the same relationship to each other.

   On some boats with vertical rudderposts, the propeller shaft cannot be pushed far enough aft to provide the necessary room to slide the transmission coupling off its shaft! The propeller hits the rudderstock and will go no farther. In this situation, remove the rudder or lift the engine off its mounts to provide the necessary space (this is an awful lot of work to change an oil seal; to avoid this problem

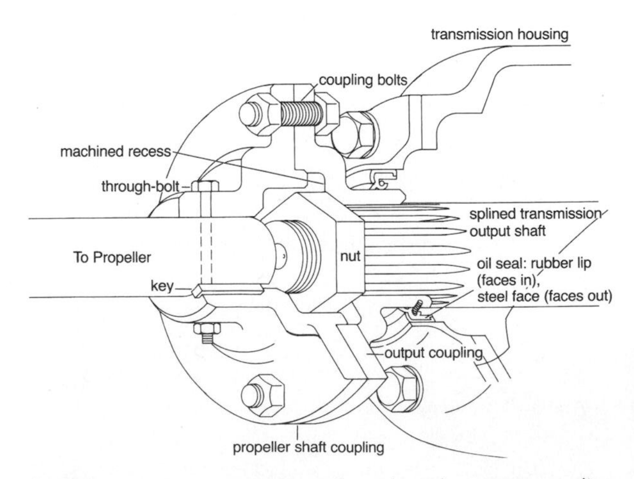

FIGURE 8-29. *A transmission oil seal and output coupling arrangement. The recesses machined into the faces of the two coupling halves assist in shaft alignment. (Jim Sollers)*

in the future, you may want to consider having the propeller shaft shortened and installing a small stub shaft in-line between the transmission and propeller shaft).

2. Remove the coupling half attached to the transmission output shaft. This coupling is held in place with a central nut, which is done up tightly on most modern boxes; on some older boxes, it is just pinched up and locked in place with a cotter pin (split pin).

   The coupling rides on either a splined shaft (one with lengthwise ridges all the way around) or a keyed shaft. (In the latter case, do not lose the key down in the bilges when removing the coupling!) The key will most likely stick in the shaft. If there is no risk of its falling out and getting lost, leave it there (tape it to the shaft for the time being); otherwise, hold a screwdriver against one end and tap gently until you can pry up the end and remove the key.

   Some couplings are friction-fit on their shafts and should be removed with a proper puller (see Figure 8-30). This is nothing more than a flat metal bar bolted to the coupling and tapped to take a bolt in its center. The bolt screws down against the transmission output shaft, forcing off the coupling.

3. Transmission oil seals are press-fit into either a separate housing or the rear transmission housing. Most seals consist of a rubber-coated steel case with a flat face on the rear end and a rubber lip on the front end (the end inside the transmission). A spring inside the seal holds this lip against the coupling face to be sealed.

   Removing a seal from its housing is not always easy. If at all possible, unbolt the housing from the transmission and take it to a convenient workbench. This is often fairly simple on older boxes and boxes with reduction gears, but may not be feasible on many modern hydraulic boxes. You may be able to dig out the seal with chisels, screwdrivers, steel hooks, or any other implement that comes to hand; it

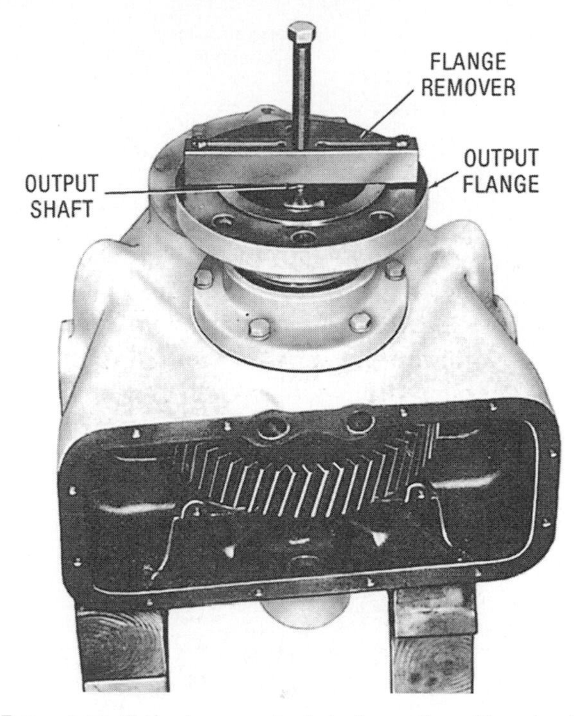

FIGURE 8-30. *Removing an output shaft coupling with a simple coupling puller. (Courtesy Allison Transmission)*

doesn't matter if the seal gets chewed up as long as the housing and shaft (if still in place) are unscratched.

4. New seals go into the housing with the rubber lip facing into the gearbox and the flat face outside. Place the seal squarely in its housing and then tap it in evenly using a block of wood and a hammer. If you force the seal in cockeyed, it will be damaged. A block of wood is necessary to maintain an even pressure over the whole seal face—hitting a seal directly will distort it. Push in the seal until its rear end is flush with the face of the transmission housing. Once in place, some seals require greasing (there will be a grease fitting on the back of the gearbox), but most need no further attention.

5. Reassemble the coupling and propeller shaft by reversing the disassembly steps. Check the alignment of the propeller shaft (see Chapter 9) anytime the coupling halves are broken loose and reassembled.

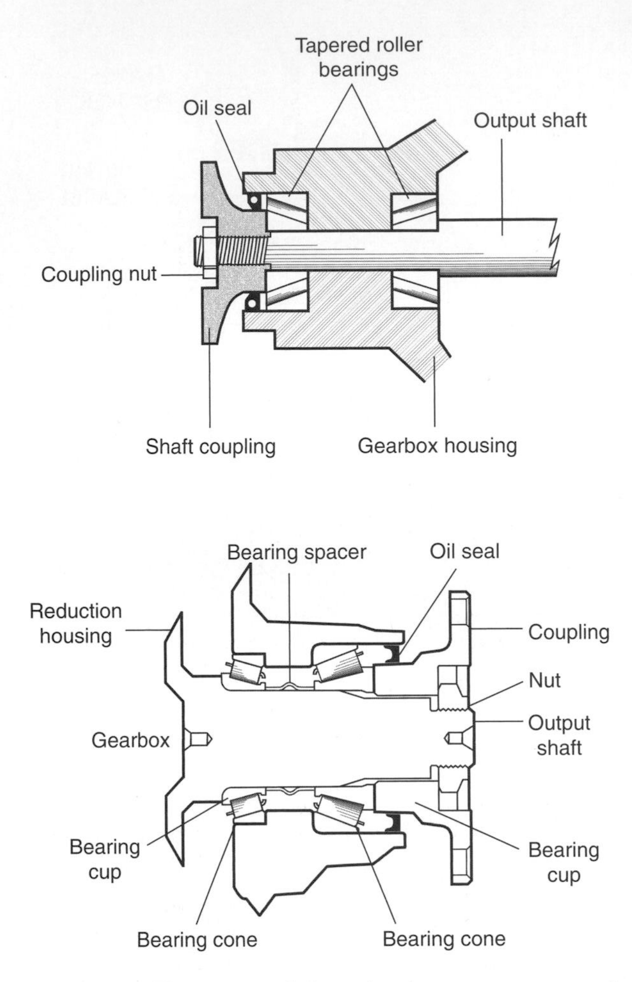

FIGURE 8-31. ***Top:*** *A typical thrust bearing arrangement and location of the engine gearbox oil seal.* ***Bottom:*** *Preloaded thrust bearings. (Fritz Seegers)*

Some transmissions have preloaded thrust bearings. The transmission output shaft, on which the coupling is mounted, turns in two sets of tapered roller bearings—one facing in each direction (see Figure 8-31). Between the two sets is a steel sleeve. When the coupling nut is pulled up, this sleeve is compressed, maintaining tension on the bearings and eliminating any play. Anytime the coupling nut is undone, use a torque wrench and note the pressure that is needed to break the nut loose. When reinstalling the nut, tighten it to the same torque plus 2 to 5 foot-pounds. In any event, the torque should be at least 160 foot-pounds on most BorgWarner boxes, but the couplings should still turn freely by hand in neutral with only minimal drag. If the transmission needs a new spacer between the thrust bearings, a special jig and procedure are necessary, and the whole transmission reduction gear will have to go to a professional.

On an older transmission in which the coupling nut is done up less tightly and restrained with a cotter pin, it is essential that you replace the nut properly. The best approach is to moderately tighten the nut, making sure everything is properly seated, then back it off an eighth of a turn or so before inserting the cotter pin. Put the transmission in neutral and turn the coupling by hand to ensure there is no binding. This type of coupling sometimes leaks oil between the shaft and the coupling; to prevent this, smear a little gasket sealer around the inside of the coupling before fitting it to the shaft.

CHAPTER 9

ENGINE SELECTION AND INSTALLATION

This chapter may not seem relevant for boatowners who already have an engine installed in their boat. However, quite commonly problems with existing installations arise from either a poor matching of the engine to the boat's needs or from poor installation practices. For either situation, these pages may throw some light on a longstanding problem.

SECTION ONE: ENGINE SELECTION

Matching the engine to its load and use are the primary considerations when installing a diesel engine in a boat.

MATCHING AN ENGINE TO ITS LOAD

Diesels are susceptible to damage from both overloading and underloading. When overloaded, generalized or localized overheating can lead to engine damage, up to and including seizure. The damage from underloading is in some ways more pernicious. It can arise from running at higher speeds with a low load (generally as a result of a mismatched propeller), or more likely, from repeated low-speed operation with little load. The latter is particularly common on auxiliary sailboats when charging a battery or running a refrigerator at anchor—it is not unusual to find a 50 hp motor carrying a 1/2 to 1 hp load.

An underloaded engine takes time to reach proper operating temperatures, and at low speeds, it also tends to run unevenly due to the difficulties of accurately metering the minute quantities of fuel needed at each injection stroke. These two factors encourage the formation of sulfuric acid in the lubricating oil (see page 51) and carbon deposits throughout the engine. The cylinder walls are likely to become glazed, and piston rings will get gummed in their grooves, resulting in blowby and a loss of compression. Valves may stick in their guides, while carbon will plug up the exhaust system. Carbon sludge will form in the oil, and if you neglect oil change procedures, the sludge will eventually plug sensitive oil passages and lead to bearing failure.

Repeated running of a diesel engine at low loads is a destructive practice, which greatly increases maintenance costs and reduces engine life.

HOW MUCH HORSEPOWER DO YOU NEED?

Does your boat have a displacement or a planing-type hull? A displacement hull is one that remains immersed at all times, whereas a planing hull develops hydrodynamic forces at speed that enable it to move up onto the surface of the water.

A displacement hull has a predetermined top speed (defined as *hull speed*), more or less irrespective of available power. This top speed is governed by specific physical properties of the waves the boat makes as it passes through the water (see the Wave

Theory sidebar), modified by the displacement-to-length (D/L) ratio of the boat. This ratio is a function of the boat's weight relative to its waterline length (I cover these issues in detail in *Nigel Calder's Cruising Handbook*). For heavier-displacement sailboats, hull speed is typically around 1.34 times the square root of the boat's waterline length. For lighter-displacement boats, it may be as high as 1.7 times the square root of the boat's waterline length.

A clean-hulled displacement craft can be driven at around two-thirds of its hull speed in smooth water by a relatively small engine, but as hull speed is approached, *wave-making drag* (resistance) increases rapidly. At this point, any additional speed can only be gained by a disproportionate increase in power (therefore fuel burned). Once hull speed is reached, it takes a great deal of additional energy to go any faster.

A planing hull, on the other hand, breaks free of the constraints imposed by the waves it generates. A certain minimum amount of power is required to come up to a plane. Thereafter, the boat's top speed is at least in part related to available power (see Figures 9-1 and 9-2).

Various formulas have been derived for determining the horsepower requirements of displacement and planing hulls. Two excellent sources are *Skene's Elements of Yacht Design*, by Francis S. Kinney, published by Dodd, Mead (a very traditional approach; see the Determining the Horsepower Requirements sidebar), and the *Propeller Handbook*, by Dave Gerr, published by International Marine (a great reference work). The formulas will enable you to take into account the effects on boat speed of a foul bottom, wave action, headwinds, and other factors.

Only in exceptional circumstances will you find that a displacement hull requires more than 1 hp per 500 pounds (fully loaded) displacement (see Figure 9-3). For many years, we had a 30 hp engine

PLANING SPEED CHART CONSTANTS

C	Type of Boat
150	average runabouts, cruisers, passenger vessels
190	high-speed runabouts, very light high-speed cruisers
210	race boat types
220	three-point hydroplanes, stepped hydroplanes
230	racing power catamarans and sea sleds

FIGURE 9-1. *Planing speed chart constants. (From Propeller Handbook by Dave Gerr; courtesy International Marine)*

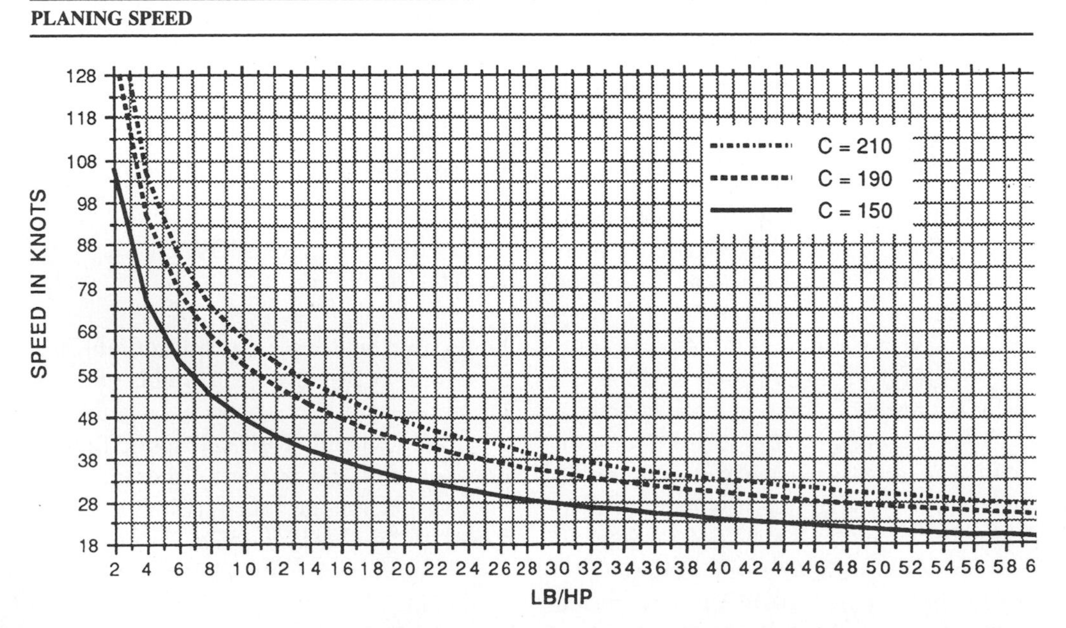

FIGURE 9-2. *This chart shows the speed attainable by planing craft as a function of available shaft horsepower. See Figure 9-1 to estimate the appropriate C value with which to enter the table. (From Propeller Handbook by Dave Gerr; courtesy International Marine)*

Wave Theory

As a boat moves through the water, it makes waves. There is a natural, physical, relationship between the speed at which waves move and the distance from crest to crest (a *wavelength*) such that:

The square root of the wavelength (in feet) × 1.34 = the wave speed (where wavelength is in feet, and speed is in knots).

This is a fact of nature that simply has to be taken into account when designing boats. Figure 9-4 shows how dramatically wavelengths increase with small increases in speed—the doubling of a wave's speed from 5 to 10 knots increases its wavelength four-fold (from 13.9 to 55.6 feet).

The bow wave of a boat is moving with the boat. Because of the physical relationship between wave speed and wavelength, the faster a boat moves, the faster the bow wave moves, and consequently, the farther apart it is from its second crest. At some point, this distance—the wavelength—will be such that there will be a crest at the bow and another at the stern. If a boat goes faster, the wave crest at the stern moves farther aft, dropping the boat's stern into the bow wave's trough so that the boat is now attempting to "climb" its bow wave. Heavy-displacement boats, which generate large bow waves, need more power than can normally be generated by sails or an engine to do this, and as a result many sailboats do not go this fast.

The boat begins to sink into the trough of the bow wave at the point when the crest of the stern-wave moves aft of the aft end of the boat's waterline—in other words, at that point when the wavelength exceeds the waterline length of the boat. Consequently, the maximum speed of many heavy-displacement boats—their *hull speed*—is generally on the order of 1.34 × the square root of the waterline length (see Figure 9-5).

VELOCITY IN KNOTS	WAVE LENGTH IN FEET
1	0.56
2	2.23
3	5.01
4	8.90
5	13.90
6	20.0
7	27.2
8	35.6
9	45.0
10	55.6
11	67.3
12	80.1
13	94.0
14	109.0
15	125.2

FIGURE 9-4. *Periods and lengths of sea waves.*

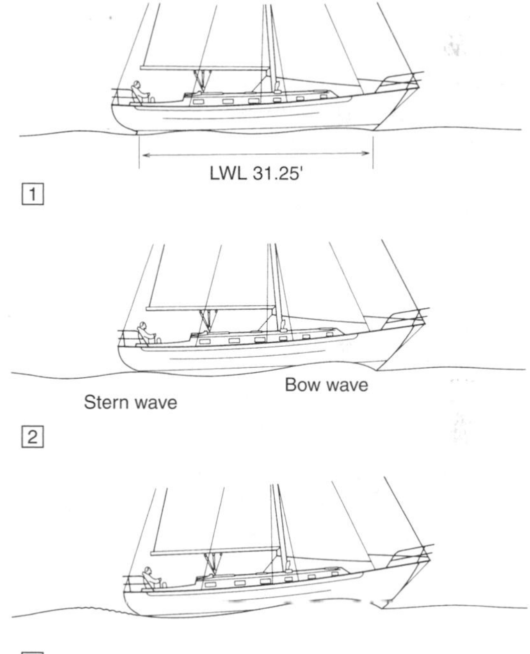

FIGURE 9-5. *Bow and stern waves. A Pacific Seacraft 40 at less then nominal hull speed (1). It takes relatively little energy to keep the boat moving at this speed. A Pacific Seacraft 40 at a little above nominal hull speed (approximately 7.8 knots; 2). The stern wave is just beginning to move aft of the stern. It takes quite a bit of energy to keep the boat moving at this speed. A Pacific Seacraft 40 at well above nominal hull speed (approximately 8.5 knots; 3). The stern wave has moved aft of the stern, which is now dropping into the trough of the bow wave. It takes a tremendous amount of energy to keep the boat moving at this speed.*

(continued)

Wave Theory (continued)

Lighter-displacement boats make smaller waves. This results in a greater potential to get somewhat ahead of the stern wave, and as such hull speed may go as high as 1.7 × the square root of the waterline length (in exceptional cases, it may reach 2.0 × the square root of waterline length).

The net result of these physical laws governing wave making is that the maximum speed of a displacement boat is substantially determined by its waterline length: the longer the waterline length, the faster the boat will go. What is important here is not the *static* waterline length, but the *sailing* waterline length. This is particularly significant on boats with long and low overhangs. When the boat is at speed or heeled, the waterline length may increase substantially and the maximum speed will rise accordingly.

Only in the most exceptional circumstances, such as when a boat surfs down the face of a wave, can a displacement hull exceed its hull speed. The closer the boat gets to its hull speed, the greater the power required for a given increase in speed. At around 75% of hull speed, the boat is extremely efficient, but beyond this point, the additional fuel burned becomes increasingly disproportionate to any increase in speed due to the rapid rise in wave-making drag. In other words, ever-increasing amounts of power are required for smaller and smaller increases in speed.

In contrast, a planing hull breaks free of its own wave formation by moving up onto the surface of the water (see Figure 9-6). The moment at which this occurs is often felt as a sudden surge in speed as the boat accelerates away from its stern wave, barely skimming the surface of the water.

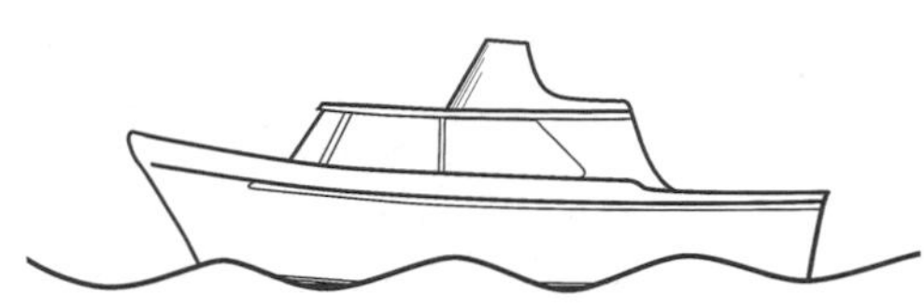

A 32-foot (waterline length) boat moving at 4 knots. There will be approximately $3\frac{1}{2}$ waves to its length.

The boat is now moving at, say, 12 knots. It has moved up onto the surface of the water and ahead of its own wave formation.

The same boat moving at $7\frac{1}{2}$ knots. There wil be approximately 1 wave to its length. This boat is moving at hull speed for displacement boats of that length.

Figure 9-6. *Planing hulls and wave formation. (Fritz Seegers)*

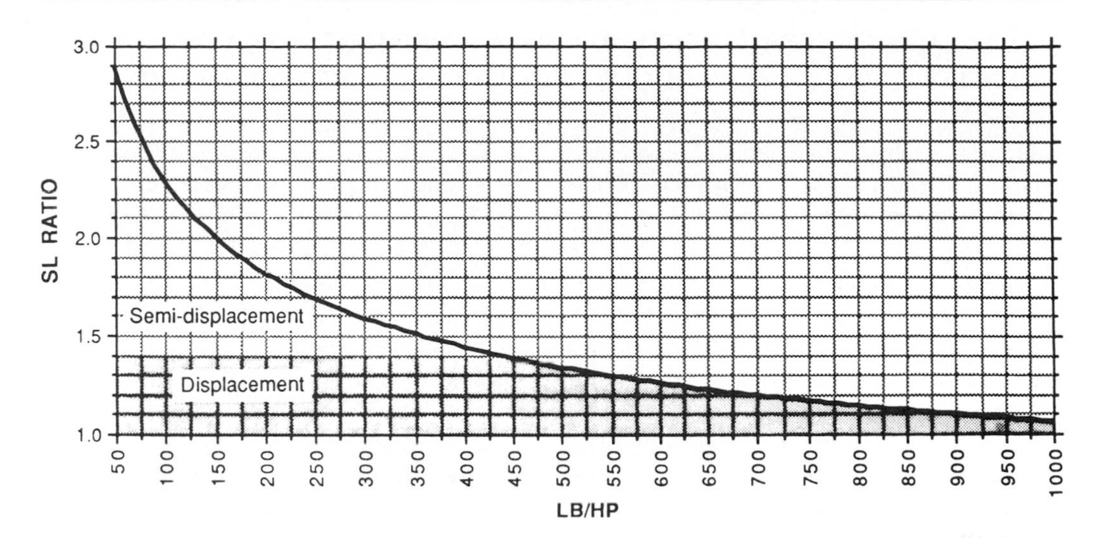

FIGURE 9-3. *This chart shows the power necessary to achieve a boat's known maximum speed-to-length ratio. It would be tempting to conclude from the chart that even a heavy-displacement hull can achieve S/L ratios of 1.5 or higher, given enough power, but in practice, such an attempt would not be feasible. For most moderate- to heavy-displacement vessels, incorporating more than 1 horsepower per 500 pounds or so of displacement in an effort to achieve S/L ratios higher than 1.3 to 1.4 is neither practical nor economical. Heavy hulls designed with planing or semiplaning underbodies may be driven to semidisplacement speeds, but only at a great cost in fuel consumption and power. (From Propeller Handbook by Dave Gerr; courtesy International Marine)*

in a 30,000-pound boat (i.e., 1 hp per 1,000 pounds) and always found it adequate, but then we had an efficient variable-pitch propeller. We were operating at the lower end of the power requirements for our boat.

BHP, SHP, and Auxiliary Equipment

The horsepower figures given in most engine manufacturers' specifications are normally the *brake horsepower* (bhp), which is measured before adding the transmission, any reduction gears, and the propeller shaft with its associated bearings and shaft seal. But the figures we have derived so far are those needed *at the propeller*, otherwise known as *shaft horsepower* (shp; see the Understanding Engine Curves sidebar for a fuller explanation of these terms). However, since power train losses are generally only 3% to 5% of the bhp, except in special circumstances, the difference between bhp and shp can be largely ignored.

The effect of belt-driven auxiliary equipment is often of more concern than power train losses. The DC loads on boats are steadily increasing from year to year as boatowners add more and more gadgets. To keep up with this burgeoning load, the tendency is to fit ever more powerful alternators—130 amp and 160 amp models are now common. At full load, these absorb up to 10 hp from an engine. (As a ballpark figure for small-frame alternators, which are typically found on boats, and taking into account all friction losses in the drive belt and other inefficiencies, you can assume 7 hp per 100 amps of alternator output with 12-volt alternators.) Engine-driven refrigeration compressors will make additional demands (generally up to 2 hp).

On large engines, the impact of such devices is not that great, but on an engine fitted to a small auxiliary sailboat, it may be considerable and will certainly need to be taken into account when sizing the engine. Every horsepower absorbed by auxiliary devices is a horsepower lost at the propeller shaft. Once an initial horsepower requirement has been calculated using either Skene or Dave Gerr's graphs, you will need to add to it an estimate for auxiliary equipment loads.

Finally, when comparing manufacturers' specifications, you must differentiate between an engine's horsepower rating in continuous duty versus intermittent duty. An intermittently rated engine is designed to be operated at full power for limited periods only. Auxiliary sailboats, which rarely use

Determining the Horsepower Requirements of an Auxiliary Sailboat Using the Formulas in *Skene's Elements of Yacht Design*

To determine the hp requirements of an auxiliary sailboat, start with the boat's waterline length (LWL)—let us assume 32 feet—and a speed/length ratio of 1.34. Hull speed is calculated as 1.34 × the square root of LWL = $1.34 \times \sqrt{32} = 7.58$ knots.

Enter the graph shown in Figure 9-7 on the bottom line at 1.34 and trace upward to the lower curve for light-displacement hulls and the upper curve for heavy-displacement hulls. Assume a heavy cruising boat of 26,000 pounds (note that this weight should include all stores normally on board). Using the upper curve, move horizontally to find the resistance in pounds for each long ton of displacement (a long ton = 2,240 pounds). For a speed/length ratio of 1.34, the resistance is 55 pounds per long ton; 26,000 pounds = 11.6 long tons, therefore total resistance at hull speed (7.58 knots) for this hull is 11.6 × 55 = 638 pounds.

Effective horsepower (ehp—the power converted to useful work), which assumes an engine and propeller are 100% efficient at transmitting the engine's power, is given by the formula: ehp = resistance × speed × 0.003, so for our example 638 × 7.58 × 0.003 = 14.5 hp.

In practice, propellers are notoriously inefficient at transmitting power. Kinney uses the following factors: folding two-blade, 10%; auxiliary two-blade, 35% to 45%; fixed three-blade, 50%.

Let us assume an average auxiliary two-bladed propeller with a 40% rating. We arrive at the following shaft horsepower (shp) to drive our boat at hull speed: 14.5/0.40 = 36.25 hp.

Kinney also adds 33% for losses in the power train and adverse conditions, to give a maximum power requirement (brake horsepower—bhp), which in this example is 48 hp (this is 542 pounds per horsepower, which is pretty much the same as the figure derived from Dave Gerr's chart using a speed/length ratio of 1.34). Note that if we had used a speed/length ratio of 1.00 with this boat, giving a top speed of 5.7 knots, the horsepower requirement would have only worked out to be 9.5. Put another way, this hull can be pushed at 5.7 knots by 9.5 hp, but will need 36 or more hp to move at 7.58 knots. This dramatically illustrates the increase in drag, and therefore fuel consumption, as hull speed is approached.

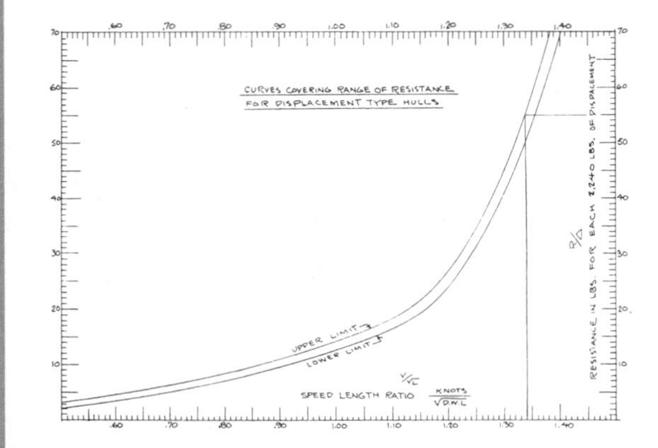

FIGURE 9-7. *This graph illustrates the dramatic rise in resistance experienced by displacement boats as they approach hull speed. (From Francis S. Kinney's Skene's Elements of Yacht Design, courtesy Dodd, Mead)*

their engines at full power for prolonged periods, can use an intermittent-duty rating for choosing a suitably sized engine. Many other boats, however (e.g., oceangoing motorsailers or sportfishing boats), will need to have an engine based on its continuous-duty rating.

SECTION TWO: PROPELLER SIZING AND SELECTION

There is a complex relationship between the available clearance under a hull for a propeller (and thus its maximum diameter—in general, the larger the propeller, the more efficient it is), the pitch of the propeller (see below), and the speed at which it turns (and thus the necessary reduction gear between it and the engine driving it—in general, the slower it turns, the more efficient it is).

Because of the complexities involved, propeller sizing is both a science and an art; that is, after the science has been completed, a certain amount of intuition needs to be factored into the equation. I found this out the hard way. On our old boat, when we re-engined and re-propped, the propeller manufacturer made the calculations and came up with a propeller size, which I ran past several experts for a second opinion. A couple of

Understanding Engine Curves (see Figure 9-8)

Brake horsepower (bhp) is the maximum horsepower delivered by the engine. It is measured by connecting the engine to a braking device, running the engine up to full speed, then progressively loading the brake in order to slow down and eventually stall the engine. The brake load is plotted against the engine speed to get the bhp curve. This test is conducted without the addition of alternators or any other auxiliary equipment.

Shaft horsepower (shp) is the power transmitted to a propeller shaft as found at the output coupling on the transmission. In other words, it accounts for the energy losses in the transmission (generally 3% to 5%) but does not include any additional energy losses in the propeller shaft and its associated bearings and shaft seal (which, if it is a long shaft with two or more bearings, may be at least as high as the transmission loss). Once again, it does not include alternator and other auxiliary loads.

Torque is the turning force transmitted to the propeller. It is measured in much the same way as bhp, resulting in the plotting of a torque curve as a function of engine speed. This curve is much flatter than a horsepower curve, and in fact it generally tapers off at around two-thirds of top engine speed. This does not mean the engine is losing power above this speed, because power is a function of torque × speed. As long as the engine speed increases faster than the torque curve tapers off, there will be a net increase in power, although the power curve will taper off (which is why the top ends of the bhp and shp curves invariably taper off).

The *propeller power curve* typically measures the amount of power absorbed by a fixed-pitch, three-blade propeller at different speeds of rotation. Whereas engine horsepower curves are convex in shape (curve upward in the center), propeller power curves are concave (curve downward in the center). The net result is that the two curves can only be matched at one engine speed. This is generally done at or around maximum engine speed. If a lower speed is chosen, the engine will be overloaded at full speed; if a higher speed is chosen, the engine will never be fully loaded.

At anything other than full power, the propeller will be absorbing considerably less power than the engine is capable of producing. This is unavoidable. This does not mean, however, that the engine is especially inefficient in terms of its fuel consumption. The engine's governor will simply cut back the fuel flow to whatever is necessary to produce the horsepower that the propeller absorbs at a given speed of rotation.

The *specific fuel consumption curve* measures the amount of fuel consumed per horsepower produced at the propeller. It is a measure of the efficiency of the installation. Generally speaking, a marine diesel is most efficient if operated at the speed that more or less corresponds to the top of its torque curve and is increasingly inefficient the slower it is run.

(continued)

Understanding Engine Curves (continued)

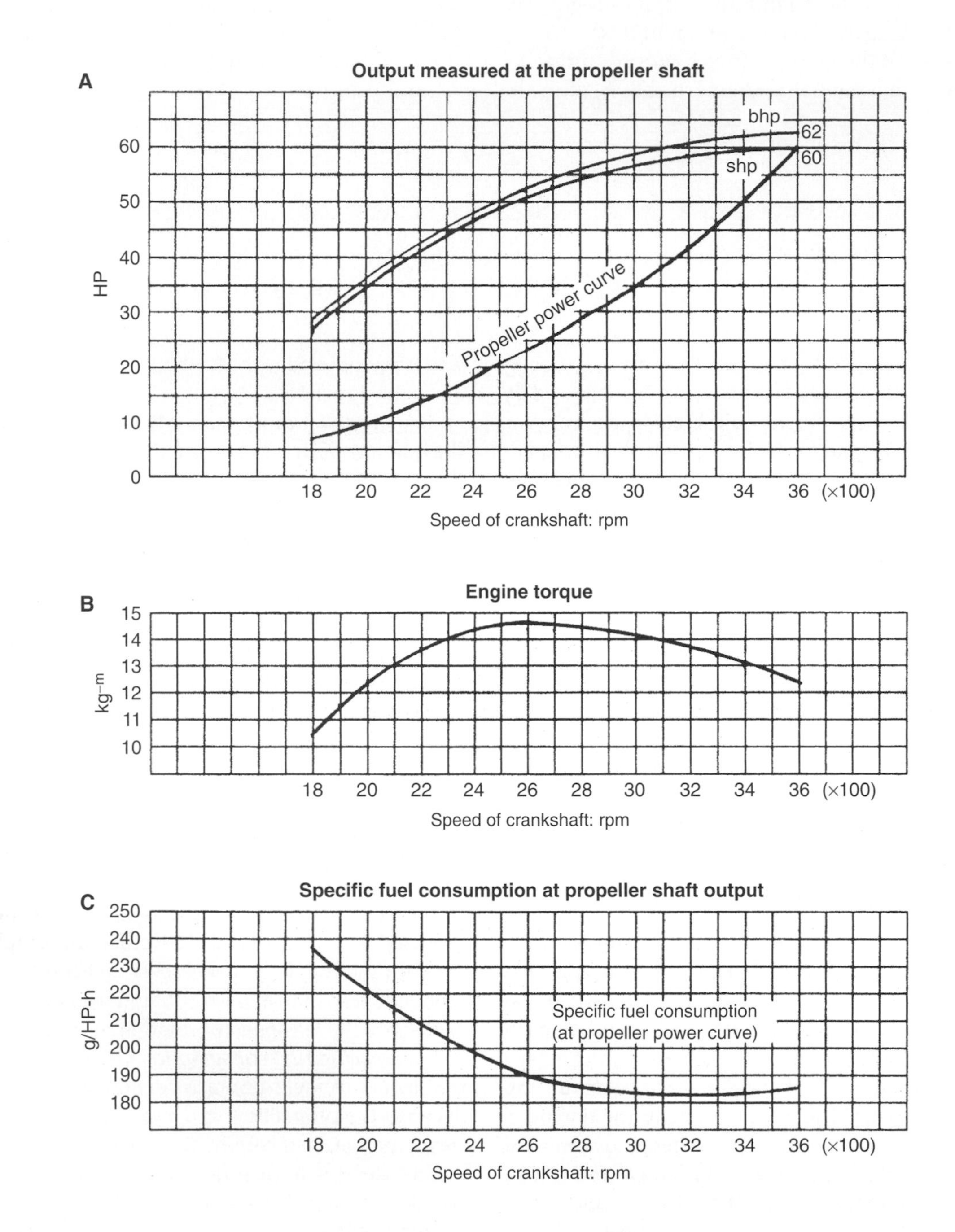

FIGURE 9-8. *Typical engine curves. (Courtesy Yanmar)*

them said it looked fine; a couple said they felt we would be over-propped. After we were all done, we duly found out that we were, indeed, somewhat over-propped, so the prop had to come off and be reworked.

Propeller Sizing

As a result of this experience, I made some calculations myself using charts, formulas, and tables developed by Dave Gerr and published in both *Propeller Handbook* and *The Nature of Boats*. These indicated that our propeller was oversized, so maybe I should have worked the numbers in the first place! The exercise is worth printing (this time, using our last boat, *Nada*, a Pacific Seacraft 40), since it demonstrates a process for anyone interested in calculating a ballpark figure for a propeller for his or her own boat, and it may help in understanding problems with an existing propeller. However, you should always bear in mind that propeller sizing is quite complex, so also seek expert advice.

Step 1: Pitch and Slip

We first need to get a handle on a couple of technical terms: pitch and slip. To understand them, we can compare a propeller to a wood screw (although this comparison is not technically accurate). The propeller can be considered to slice its way through the water much as a screw penetrates wood. In theory, the more extreme the angle at which the blades cut the water, the farther the propeller will move at each complete revolution. This theoretical distance is known as *pitch*. If a propeller were 100% effective and if it had, for example, a 12-inch pitch, it would move its boat forward 12 inches for each revolution it made. Of course, it doesn't do this. The amount by which it fails to achieve this theoretical result is known as *slip*. If the boat moves forward 9 inches, there is 25% slip.

Nada is a moderately heavy-displacement (24,280 pounds) cutter with a hull speed of around 7.3 knots. One knot equals 101.3 feet per minute, so at 7.3 knots, *Nada* travels 739.5 feet per minute (101.3 × 7.3 = 739.5 feet). The first thing we want to know is what pitch propeller is needed to produce this speed.

We chose a Yanmar with a maximum rated speed of 3,600 rpm and a reduction gear ratio of 2.62:1 (3.06:1 in reverse). At full engine speed, the shaft rotates at 3,600/2.62 = 1,374 rpm. We could use this number to determine pitch, but to provide a bit of a cushion for adverse conditions, we want the boat to reach rated speed at 95% of rated engine speed; that is, at a shaft speed of 1,374 × 0.95 = 1,305 rpm. (If you look at Figure 9-8, you will see that a typical marine diesel engine is up to its maximum rated hp at around 95% of its maximum speed.)

Without slip, if the propeller is turning at 1,305 rpm, a pitch of 739.5/1,305 = 0.56 feet = 6.8 inches will produce a boat speed of 7.3 knots. So far, so good, but now we get to the tricky part. To what extent will *Nada*'s moderately heavy displacement induce slip? There are various tables that provide a guide, but at the end of the day there is no substitute for experience when making this judgment call. Better still is a direct comparison with the performance of sister ships.

Based on my reading of several textbooks, 40% to 45% seemed a reasonable amount to factor in for slip, which means the propeller is assumed to be 55% to 60% effective at full speed. Let's split the difference at 57.5%. The theoretical 6.8-inch pitch was increased to a real-world pitch of 6.8/0.575 = 11.8 inches (rounded up to 12 inches).

Step 2: What Diameter Should This Propeller Be?

Now we need to determine the diameter of the propeller. Larger-diameter propellers are more efficient than smaller-diameter ones, but obviously there is a limiting factor in terms of the size of aperture available—the clearance beneath the hull. *Nada* has a 23-inch aperture. It is desirable to maintain 10% to 15% of this (i.e., 2.3 to 3.45 inches) as clearance between the tips of the propeller blades and the boat to reduce turbulence (see Figure 9-9). Given that *Nada*'s propeller is mounted in an aperture, this clearance is needed both at the top and bottom of the aperture, leaving us with a maximum propeller diameter of 23 inches – 4.6 inches = 18.4 inches. Would this size propeller be too big for the engine to handle (over-propped), or too small to absorb the engine's full output (under-propped)?

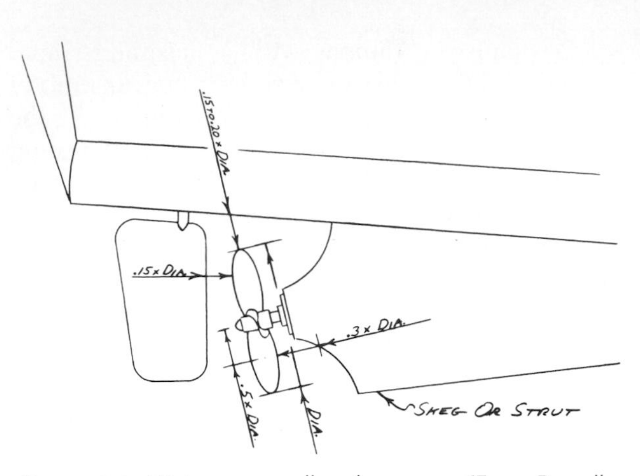

FIGURE 9-9. *Minimum propeller clearances. (From Propeller Handbook by Dave Gerr, courtesy International Marine)*

Dave Gerr has developed a formula and a nomograph that show the relationship among maximum shaft speed, shaft horsepower, and propeller diameter. The formula is:

$$D = 632.7 \times shp^{0.2}/rpm^{0.6}$$

where D = propeller diameter, shp = shaft hp at the propeller, and rpm = maximum shaft rpm at the propeller.

This is a bit complicated! The nomograph is much easier to use. See page 406 of *The Nature of Boats* and also Figure 9-10.

We already know our maximum shaft speed (1,374 rpm). The engine is rated at 50 hp, but with the loss in the transmission (approximately 5%), we have a maximum shp rating of 50 × 0.95 = 47.5 hp. Our high-output alternator frequently ran at near full output, absorbing up to 7 hp, so I decided to average its draw at 3.5 hp. Taking these factors into consideration, I chose to downgrade the shp rating to 44 hp as the basis for determining propeller diameter.

If we enter Dave's nomograph with a shaft speed of 1,374 rpm and a shp of 44, it tells us that we should use a 17.7-inch-diameter propeller; 18 inches is close enough (this is not rocket science), so we end up with an 18-inch-diameter × a 12-inch-pitch propeller.

If the recommended diameter is too large for the aperture, a lower gear ratio is needed on the transmission (typically, engine manufacturers offer a choice of two or three different transmission ratios). This raises the shaft speed, which results in a smaller-diameter propeller. If the recommended propeller is considerably smaller than the aperture will accommodate, it is desirable to use a higher gear ratio; this lowers the shaft speed and results in a larger-diameter (more efficient) propeller.

STEP 3: CALCULATE THE BLADE AREA

Finally, we need to calculate the blade area, which determines *blade loading*—the amount of thrust developed by each square inch of the blades. The higher the engine output and the lower the blade area, the higher the blade loading. Above a certain point, this blade loading results in cavitation, which in turn may damage the propeller. The forward face of a propeller operates something like an airplane wing, reducing pressure at its surface. The greater the blade loading, the greater the reduction in pressure. When you reduce pressure, you lower the boiling point of water. With a sufficient reduction in pressure, water boils off on the face of the propeller, and the steam bubbles implode, eroding the surface of the metal and resulting in cavitation erosion. In other words, blade loading must be kept below a certain level.

Calculations of blade loading can get quite complex, but luckily Dave has developed another formula and a graph that enable fair estimates to be made quite easily. The formula determines the blade area needed to absorb the energy produced by the engine:

$$\text{blade area} = 100 \times \text{shp}/\text{hull speed} \times \sqrt{\text{hull speed}}$$

Looking at *Nada*, we have the following data: shp = 44, top speed = 7.3 knots, and $\sqrt{\text{top speed}}$ = 2.7 knots. Therefore,

$$\text{blade area} = 100 \times 44/7.3 \times 2.7 = 223 \text{ square inches}$$

Looking at Dave's graph of blade area against propeller diameter (see left illustration on page 246 of *The Nature of Boats* and also Figure 9-11), we see that *Nada* falls between a three- and a four-blade propeller. Three-blade propellers are widely available and relatively cheap, so this is what we used. We installed the propeller (18-inch diameter × 12-inch pitch) knowing that at full engine speed there was likely to be a certain amount of cavitation.

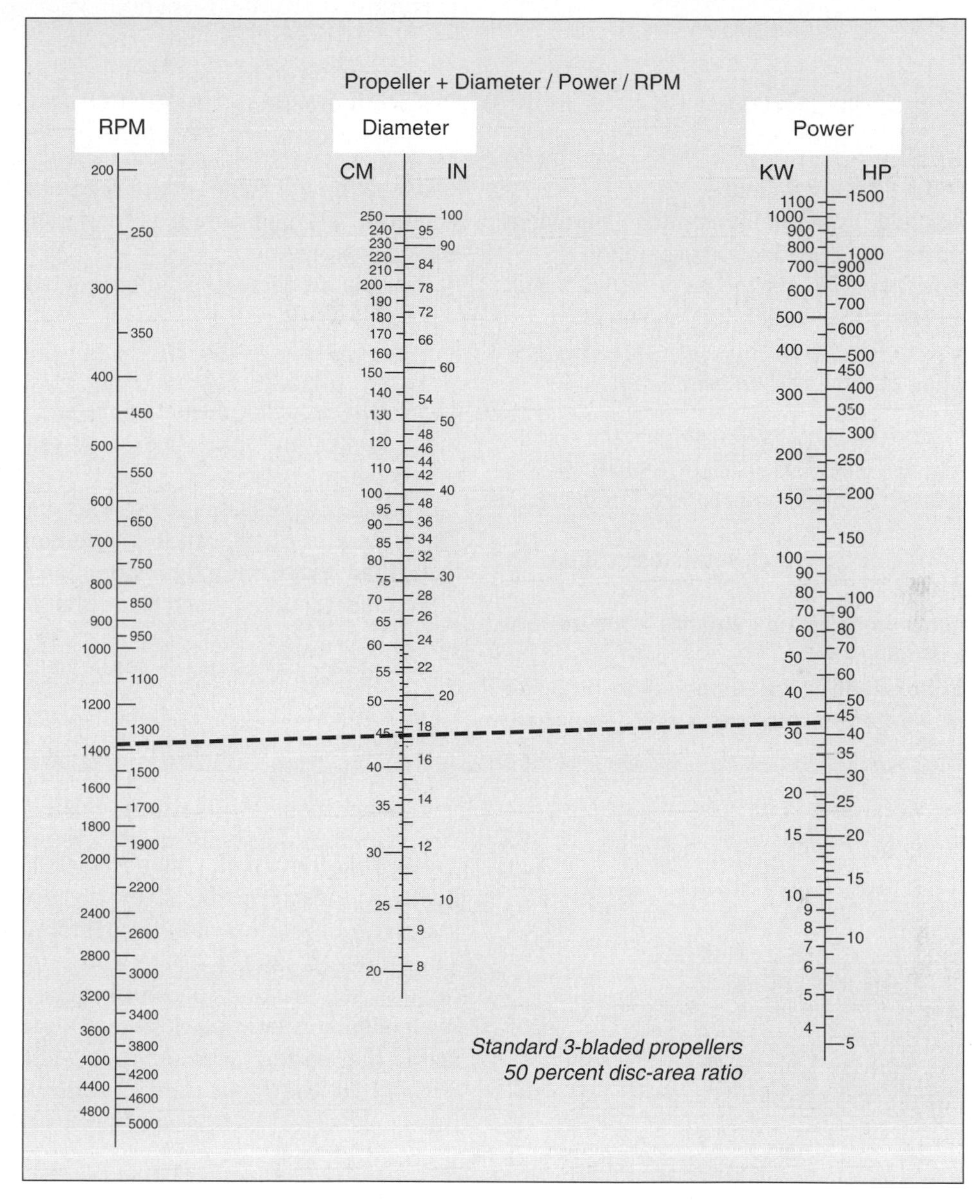

FIGURE 9-10. *Nomograph indicating shaft speed, propeller diameter, and engine shaft horsepower. (Illustration by Kim Downing © SAIL Publications from data supplied by Dave Gerr)*

This was not a problem in normal use since the engine was rarely used above 80% of rated rpm.

PROPELLER SELECTION

The past couple of decades have seen the introduction of a wide range of new propellers, including feathering, folding, and controllable pitch, providing a wide range of choices with different trade-offs for each.

STANDARD PROPELLERS

For sailboats, standard propellers have a fixed pitch. The blade shape is optimized to produce maximum thrust in forward at or close to full engine speed. At all other speeds, the propeller is increasingly inefficient. Sometimes the propeller is deliberately optimized for

Useful Numbers and Formulas for Sizing Propellers

1 knot = 101.3 feet per minute

Shaft speed for determining pitch = maximum engine speed × 0.95/reduction gear ratio

Pitch without slip (in inches) = 101.3 × hull speed × 12/shaft speed for determining pitch

Typical auxiliary sailboat propeller effectiveness as a fraction of pitch without slip:

- three-blade propeller—0.5 to 0.6
- fixed two-blade propeller—0.35 to 0.45
- folding two-blade propeller—0.1 to 0.35

Real-world pitch = pitch without slip/propeller effectiveness

Minimum propeller tip clearance = 10% to 15% of propeller diameter

Shaft horsepower for determining propeller diameter = rated horsepower × 0.95 – auxiliary loads that do not come with the engine (the most significant are high-output alternators and refrigeration compressors)

Formula for determining propeller diameter:

$$D = 632.7 \times shp^{0.2}/rpm^{0.6}$$

where D = propeller diameter, shp = shaft hp at the propeller, and rpm = maximum shaft rpm at the propeller.

Note: It is much easier to use the nomograph in *The Nature of Boats*, which is based on this formula, than it is to make this calculation.

Formula for determining blade area:

$$\text{blade area} = 100 \times \text{shp}/\text{hull speed} \times \sqrt{\text{hull speed}}$$

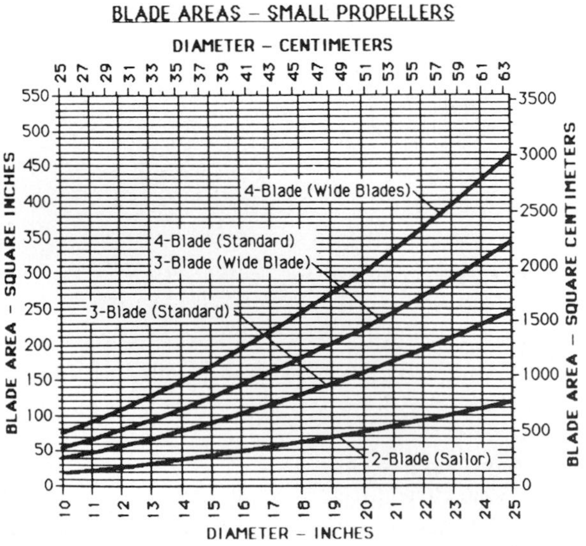

Figure 9-11. *Propeller blade areas as a function of diameter. (From Propeller Handbook by Dave Gerr, courtesy International Marine)*

a lower engine speed, which is logical on many sailboats as most cruising is by preference done at about three-quarters of the maximum engine speed. The trade-off here is that if an attempt is made to drive the engine to its maximum speed, it may overheat, will likely emit black smoke, and may get damaged; the engine warranty may also be voided. Given that the blades are optimized for forward thrust, the shape is especially inefficient in reverse.

Folding Propellers

Folding propellers have two significant benefits over other propeller types. They have the least amount of drag when under sail, and the least likelihood of any propeller type on the market of snagging seaweed, fishing lines, lobster-pot warps, and other debris.

The old style has two hinged blades. When sailing, the blades are held in the closed position by the flow of water over the propeller. When the engine is cranked and put in forward, the centrifugal force opens the blades partway, and then the developed thrust drives them all the way open, sometimes with considerable force. The net result is a fixed-pitch propeller. In reverse, any thrust developed tends to close the blades. This tendency is counteracted by centrifugal force, but efficiency is often minimal.

The most common problem with this type of propeller is a failure to open at all because of weeds or barnacles in the hinge. Partial opening of one or another blade will result in severe vibration. These propellers must be kept clean. Thrust in reverse is minimal, and even in forward the propellers are typically very inefficient (often converting considerably less than 50% of the available shaft horsepower into thrust).

The new style of folding propellers—e.g., Flex-O-Fold (www.flexofold.com), Gori (www.gori-propeller.com), Brunton's Varifold (www.bruntons-propellers.com or www.varipropusa.com), and Volvo (www.volvopenta.com)—have a gear on the base of each propeller blade that is engaged with the other blade(s). This ensures that the blades open and close in tandem (see Figures 9-12A to 9-12F). The design is such that these propellers open better and are far more efficient in both forward and reverse

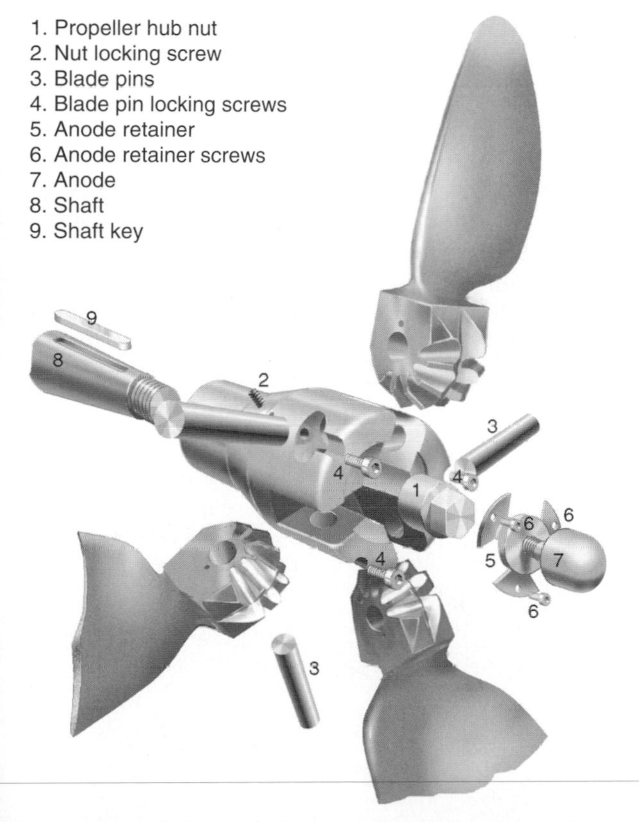

FIGURE 9-12A. *A Varifold folding propeller. (Brunton's Propellers)*

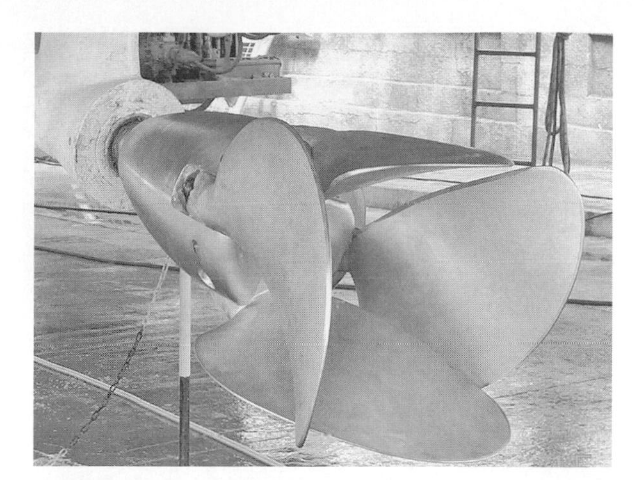

FIGURE 9-12B. *A Varifold propeller in the folded position.*

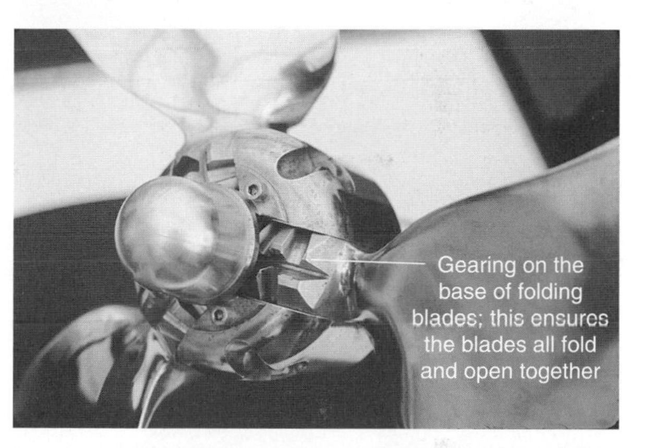

FIGURE 9-12C. *The geared hub of a contemporary folding propeller.*

than traditional folding propellers. In fact they are as efficient as any other propeller on the market and more efficient than many. (In tests conducted at the University of Berlin in Germany, the Flex-O-Fold was as efficient in forward gear as the best fixed-blade propellers and substantially more efficient than many feathering propellers). The blades open to a preset position and as such have a fixed pitch. They are especially suitable for high-speed multihulls. Some manufacturers offer these props in three- and four-blade versions and very large diameters for big boats and high-horsepower engines.

FEATHERING PROPELLERS

As with the new-style folding propellers, a feathering propeller has a gear on the base of each propeller blade, but in this case it engages a central gear mounted on the propeller shaft. The whole unit is

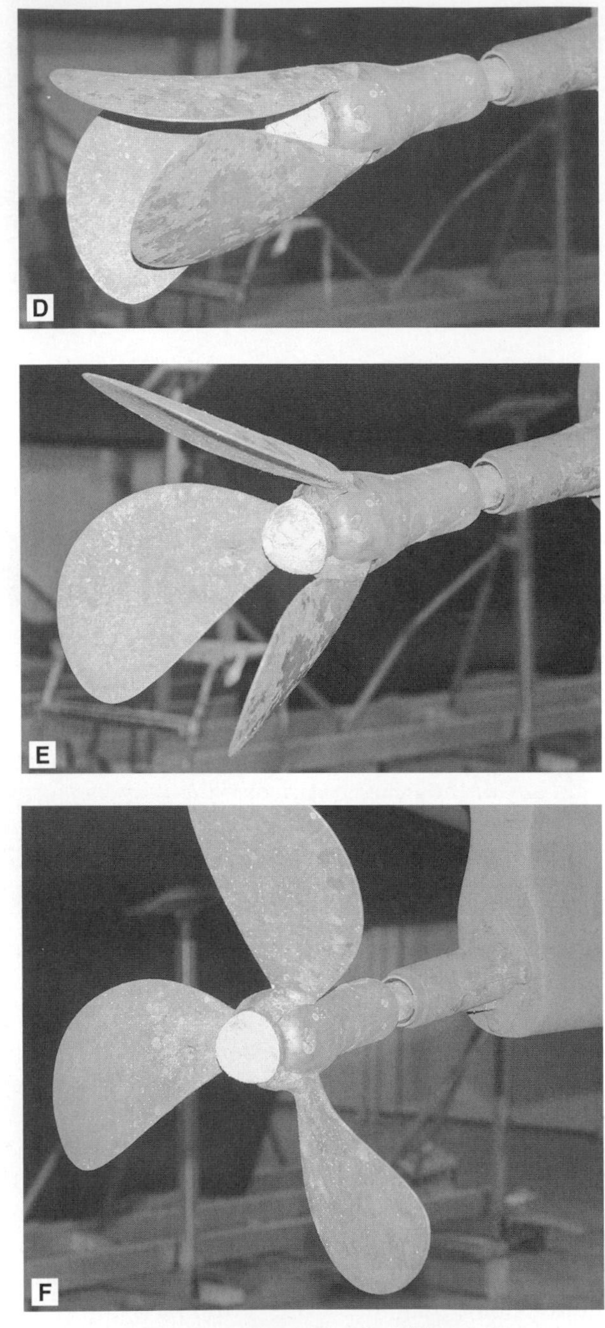

FIGURES 9-12D, 9-12E, and 9-12F. *Flex-O-Fold propellers.*

enclosed in a case (see Figure 9-13; for examples, go to www.pyiinc.com, www.martec-props.com, www. bomon. com, www.varipropusa.com, and www. peluke. com).

When the engine is cranked and put in gear, the propeller shaft and the gear mounted on it turn while the case and blades tend to lag due to inertia. The blades are in a feathered position. The rotating gear on the propeller shaft turns the propeller blades inside the case. The blades have a preset stop, so that they cannot rotate beyond a certain point. This preset stop determines the pitch of the blades. Once it is reached, the propeller shaft spins the whole unit including the case. When the engine is shut down, water pressure on the blades forces them back to the feathered position.

When put in reverse, the propeller shaft once again drives the blades to their full pitch before spinning the whole unit. And unlike a standard fixed-pitch propeller, a feathering propeller has the same efficiency in reverse as in forward. This is because the blade faces are reversed; as a result, considerably more thrust is developed than by a standard propeller (note, however, that some feathering propellers have flat blades with little camber; these are significantly less efficient in forward than most standard propellers and new-style folding propellers).

On many older models, the blade pitch can be adjusted by altering the stops, but to do this, the case must be disassembled and on some propellers (notably Luke and Prowell) the stops have to be machined down or built up by welding in additional material. On others, the stops are adjustable mechanically. Newer feathering propellers tend to have an external pitch-adjustment mechanism (the pitch can be changed with no disassembly). A couple of brands offer independently adjustable forward and reverse pitches, which is very useful because prop walk (the "paddle-wheel effect") in reverse can be greatly reduced or even eliminated by flattening the reverse pitch without compromising the forward pitch. Note that in all cases these pitch-adjustment capabilities do not in any sense equate with a controllable-pitch propeller (see below). It is simply a method for fine-tuning the pitch(es) in the event the propeller was not properly matched to the boat and engine in the first place—or as the boat gains weight and the engine loses power over the years!

Maintenance is a matter of checking sacrificial zinc anodes regularly (see Figure 9-14) and renewing the grease every one or two years. To renew the grease some units must be taken apart (e.g., older Max-Props), whereas on others a grease fitting is simply screwed into the case (e.g., Luke, Variprop). In the former case, make a careful note of where everything goes and put it all back the same way. A waterproof bearing grease such as Lubriplate Marine-Lube "A"

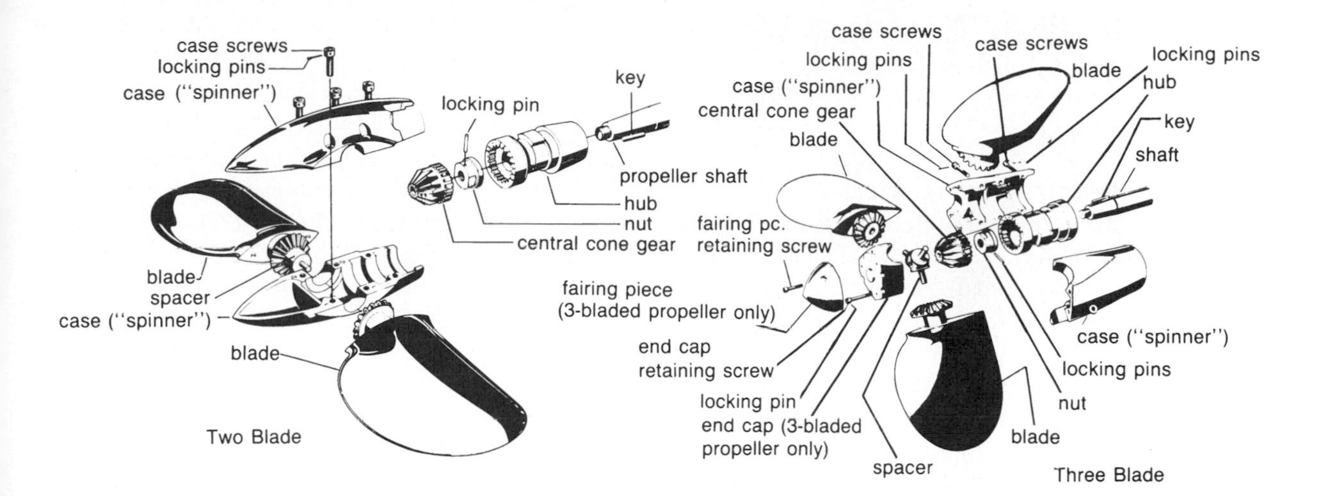

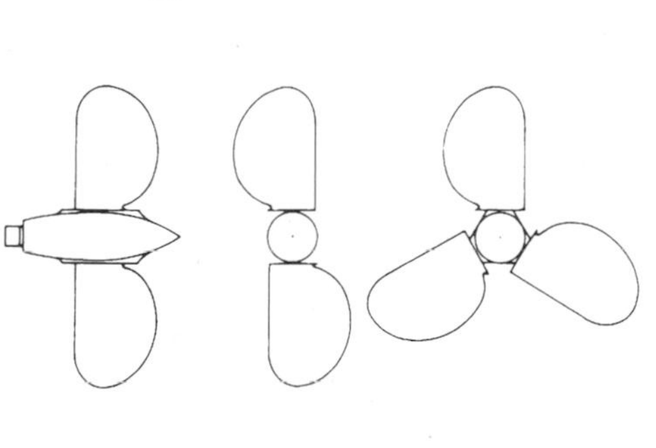

FIGURE 9-13. ***Top:*** *Fully feathering propellers.* ***Bottom:*** *The operation of Max-Prop propellers. Under sail, the Max-Prop feathers to a low drag shape. In forward, the torque of the prop shaft acts to force the blades open to a preset pitch at any throttle setting. In reverse, as in forward, the torque of the shaft will rotate the blades 180 degrees, presenting the same leading edge and pitch in reverse. (Courtesy Max-Prop)*

works well (lighter Teflon greases will be washed away). Some manufacturers insist that you use their own proprietary grease; take heed and use it—there is a good reason.

If the propeller blades are stiff to rotate by hand, or have hard spots, it may simply be the result of too much grease, so take a little out. If that doesn't work, check for small burrs on the gears or minor damage on any of the bearing surfaces (the propeller blade bases, bearing surfaces in the case, and the hub). Remove the cone gear and reassemble the unit. Now check each individual blade for stiffness and test the rotation of the case around the hub; you should be able to isolate problem areas. Dressing up with emery cloth or a fine file will solve most problems.

FIGURE 9-14. *This zinc is past its replacement date! Note that galvanic interaction between the stainless steel fasteners and the zinc has caused the zinc to become loose. This will interfere with the electrical connection (metal to metal) between the zinc and the propeller, rendering the zinc ineffective regardless of how much of it is left.*

Be sure to replace the zinc anode tailcones when they are about two-thirds gone—you are protecting a lot of very expensive machinery inside these props!

Controllable-Pitch Propellers

A traditional controllable-pitch propeller has the propeller hub bolted to the boat with a mechanism to move the propeller shaft forward and aft. Moving the shaft rotates the propeller blades in the hub, changing their pitch. Propeller pitch can be matched to engine speed to maximize the thrust and efficiency over the full speed range as opposed to at a single point on the engine's speed/power curve. Depending on the model, it is sometimes possible to align the propeller blades with the water flow—i.e., fully feather the blades—when sailing. This kind of propeller is rarely found on recreational boats.

Brunton Autoprop

The Brunton Autoprop is a feathering propeller that uses some very clever engineering to perform these pitch-adjusting functions automatically (see Figure 9-15). The blades are designed and weighted so they increase the pitch at lower engine speeds, resulting in increased speed for a given engine speed and lower fuel consumption. The blades also reverse when the engine is put in reverse, providing the same thrust in reverse as in forward. Vibration can be a more common problem than with the best modern feathering or folding propellers. Under sail, the water flow past the boat causes the blades to feather. It should be noted that the manner in which the blades are weighted results in the lower one hanging down somewhat, causing more drag than with a fully folding or feathering propeller (see Figure 9-16).

The Autoprop is probably best suited for motor-sailer applications or for boats that make long and frequent passages under power. Although it does work as advertised, it also takes a little getting used to, especially when going into reverse. Initially, nothing seems to happen, then suddenly, the propeller kicks in with considerably more thrust than you get from a typical propeller.

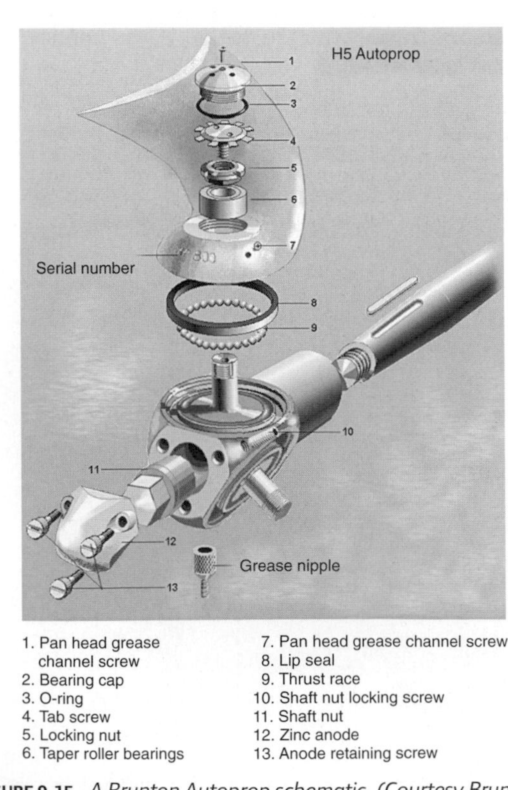

Figure 9-15. *A Brunton Autoprop schematic. (Courtesy Brunton's Propellers)*

Figure 9-16. ***Top:*** *What an Autoprop looks like in practice.* ***Bottom:*** *The way the blades are weighted causes one to hang down when under sail, creating a little more resistance than a feathering propeller. (Courtesy Brunton's Propellers)*

SECTION THREE: CONNECTING THE TRANSMISSION TO THE PROPELLER

Note that this section does not apply to inboard/outboards, saildrives, or the Volvo Penta IPS system.

COUPLINGS

Couplings should always be keyed to their shafts and then pinned or through-bolted so that they cannot slip off (see Figure 9-17). The practice of locking a coupling with setscrews is not very seaworthy. If a coupling is held this way, be sure the setscrews seat in good-sized dimples in the shaft, and preferably are tapped into the shaft a thread or two so that there is no risk of slipping. *Should they slip, the propeller and shaft are liable to pull straight out of the boat in reverse, leaving the ocean pouring in through the open shaft hole!* (I hear of one or two instances of this every year.) Just for insurance, it is a good idea to place a stainless steel hose clamp (Jubilee clip) around the shaft in front of the shaft seal; if the coupling should ever work loose, the hose clamp will stop the shaft from leaving the boat. (With high-speed shafts, use either two hose clamps with the screws on opposite sides of the shaft to provide some balance, a specially made collar bolted around the shaft, or a propeller shaft zinc).

Of course, if you have to remove a coupling (e.g., for shaft removal), you'll probably find that it's frozen

FIGURE 9-17. *Shaft couplings should be keyed (as shown here), although many are not, and preferably through-bolted, as opposed to being held by a setscrew, as in this case. At least this setscrew is lock-wired in place so it cannot vibrate loose.*

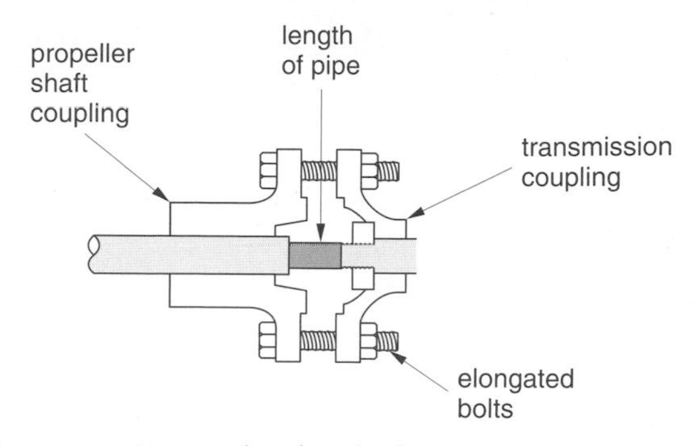

FIGURE 9-18. *Using a short length of pipe to remove a coupling from its propeller shaft.*

in place. You can remove a stubborn coupling by making a coupling puller as in Figure 8-30 or by: (1) separating the coupling halves, (2) inserting a length of pipe on the end of the shaft (it must have a diameter less than the size of the hole in the coupling), and (3) pulling the coupling halves together again with elongated bolts (see Figures 9-18, 9-19, and 9-20). Make sure to pull up the bolts evenly to avoid

FIGURES 9-19 and 9-20. *In this case, an appropriately sized socket is being used as a spacer to remove a stubborn coupling.*

distorting the coupling halves. Heating the coupling with a blowtorch may help loosen it, but given the awkward access to many couplings, you must be careful not to wave the blowtorch around and accidentally scorch surrounding surfaces. If the coupling won't come off, you'll have to cut the propeller shaft and replace both the shaft and coupling (you may be able to salvage the coupling by using a bench press to drive out the shaft).

ENGINE ALIGNMENT

Most engines use conventional couplings, either solid or flexible, between the transmission and the propeller shaft. Accurate alignment is critical to smooth, vibration-free running and a long life for the transmission bearings, transmission oil seals, and the stern (Cutless) bearing. Engine alignment should be checked periodically (it is traditional to recommend once a year, but in practice modern hulls are stable enough that in most cases this can be extended to every couple of years). When checking the alignment, the boat must always be in the water (the hull may well have a different shape there than on land; it is preferable to wait a few days to a week after launching a boat to give the hull time to settle, especially a wooden hull). However, alignment cannot be checked unless the propeller shaft is straight and the two coupling halves are exactly centered on and square to their shafts. Therefore, *a coupling should be fitted to its shaft and, if necessary, machined to a true fit in a lathe before putting the shaft in the boat.* The flanges on the coupling halves should also be exactly the same diameter.

FLEXIBLE FEET AND COUPLINGS

Almost all engines today are mounted with flexible feet and couplings. *They are not substitutes for accurate engine alignment.* Modern, lightweight hulls tend to flex in a seaway, whereas engines are necessarily extremely rigid. The principal reason for flexible feet is to absorb hull movements and lessen hull-transmitted engine vibrations, not to compensate for inadequate alignment. Flexible feet slowly compress over time, while the rubber is subject to softening, especially if it becomes soaked with oil or diesel fuel. Inspect the feet annually and replace them if deteriorated (Yanmar recommends every two years, although I know no one who does this—the feet are expensive!). Always check the feet and replace them if necessary before checking the alignment.

CHECKING AND ADJUSTING ENGINE ALIGNMENT

To check alignment, undo the coupling bolts and separate the coupling halves (first mark the two halves so you can put them back in the same relationship). If there is a flexible insert, remove it. In cases where a long run of propeller shaft is unsupported by a bearing, the shaft will sag under its own weight and the weight of its coupling half. In this case, the correct procedure is to calculate half the weight of the protruding shaft, add to it the weight of the coupling, and then pull up on the shaft by this amount with a spring scale (like those used for weighing fish; see Figure 9-21). In practice, smaller shafts can generally be flexed up and down by hand to get a pretty good idea of the centerpoint, then supported with an appropriately sized block of wood. A notch in the wood will hold the shaft and allow it to be rotated.

Preliminary Check

There should be a machined step on one coupling half that fits closely into a recess on the other. Bring the two halves together—the step should slip into the recess cleanly and without snagging at any point. If it does not, the shafts are seriously misaligned, and the engine must be moved

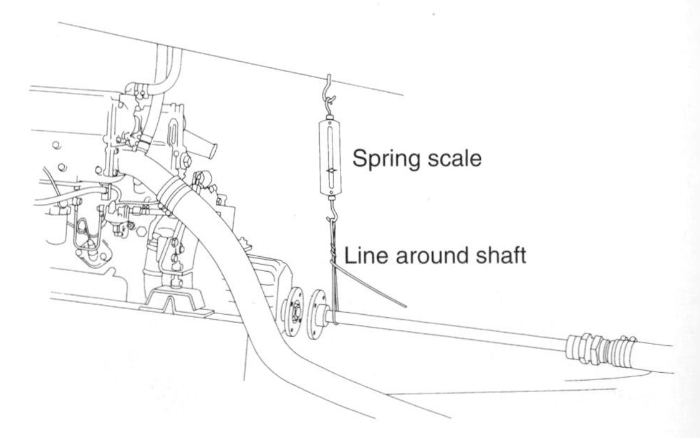

FIGURE 9-21. *Using a spring scale to eliminate shaft droop during engine alignment. The scale is tensioned until it reads one-half the free-hanging weight of the shaft plus the weight of the coupling. (Jim Sollers)*

around until the two halves come together (see Figures 9-22A and 9-22B). When the halves come together, the circumferences of the couplings should be flush with one another (see Figure 9-22C); if one has a larger diameter than the other, this eliminates a couple of useful checks (see below) but is not insuperable.

Assuming the couplings come together cleanly, bring the faces into contact and then try to insert a feeler (thickness) gauge into the gap between the

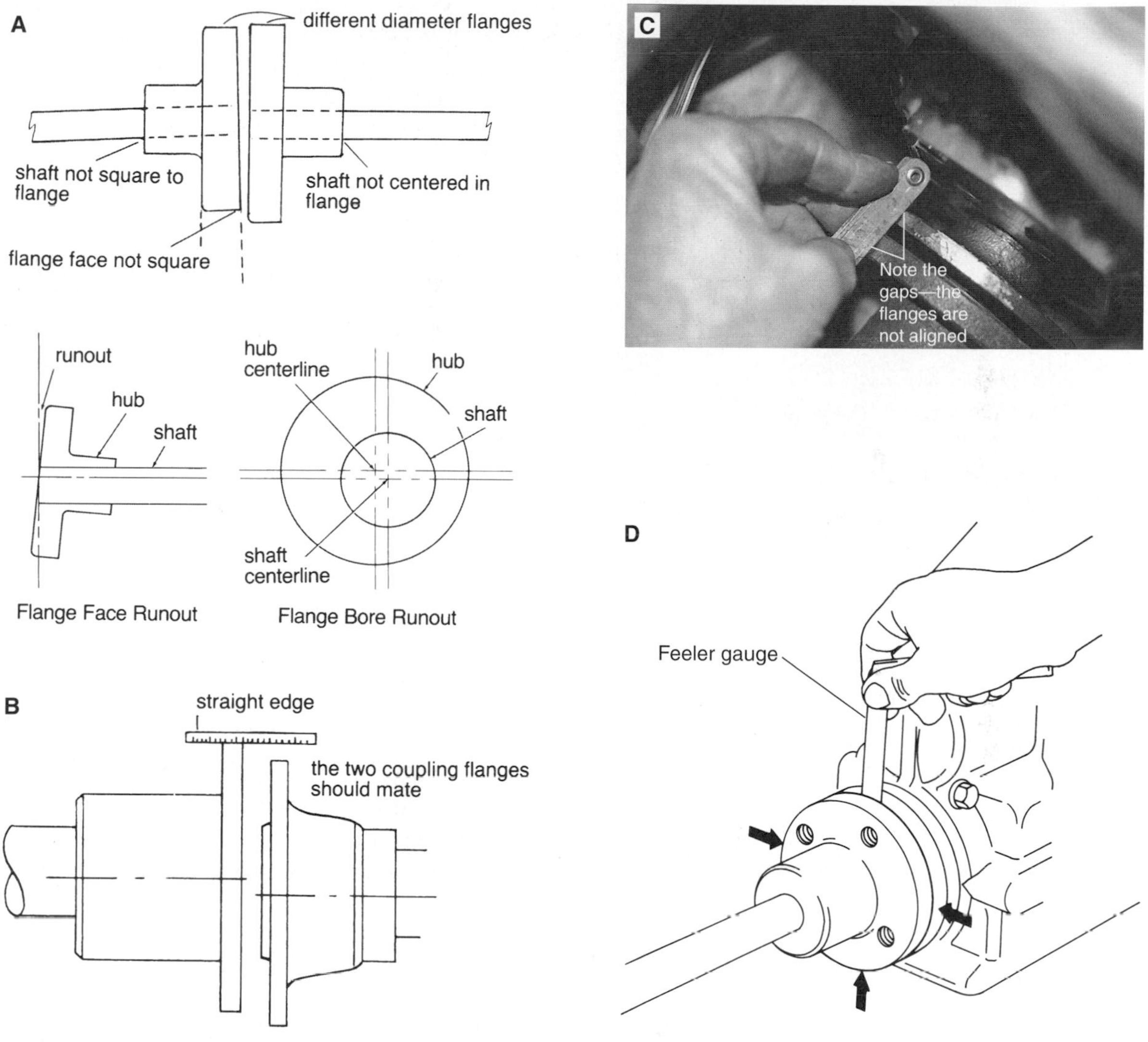

FIGURES 9-22A to 9-22H. *The shaft and coupling problems shown here (A) will make accurate engine alignment impossible. For effective alignment, both shafts must be square to their couplings; the coupling faces must be square; and the coupling diameters must be exactly the same. Bore alignment (B, C). The machined step on one coupling half must slip easily into the recess on the other. In this example, the step on one coupling half will not slip cleanly into the other half, and the flanges are clearly not lined up. The engine must be moved around until the two halves come together cleanly. Once the coupling flanges match and the machined step fits into its recess, use a feeler gauge at four points around the circumference of the coupling to measure the misalignment (D). If needed, raise engine feet to adjust alignment (E). In F, the widest gap is a little to one side of the top of the coupling. The engine front will need to come up and also move across to the left. After getting the top and bottom flange gaps equal, the flange halves are likely to be at different heights (G); in this case, the entire engine will have to be lowered. To shift the engine sideways, knock the slotted end of the feet with a mallet (H). (Courtesy Caterpillar and Jim Sollers)*

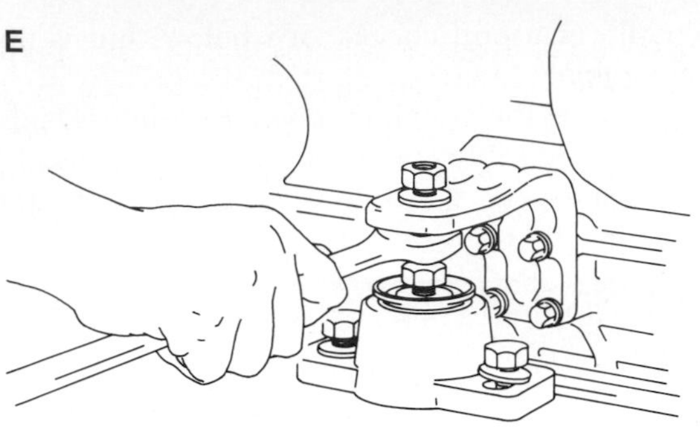

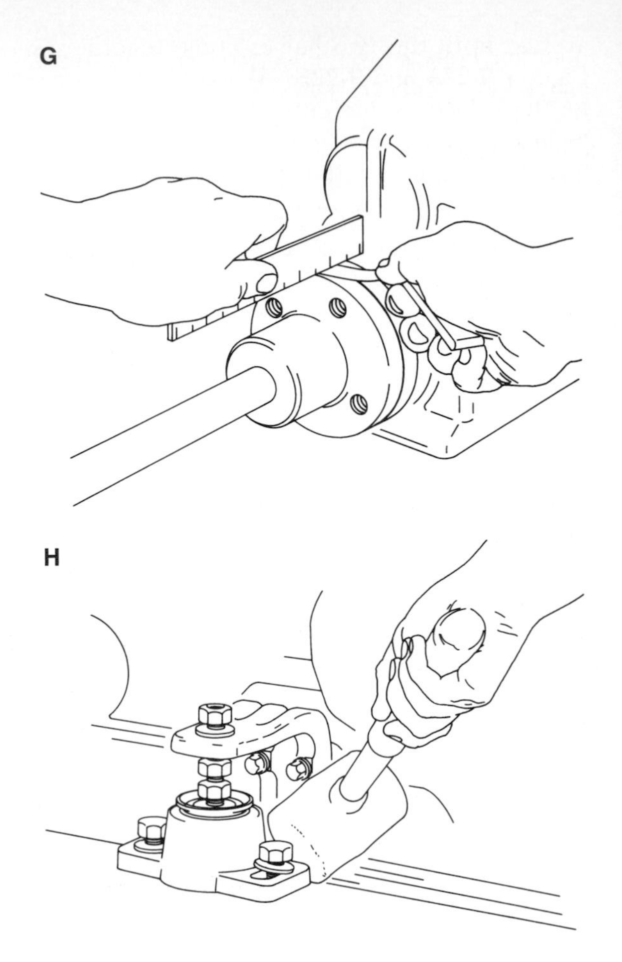

FIGURES 9-22A to 9-22H. *(continued)*

coupling faces. Start with a very thin feeler gauge. If it slides in, work up to thicker gauges until some resistance is felt when pushing the gauge in. Be careful not to push the faces apart or you will completely mess up the readings. Note the width of the gap. Repeat this procedure at the bottom of the faces and on both sides (see Figure 9-22D). The difference in the gap from any one point to another should not exceed 0.001 inch (one thousandth of an inch) per inch of coupling diameter, or 0.01 mm (one hundredth of a mm) per cm of diameter (e.g., 0.003 inch on a 3-inch-diameter coupling, or 0.06 mm on a 6 cm diameter coupling). Note that if alignment is spot on, you will be unable to get a feeler gauge in anywhere, but almost always the faces will only make contact at one point, and the feeler gauge will slide in at all other points.

If the gap exceeds tolerable limits at any point, turn the propeller shaft coupling through 180 degrees while holding the transmission coupling stationary, then measure the clearances again. If the widest gap is still in the same place, the engine alignment needs correcting. If the widest gap has also rotated 180 degrees, either the propeller shaft is bent, or its coupling is not squarely on the shaft, or both. The shaft and the coupling should be removed from the boat and trued up by a machine shop.

To adjust alignment, the engine must be moved up and down and from side to side until the clearances all around the coupling are in tolerance. Some engines have adjustable feet (see Figure 9-22E), which greatly simplifies things; others need thin strips of metal (shims) placed under the feet until acceptable measurements are reached. Make sure that all feet take an equal load, otherwise when the mounting bolts are tightened you risk distorting the engine block and causing serious damage.

Initial Adjustment

If the gap between the coupling faces is wider at the top of the couplings than at the bottom (see Figure 9-22F), raise the front feet until the gaps at the top and bottom are equal. This has the effect of pivoting the engine around the rear feet and dropping the height of the output coupling on the transmission with respect to the propeller shaft coupling. Now, even if the top and bottom gaps between the coupling faces are the same, the transmission coupling may well be lower than the propeller shaft coupling. All four engine feet will need to be raised an equal amount until the couplings are level with one another. Check the level by laying a straightedge across the top and bottom of the couplings (see Figure 9-22G); if the couplings have different diameters, this cannot be done.

If, on the other hand, the gap between the coupling faces is wider at the bottom than the top, you will have to lower the front feet and then probably the whole engine as well.

Side-to-Side Alignment

Once the gaps between the top and bottom coupling faces are equal, and the couplings are the same height, check the side gaps. If one is greater than the other, move the front of the engine toward the side that has the greater gap until the gaps are the same. Then place a straightedge across the sides of the couplings to make sure they are in alignment. If they are not, you will have to move the entire engine sideways. This is always the most difficult part of engine alignment because there are almost never any jacking screws for sideways movement. So the engine has to be levered across in a crude, and somewhat uncontrolled, fashion—all too often suddenly moving too far!

With a bit of luck, the engine feet will have two hold-down bolts on each foot: one through a close-fitting hole, and the other through a slot. To shift the engine sideways, loosen the bolts and knock the slotted end of the foot using a mallet, or lever it, so that the foot pivots around the other bolt (see Figure 9-22H).

Now it's time to go back and run the feeler gauges all the way around the coupling faces once again. If the clearance is within tolerance, and a straightedge across the couplings at top and bottom and both sides shows the coupling sides to be aligned, bolt down the engine and check the alignment once again. Unfortunately, tightening the bolts often throws out the alignment, in which case more adjustments will be needed. If this happens, loosen the mounting nuts one at a time and check the alignment after each is loosened; you may find it is just one nut that is causing the problem.

Engine alignment can be a time-consuming and frustrating business. Patience is the order of the day.

Constant-Velocity Joints

Back in the 1950s, constant-velocity joints (CVJs) were developed for front-wheel-drive cars. They were a special refinement of a universal joint, allowing a limited amount of shaft play in all directions. CVJs have since been adapted for marine installations, notably by Aquadrive (see Figure 9-23—www.aquadrive.net; see also the Python-Drive from PYI Inc.).

CVJs are used in pairs; one joint has a short, splined shaft that slides into a splined collar on the other joint. The entire unit is bolted between the transmission and propeller shaft, and according to manufacturers, will permit misalignment of up to ½ inch or 13 mm! Since reverse thrust of the propeller would pull the two sliding shafts apart, the unit is combined with a thrust bearing. The propeller shaft is locked into this bearing, which in turn is fastened to a hull-bonded bulkhead that absorbs all forward and reverse thrust from the propeller, leaving the CVJs to cope solely with misalignment. The engine can now be soft mounted, eliminating much hull-transmitted vibration. Thus CVJs not only eliminate alignment issues, but also provide a quieter and smoother installation.

CVJs require no maintenance; the various bearings are packed in grease and sealed in rubber boots. However, because all the main components are steel, watch carefully for corrosion, especially on boats with wet bilges. Should the rubber boots ever get damaged, they will need immediate replacement. Despite the tolerance of CVJs for extreme misalignment, their life expectancy will be increased if alignment is kept fairly accurate. Ironically, however, perfect alignment can lead to accelerated wear, so a very small amount of misalignment is desirable

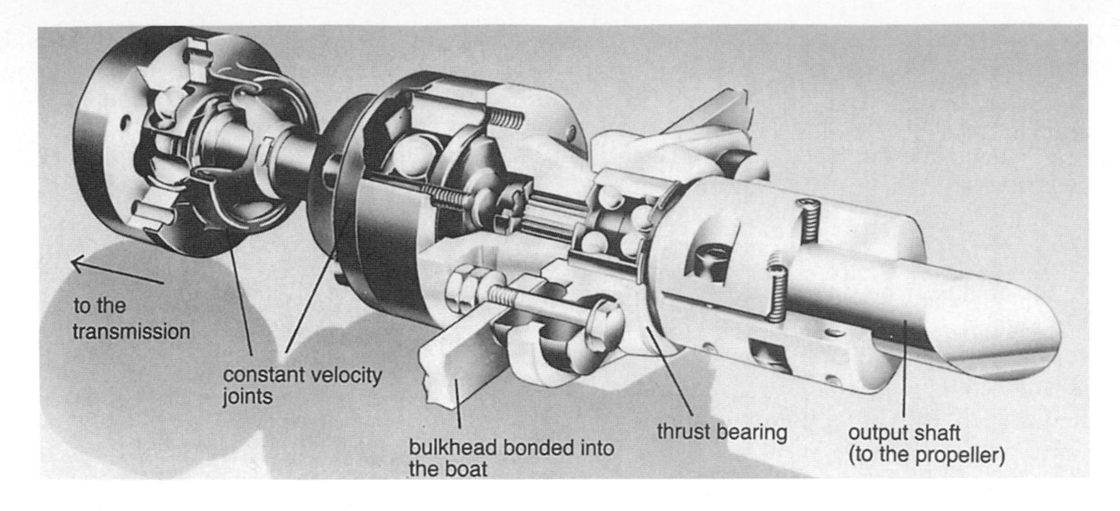

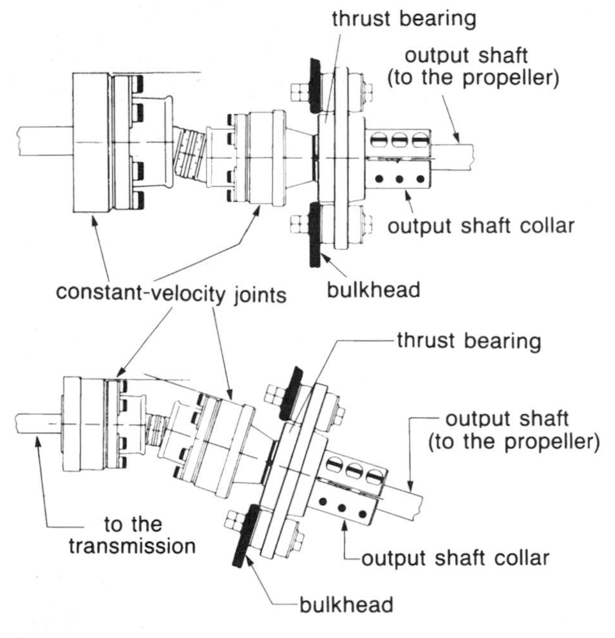

FIGURE 9-23. *Top: Constant-velocity joints, such as this Aquadrive unit, compensate for shaft misalignment. Bottom: How constant-velocity joints handle different types of misalignment. (Courtesy Aquadrive)*

(my understanding is that with perfect alignment the whole assembly is likely to "chatter").

SHAFT SEALS

Although most boats still have a traditional stuffing box, or packing gland, mechanical seals and lip-type seals increasingly are being used.

STUFFING BOXES (PACKING GLANDS)

In all forms of a stuffing box, a chamber is formed around a shaft. Rings of greased flax or something similar *(packing)* are pushed in, and some form of a cap is tightened down to compress this material around the shaft, creating a close fit that excludes almost all water from the shaft/packing interface (see Figure 9-24). Some stuffing boxes have grease fittings, and in this case, a bronze spacer ring is generally incorporated between the second and third rings of packing and directly below the grease fitting, allowing the grease to be distributed around the stuffing box (see Figure 9-25). A shot of grease (or one turn on the grease cup) should be put in about every 8 hours of engine-running time.

At one time most stuffing boxes were bolted to the deadwood on a boat (rigidly mounted, often incorporating a stern bearing—see Figure 9-26), but today the majority are flexibly mounted in a

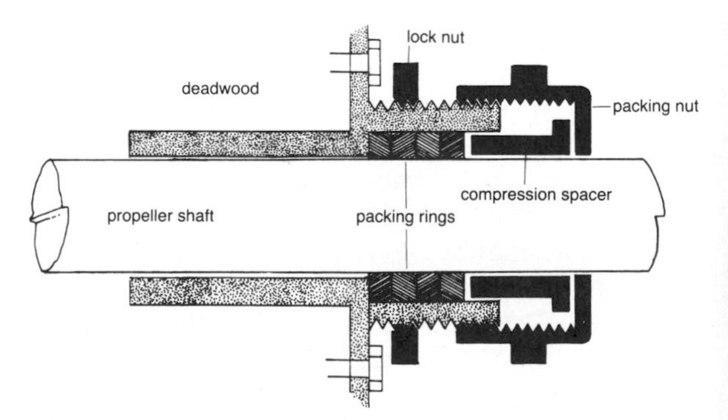

FIGURE 9-24. *A cross section of a rigid stuffing box. (Courtesy Ocean Navigator)*

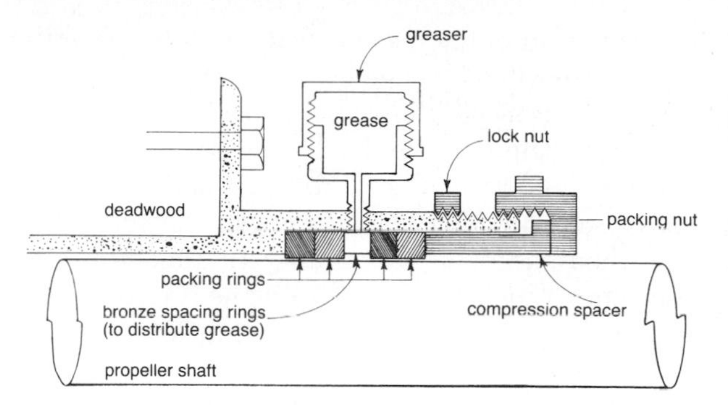

FIGURE 9-25. *A cross section of a stuffing box equipped with a greaser. The screw-down cap-style cup greaser could be replaced with a standard grease nipple or a remote greaser. (Courtesy Ocean Navigator)*

FIGURE 9-27. *A flexible stuffing box. Note the use of double hose clamps, with all 300-series stainless steel construction—a quality installation.*

length of hose that in turn is fastened to the inner end of the shaft log (stern tube; see Figures 9-27 and 9-28). *If this hose fails, water may pour into the boat at an alarming rate.* Note that ABYC standards require shaft seals "to be constructed so that, if a failure occurs, no more than two gallons of water per minute can enter the hull with the shaft continuing to operate at low speed." This requirement can most easily be met by minimizing the gap between the shaft log and the propeller shaft.

The hose on a flexible shaft seal must have two all-stainless-steel hose clamps at each end, with 300-grade stainless steel screws (see page 63). Inspect the hose annually for any signs of cracking or bulging, and undo the clamps a turn or two to make sure that galvanic corrosion is not eating away the band where it contacts the worm screw. If it is necessary to replace the hose, the propeller shaft coupling will have to be broken loose and the coupling removed from the shaft (see above).

Stuffing Box Blues

The concept of a stuffing box is simple enough. Done right, on a properly installed and aligned engine, the packing will not leak when a shaft is at rest, but when the shaft is in motion, it will allow the occasional drip. Done wrong, the stuffing box will leak continuously, or else the shaft will be permanently damaged.

Note the emphasis on the *engine installation and alignment*. It is an unfortunate fact of life that many

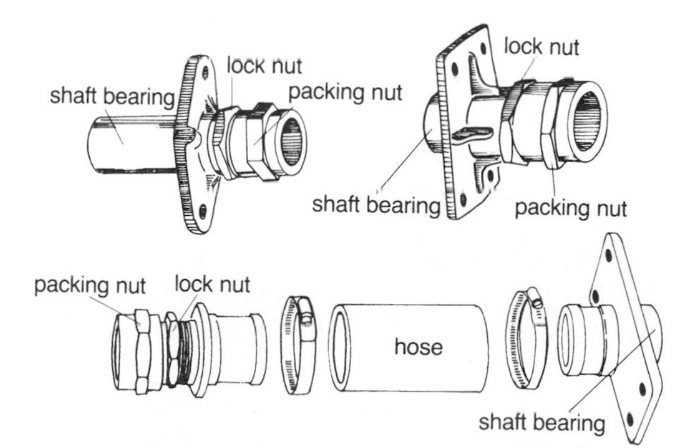

FIGURE 9-26. *Rigid stuffing boxes (top) and a flexible stuffing box (bottom). (Courtesy Wilcox Crittenden, adapted by Jim Sollers)*

FIGURE 9-28. *A stuffing box with single hose clamps in which the screws are 400-series stainless steel that is corroding. Meantime, the packing nut has corroded to the shaft. One way or another, when the engine is next put in gear, something is likely to tear loose, allowing large amounts of water to enter the boat!*

engines are not properly aligned, or else they suffer from other installation problems such as a misaligned strut or stern (Cutless) bearing (see below). In these cases, the shaft whips around inside the stuffing box, beating the packing back against the sides of the box and allowing more and more water into the boat, both with the shaft turning and with it at rest. Regardless of the skill of the mechanic who fitted the packing, the stuffing box will need constant attention and will still never give satisfactory service.

Even on a properly installed engine, a stuffing box needs some maintenance, including adjustment, greasing, and periodic replacement of the packing, which otherwise hardens over time and may then damage the shaft. Access to many stuffing boxes is poor, to say the least, making such maintenance an onerous chore—it is not uncommon to have to hang upside down in a cockpit locker with insufficient room to swing the necessary wrenches. In extreme cases, notably some V-drives and boats with very short propeller shafts, the packing cannot be reached or replaced without removing either the propeller shaft coupling or the engine (see Figure 9-29; this is in clear violation of ABYC standards, which call for sufficient clearance "along the shaft line to permit replacement of the packing without uncoupling the shaft or moving the engine").

FIGURE 9-29. *There is not enough room in this installation to back off the packing nut; as a result the packing has not been changed in ten years! The stuffing box has been leaking profusely, corroding the coupling and everything else in range of the saltwater spray produced when the shaft is turning. The transmission controls were close to failing, and an electric water heater was almost corroded through. The leak was such that a dead battery, or a failed bilge pump, could have resulted in the loss of the boat in just a few days.*

Poor access and difficulties in packing adjustment and replacement often lead to neglect. Add a little engine misalignment and vibration, and the stuffing box is soon dripping when the engine is at rest as well as when the shaft is turning. In time this drip becomes a trickle. When the engine is running, salt water is sprayed all over the back of the engine room; when the boat is left unattended, the automatic bilge pump and its battery become the sole line of defense against a sinking—the loss of the pump can result in the loss of the vessel. Small wonder then that many owners regard their stuffing box with hostility, little realizing that in most instances the engine installation is the source of their problems rather than the stuffing box. It's the old story of shooting the messenger rather than heeding the message!

Packing Adjustment and Replacement

A stuffing box is meant to leak. When the shaft is turning, two or three drops a minute are needed to keep the shaft lubricated. If the leak is worse than this, tighten down the nut or clamp plate to compress the packing a little more. If a greaser is fitted, pump in a little grease first. Tighten down the nuts no more than a quarter turn at a time. With a clamp plate, tighten the two nuts evenly.

Start the engine and put the transmission in gear for a minute or so. *Shut down the engine.* Feel the stuffing box and adjacent shaft; *if they are hot, the packing is too tight.* A little warmth is acceptable for a short while as the packing beds in, but any real heat is unacceptable. It is quite possible (and common) to score grooves in shafts by overtightening the packing, in which case the shaft will never seal and will have to be replaced. (Sometimes it can be turned end for end to place a different section in the stuffing box.)

If the shaft cannot be sealed without heating, replace the packing. You should, in any case, renew the packing every few years, since old packing hardens and will score a shaft when tightened. The hardest part of the job generally is getting the old packing out. *It is essential to remove all traces of the old packing or the new wraps will never seat properly.* With a deep, awkwardly placed stuffing box, it is next to impossible to pick out the inner wraps of packing with screwdrivers and ice picks; you need a special tool consisting of a corkscrew on a flexible

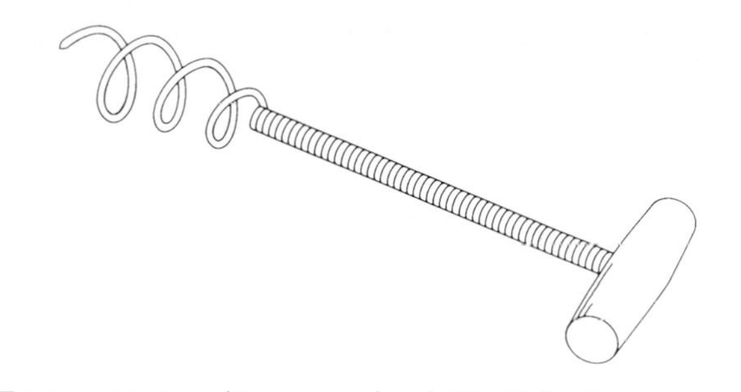

FIGURE 9-30. *A packing removal tool. (Jim Sollers)*

shaft (see Figure 9-30; available from West Marine and other chandleries—www.westmarine.com). Unless this tool is on hand, it is not advisable to start digging into the packing, especially if the boat is in the water. Appreciable quantities of seawater may start to come into the boat as the packing is removed, in which case speed is of the essence.

Packing itself comes as a square-sided rope in different sizes: 3/16 inch (5 mm), 1/4 inch (6 mm), 3/8 inch (10 mm), etc. It is important to match the packing to the gap between the shaft and cylinder wall. You can buy packing as pre-formed rings to match the stuffing box (the best option for most people) or by the roll. When cutting rings off a roll, make about five tight wraps around the propeller shaft at some convenient point and then cut *diagonally* across the wraps with a very sharp knife.

To fit new rings of packing, grease each with a Teflon-based waterproof grease before installation, and tamp down each ring before putting in the next. I use some short pieces of pipe slit lengthwise, slipped around the shaft, and pulled down with the packing nut or clamp plate to *gently* pinch up the inner wraps (not too tight, or the shaft will burn). Stagger the joints from one wrap to another by about 120 degrees. Don't forget the greasing spacer (if one is fitted) between the second and third wraps.

Dripless Packing

Drip-free or dripless packing (available from West Marine and others) has a slightly crumbly, clay-like consistency that can be rolled between the palms of the hand until it is an appropriately sized diameter to go in the stuffing box. (It tends to break up as you push it in, but that's OK.)

Dripless packing is sandwiched between outer and inner rings of conventional packing, which keeps it from working its way out of the ends of a stuffing box. After the first ring of packing is in, massage the dripless material into an appropriate shape, daub it in a special (PTFE-based) grease, then push it into place. Add the outer ring of packing, and tighten the packing nut just enough to stop all drips. That's it! Since it's moldable, dripless packing will conform to even corroded and grooved shafts. It will not eliminate drips in cases of severe engine misalignment or shaft vibration, but it does work exceptionally well in most situations. Once installed, it almost never needs renewing and very rarely needs adjustment. Typically, problems arise from improper installation (not reading the instructions!) and from overtightening the packing nut, causing the stuffing box to run hot. In this case, simply undoing the packing nut may not relieve enough pressure to stop the heating, and it may be necessary to pick out the packing and put it back in.

MECHANICAL SEALS

Mechanical seals are not an adaptation of the traditional stuffing box, but instead are a complete replacement. A mechanical seal has two components: a boot or bellows, which is attached to the inner end of a shaft log; and a sealing ring *(rotor)*, which is attached to the shaft and rotates with it (see Figure 9-31). The face of the boot has a hard machined surface (a *stationary flange*); the sealing ring sometimes consists of a stainless steel unit, but other times consists of a second machined surface embedded in another boot. One or the other of the two components either contains a spring or is constructed so that it resists compression. On installation the two components are pushed together against the spring or boot resistance, and then the sealing ring is clamped to the shaft so the pressure is maintained. As the shaft rotates, the machined faces of the stationary flange and the sealing ring form a close-enough fit to keep water out of the boat.

On a properly installed and aligned engine, such a seal will eliminate all leaks. Once installed there is no adjustment or maintenance beyond a routine inspection of the hoses, clamps, and sealing faces. Since no part of the seal is bearing on the propeller shaft, there is no risk of overheating or scoring the shaft. What is more, the seal will be effective regardless of the surface condition of the shaft (i.e.,

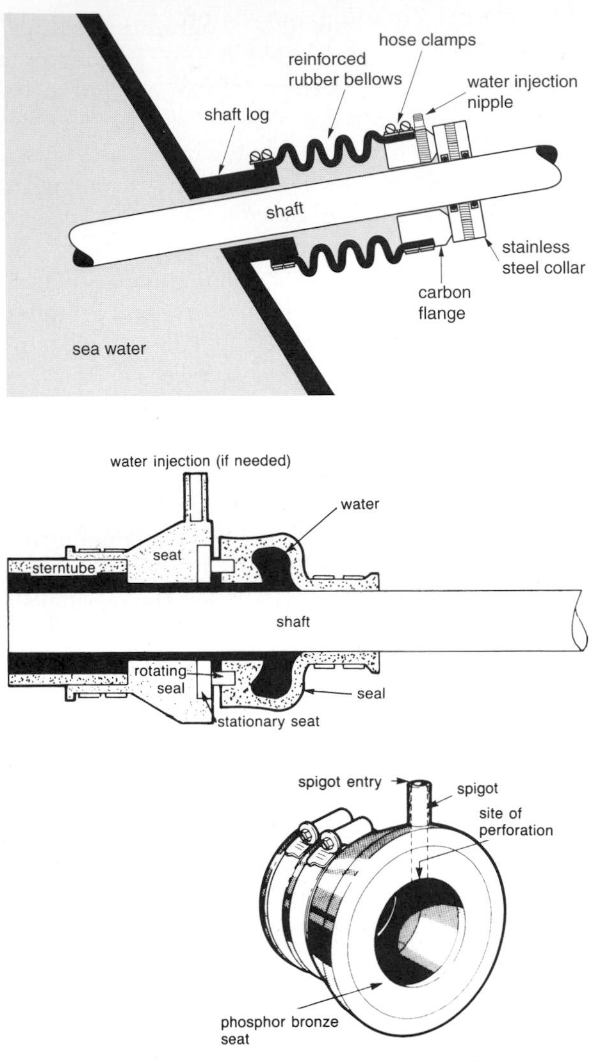

FIGURE 9-31. *Two types of mechanical seals. Mechanical seals all work on the same principle—two sealing surfaces are held in contact by some sort of spring pressure. A PSS shaft seal (top) and a Crane shaft seal (bottom). (Courtesy Ocean Navigator and Halyard Marine)*

the seal will work even if the shaft is corroded or otherwise damaged).

Problem Areas

Many face seals work perfectly—after five or ten years, the owner hasn't seen a drip and is ecstatic. But then there are the ones with problems. Generally the problems have not stemmed from an outright failure of the seal itself, but have resulted from unforeseen operating conditions or improper engine installation and alignment. For example, a one-cylinder diesel engine placed on flexible mounts will jiggle all over the place. When the throttle is pushed forward, the force of the prop may push the engine forward as much as $^3/_8$ inch, which may allow the seal faces to open up and spray water everywhere. In addition, there is the movement that occurs when a boat is under sail. As the boat comes about, the heavy mass of the engine may cause it to sway from one side to the other, moving one half of the seal sideways, potentially resulting in a severe leak until the boat rights itself.

As a result, some installations require a greater initial compression than others. As far as sideways movement is concerned, some seals have a broad seating surface that allows the seal to oscillate without loss of contact, while others have either a lip or a chamfer built into the face of the stationary seal and sealing ring so that the two are held in a positive mechanical alignment.

The nature of the boots or bellows has come under close scrutiny over the years. Some early bellows were susceptible to softening caused by diesel, oil, and gas, reducing their operating life. These were weak in compressive force and got weaker over time. Some seals containing a spring had failures when the spring came through the bellows. Any kind of a bellows failure is, of course, potentially catastrophic. Nowadays, newer, more chemically resistant and resilient materials and expensive high-tech construction methods are mostly used.

Regardless of the engineering that has gone into a mechanical seal, its success and life expectancy in any given installation will depend to a considerable extent upon the conscientiousness of the seal installer. Shafts must be deburred before installation, and sealing rings must be properly secured to the propeller shaft, with the fasteners locked against vibration. If a bellows is attached cockeyed to a shaft log, or the seal is inserted crookedly into the other end of the bellows, the seal faces will be subjected to uneven pressure; even if the seal does not leak, it will wear rapidly. And finally, if provision is not made to lubricate a seal, it will eventually burn up.

Two situations can typically cause a loss of lubrication. The first is high boat speed, which creates a vacuum that sucks the water out of the stern tube. The second is a haulout, which allows the water to drain out of the shaft log. To combat both situations, most rotating seals include an option for water injection using water from the engine's raw-water circuit. If water injection is not used, there may be an air-vent valve that will release trapped air after a haulout, allowing the stern tube to refill with water. If not, *it is essential*, once a boat is back in the water, *to pull back and hold the boot until water spurts out of the seal face* (see Figure 9-32).

Failures

Properly installed mechanical seals will perform as advertised, even under adverse conditions. But don't forget that mechanical seals are *two-part* seals (the boot and sealing ring) in which the two parts are not, and cannot be, fastened together. If the two halves become separated, or the seal between them is damaged, the seal is likely to leak far more rapidly than a traditional stuffing box. This will happen if the sealing ring fastened to the propeller shaft works loose. For this reason, some owners like to put a hose clamp or propeller shaft zinc around the shaft behind the sealing ring as a security precaution.

If a seal fails, at the very least the propeller shaft coupling must come off in order to replace it. On some boats, the propeller also has to come off to create sufficient longitudinal shaft movement to get off the coupling, and occasionally the engine has to be moved off its mounts. In the meantime, if the boat is in the water, the flow rate from the failed seal into the boat is determined by the clearance between the shaft log and the propeller shaft. On those boats in which a rubber sleeve bearing is inserted in the shaft log (mostly boats with a propeller in an aperture in the rudder), the flow rate is almost always relatively low, such that a bilge pump can keep up with it. But on those boats in which the rubber sleeve bearing is installed in a separate strut (the majority of modern boats), the flow rate may be quite high (see below for more on rubber sleeve bearings). In the latter case, it is worth ensuring that the shaft log makes a close-enough fit around the propeller shaft to keep the maximum flow rate down to a level that the bilge pump can handle.

FIGURE 9-32. *After launching, mechanical seals should be pulled back to allow the seal to flood.*

I once found the bilge pump on my last boat constantly kicking on. An investigation revealed that the sealing ring on my shaft seal had worked loose, and the two halves of the seal had separated. Water had been flowing into the boat since the last time I used it, two weeks previously! The boat had not sunk because the shaft log on our boat was a moderately close fit around the shaft, which kept the flow rate relatively low, and we had a large battery bank, which was able to support the bilge pump for two weeks. Further investigation revealed that the locking screws had been left out of the sealing ring when it was installed. After locking the ring, I added a propeller shaft zinc behind it as a backup.

Lip-Type Seals

It is this nagging fear of a major leak as much as any other factor that causes many boatowners to remain wary of mechanical seals. The various lip-type seals on the market attempt to address these concerns by seeking to eliminate the leaks of a stuffing box without in any way increasing the risk of a catastrophic failure. This is done, essentially, by replacing the packing in a conventional stuffing box with a nitrile lip seal of the type universally used to seal crankshafts and output shafts in engines and transmissions.

Lubrication

The simplest units are water lubricated and closely resemble a flexible stuffing box. One end of the seal body is fastened to the shaft log with a length of hose; the other end has a recess into which the seal is slipped. In order to keep the seal lubricated, provision is made to maintain a water supply from the engine's raw-water circuit. Experience has shown that unless corrected for, shaft vibration will cause

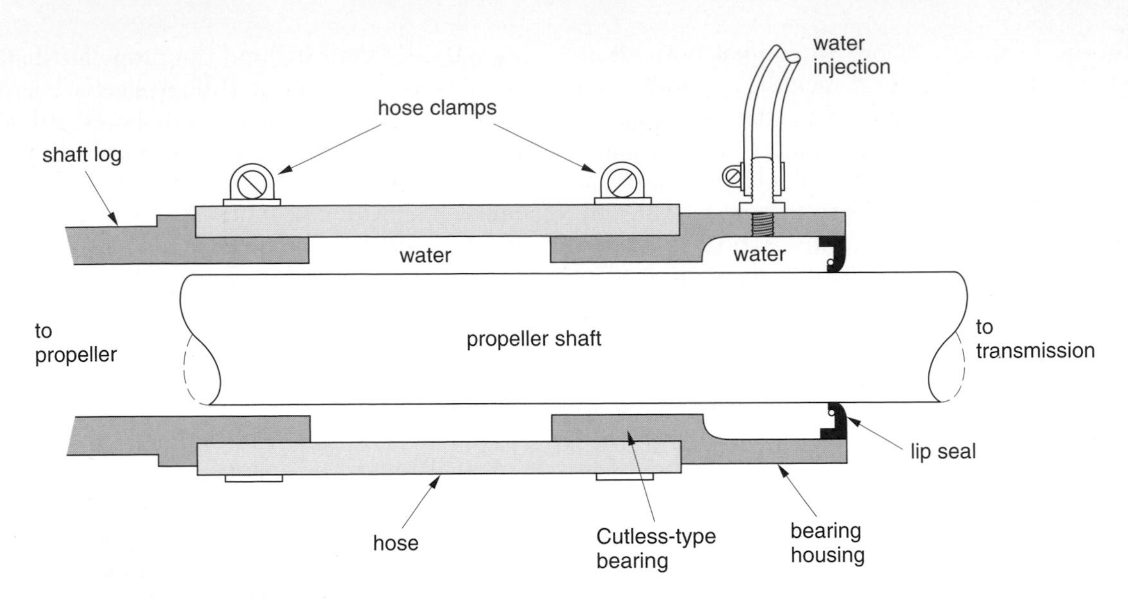

Figure 9-33. *A water-lubricated lip seal.*

such a seal to leak, so a rubber sleeve–type bearing is built into the inside of the seal body, forming a close fit around the propeller shaft. This keeps the seal in a constant alignment with the shaft, regardless of engine misalignment and other problems (see Figure 9-33).

Other seals are oil lubricated. These are physically similar in appearance, but instead of a single seal at the inboard end, they have seals at both ends of a tube (see Figure 9-34). Lubrication is provided by a gravity-feed tank of oil set above the seal. There seems little doubt that the oil-lubricated seals last longer than the water lubricated, but on the other hand, the oil reservoir must be checked regularly, and if it is allowed to run dry, the seal will heat up and fail. Unlike the water-lubricated seals, these shaft seals require a minimal level of operator attention and maintenance.

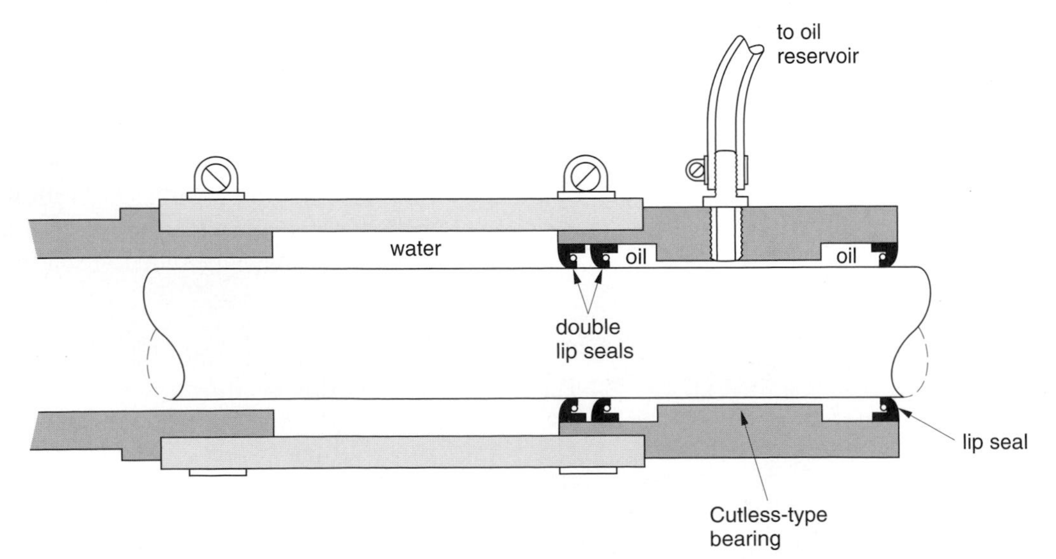

Figure 9-34. *An oil-lubricated lip seal.*

Failures

The number one reason for lip-seal failures is improper installation. The seals are generally installed over the inboard end of the propeller shaft and then slid into place toward the outboard end. The lip on the seal is facing aft. *If the end of the shaft is not adequately chamfered during installation, there is a good possibility that the lip will be rolled under. If the seal is put in at all cockeyed, it will probably leak.* Neither of these situations is easily detected until the boat is in the water, at which point you have to haul the boat back out and remove the propeller shaft coupling to repair the seal—an expensive and time-consuming business.

Even if properly installed, a lip-type shaft seal, which has the seal bearing directly on the propeller shaft, will leak if the shaft is not in an unblemished condition and polished. One way to get around this is to give the seal a machined inner sleeve, which is rigidly fastened and sealed to the propeller shaft, and then enclose this in an oil-filled outer sleeve containing a bearing surface and lip seals. The outer sleeve is fastened with a bellows to the stern tube. The inner sleeve now becomes the seal-seating surface, so the condition of the propeller shaft is not critical.

In the event of complete seal failure, the rate of leakage is not likely to exceed that of a badly leaking stuffing box—the potential for a major failure is less than that of a mechanical seal. Replacement once again requires removal of the propeller shaft coupling, which on some boats may also require removing the propeller as well as moving the engine.

Struts and Bearings

Struts

In the old days, the propeller shaft invariably emerged out the back of a hull through a shaft log mounted in the deadwood, with the propeller mounted immediately aft of the deadwood (see Figure 9-35). With today's flatter-bottomed hulls, the propeller shaft must extend some way beyond the shaft log before there is adequate hull clearance to mount the propeller. This run of exposed shaft needs a strut to support it (see Figure 9-36). Some are in the shape of an I, and some a V. The V-shaped strut is inherently stronger. On sailboats, an I-shaped strut is almost always used because it minimizes drag.

All too often these struts are inadequately mounted. The stresses from a fouled propeller or bent shaft will work them loose. If the strut is through-bolted, with accessible nuts on the inside of the hull, check the fasteners annually and tighten as necessary. If the bolts are a loose fit in the hull due to elongated holes, a good dollop of bedding compound combined with a good-sized, well-bedded backing block will tighten things up. Many struts,

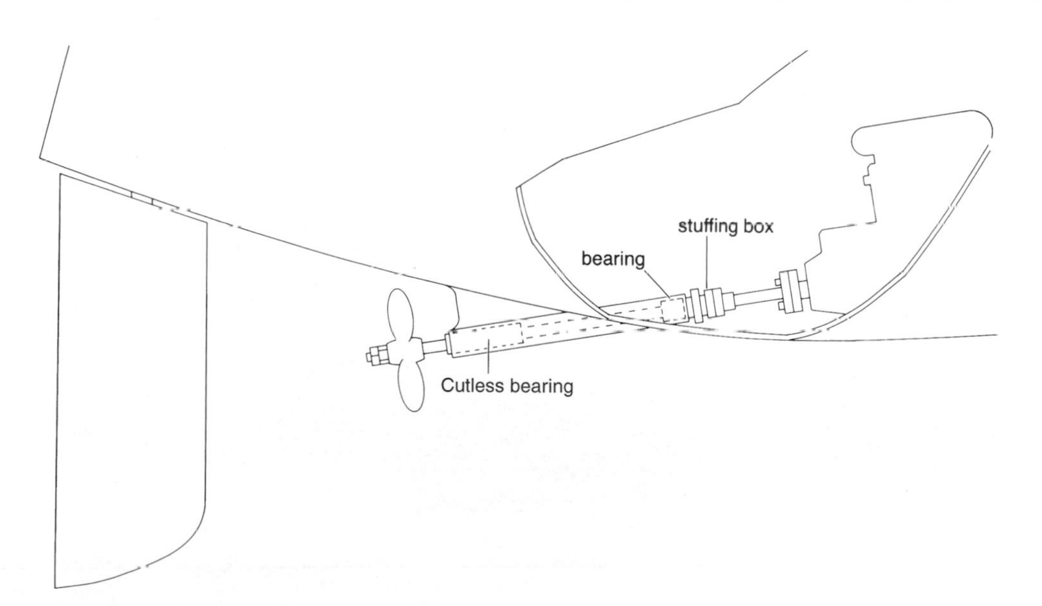

Figure 9-35. *A propeller shaft supported in a stern tube. (Jim Sollers)*

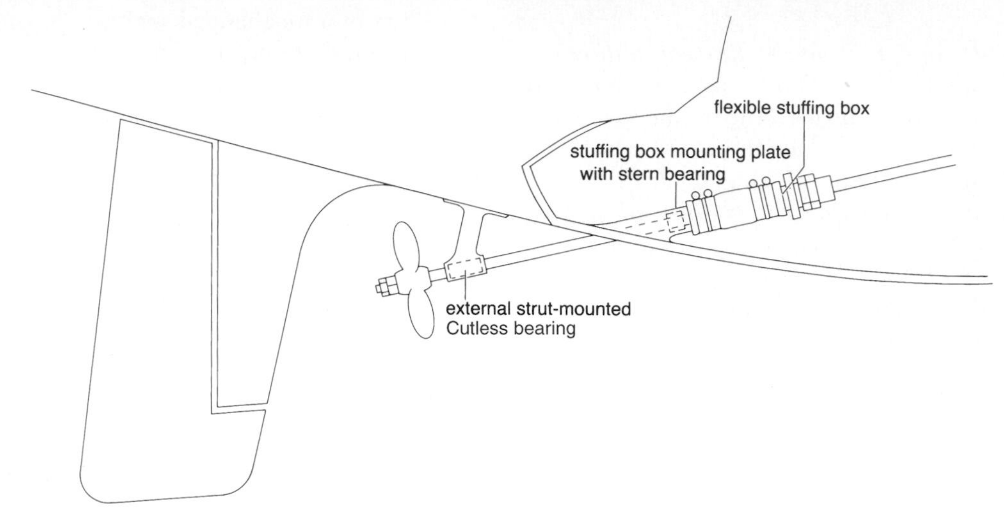

FIGURE 9-36. *A propeller shaft supported by an external strut. (Jim Sollers)*

however, are glassed in place. If the strut develops any play, some extensive repair work will be necessary (in a bind, it may be possible to drill through the strut and hull and add a backing block and bolts).

RUBBER SLEEVE (CUTLESS) BEARINGS

Most stern bearings consist of a metal or fiberglass pipe with a ribbed rubber insert in which the shaft rides (see Figures 9-37 and 9-38). Water circulates up the grooves to lubricate the bearing. Externally mounted bearings in a strut need no additional lubrication, but some bearings installed in shaft logs and deadwoods have an additional lubrication channel into the stern tube. In place of the ribbed rubber, various plastics are sometimes used.

With a properly aligned engine, most rubber sleeve bearings will last for years (unless operated in silty water, which accelerates wear), as long as they are lubricated adequately. They require no maintenance. At the annual haulout, flex the propeller shaft at the propeller; if there is more than minimal movement (1/16 inch of clearance between the shaft and bearing

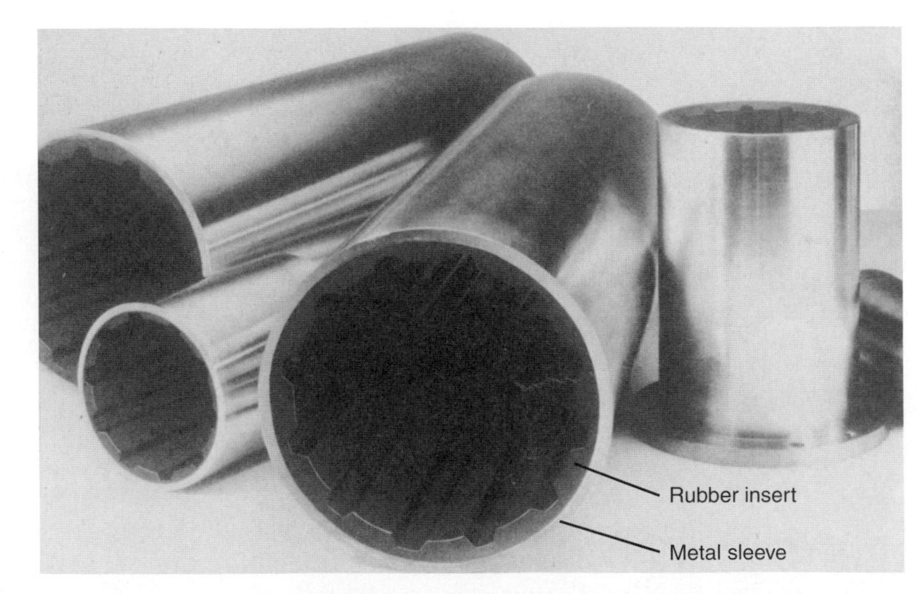

FIGURE 9-37. *Cutless bearings. (Courtesy BF Goodrich)*

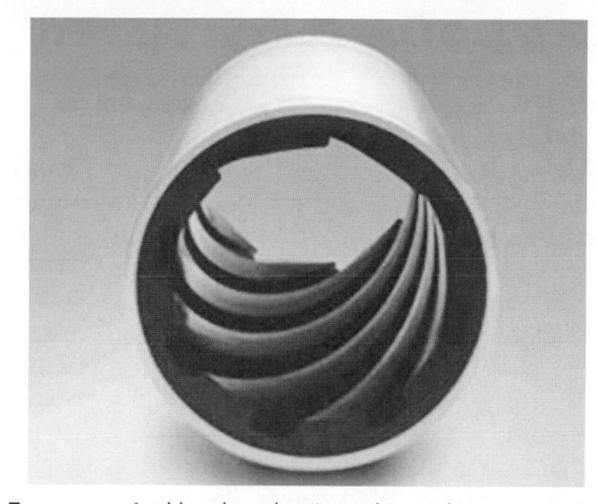

FIGURE 9-38. *A rubber sleeve bearing with curved grooves, which is reputed to help the water flow through the bearing. (Courtesy PYI)*

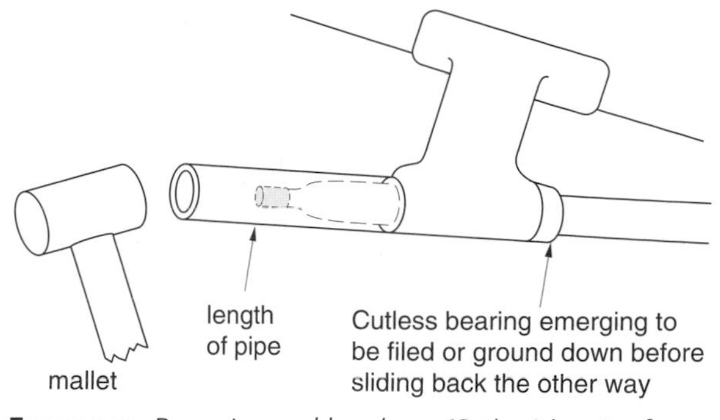

FIGURE 9-39. *Removing a rubber sleeve (Cutless) bearing from a strut, with the propeller shaft still in place.*

per inch of shaft diameter, or 2 mm per cm of shaft diameter), the bearing needs replacing. If not renewed, a worn rubber sleeve bearing will allow excessive shaft vibration, which will rapidly wear stuffing box and transmission bearings and oil seals. A rubber sleeve bearing worn on only one side is a sure sign of engine misalignment.

Replacement

Most bearings are a simple sliding fit in the deadwood or strut; some are then locked in place with setscrews but others are not (the ABYC says they should be, with at least two setscrews). Once any setscrews are loosened, a strut-mounted bearing can sometimes be replaced with the propeller shaft still in place. The problem is that there generally is not enough room between the hull and strut to maneuver the tools needed to knock the bearing aft. However, by sliding a piece of pipe (with an outside diameter a little less than that of the bearing) up the propeller shaft, you can knock the bearing up the shaft into the space between the strut and hull (see Figure 9-39). Then file or grind down the bearing's outer case until it breaks apart or slides back easily through the strut and off the shaft.

If the bearing won't budge, you can bring greater pressure to bear in a more controlled fashion by making a homemade bearing puller as shown in Figure 9-40. Cut a slot in a metal plate allowing it to slide over the propeller shaft with sufficient clearance for the bearing to pass through it. Take a second plate and drill a hole just large enough to clear the propeller shaft. Find a piece of pipe that fits around the shaft but which has an outside diameter less than that of the bearing. Fasten the two plates together with two long bolts and tighten the bolts to drive the bearing out of the strut. Once out of the strut, file or grind down the bearing case as above until it breaks apart or slides back through the strut.

In almost all other cases, the propeller shaft must come out to renew a rubber sleeve bearing. A strut-mounted bearing can generally be knocked out from the inner side of the strut, but care must be taken not to damage the strut or its mounting. To

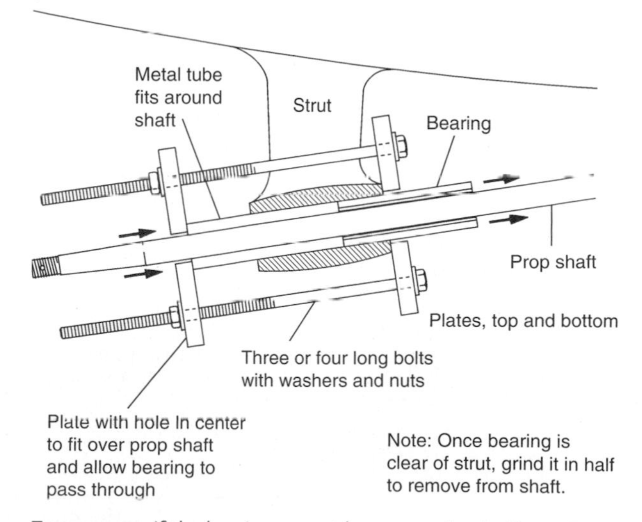

FIGURE 9-40. *If the bearing cannot be removed as in Figure 9-39, greater pressure can be brought to bear by making a tool as shown. (Jim Sollers)*

drive out a hull-mounted bearing (i.e., in the deadwood), the stuffing box has to be removed from the inner end of the stern tube.

Not infrequently, the bearing refuses to budge. In this case, given a strut, a simple puller can be made as shown in Figure 9-41. In the case of a hull-mounted bearing, or in the absence of a puller, make two longitudinal slits in the bearing with a hacksaw blade so you can pry out a section, allowing the rest to flex inward. Take care not to cut through the bearing into the surrounding strut or stern tube. Then lightly grease a new bearing and push it in using a block of wood and *gentle* hammer taps. You don't want the bearing to be a tight (interference) fit because it may distort. If a metal-shelled bearing is packed in ice before installation, it will shrink a few thousandths of an inch—enough to ease the installation.

Any retaining setscrews *must tighten into dimples in the bearing case* and should be locked in place with Loctite or something similar. Setscrews must not press on the bearing case since this could distort it, causing friction between the rubber bearing and the propeller shaft. The shaft itself should *not* be a tight fit in the bearing sleeve.

Most bearing cases are made of naval brass or stainless steel. Fiberglass-epoxy (FE) cases are becoming more common, however. This material works just as well as the metals and is especially recommended for steel and aluminum hulls, where it will eliminate the risk of galvanic corrosion. Its only disadvantage is that if the bearing is a tight fit, there is a greater risk of damaging the bearing when knocking it in or out.

FIGURE 9-41. *A homemade tool for rubber sleeve (Cutless) bearing removal. The bearing is drawn out by tightening the nuts.*

SECTION FOUR: AUXILIARY SYSTEMS

VENTILATION

A diesel engine requires a large volume of clean and cool air (see Chapter 2). As air temperatures rise, the weight of air per cubic foot falls, and the engine pulls in correspondingly less oxygen at each cycle, causing a loss in efficiency and power.

Figure 9-42 gives an approximate idea of the decrease in the weight of air as the temperature rises. It is not uncommon for an engine room in the tropics to be as hot as 120°F (49°C), with turbocharging inlet-air temperatures considerably higher. Figure 9-43 shows the decrease in the engine's rated output as inlet air temperatures rise (as determined by the Diesel Engine Manufacturers Association; the rating starts at 90°F/32°C). In excessively hot engine rooms, it will be necessary to duct air from outside directly to the air-inlet manifold. If such ducting is installed, its opening must be situated in a way that prevents water from entering. Ducting also must be as far from the exhaust as possible so that the engine does not suck in spent gases.

It is also important, particularly on auxiliary sailboat engines housed in an insulated box or in a sealed engine room, to ensure an adequate flow of air to the engine. Otherwise, at higher loadings, the engine is likely to be partially strangled, resulting in

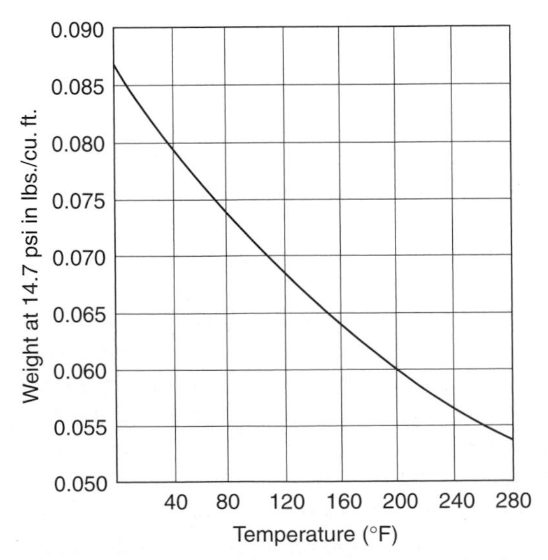

FIGURE 9-42. *The effect of temperature on the weight of air.*

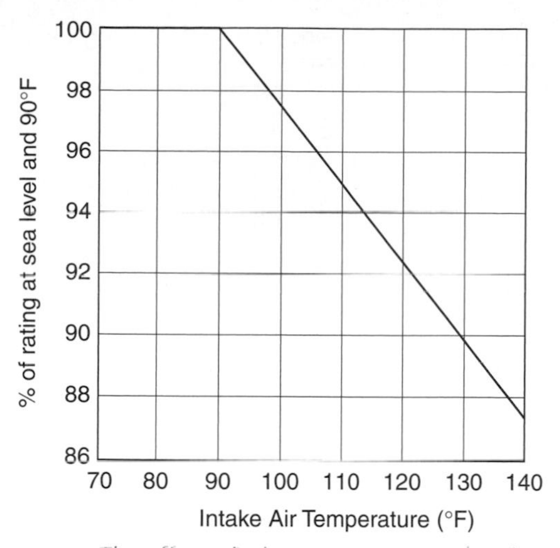

FIGURE 9-43. *The effect of air temperature on diesel engine performance.*

a loss of power, overheating, and perhaps black smoke from the exhaust.

FUEL TANKS

Diesel tanks can be made from a wide range of materials; I discuss the advantages and disadvantages of various materials at some length in *Boatowner's Mechanical and Electrical Manual*. My preference is for high-quality, cross-linked polyethylene because it will never corrode (as will all metal tanks eventually).

Tanks need to be baffled at regular intervals to prevent the fuel from sloshing around; no compartment should contain more than 25 gallons. Access hatches for cleaning are needed in every compartment. Since contamination of the fuel supply is the number one cause of problems with diesel engines, have a small sump built into the base of the tank and provide some way to drain it or pump it out.

All connections should be made through the top of a tank, especially fuel line connections. If a fuel line ruptures, this will prevent the entire contents of the tank from draining into the bilges. If the tank needs to be emptied, pump it out from above.

The fuel fill line is best extended inside the tank since this reduces foaming when filling. The fuel suction line should extend to no lower than 2 inches from the bottom of the tank to avoid sucking up contaminants. Ideally, a sampling line should run to within $^1/_8$ inch of the lowest point in the tank. Figure 9-44 illustrates a proper fuel tank design and installation.

Fuel density decreases as temperatures increase. Fuel temperatures above 90°F (32°C; the rating point for American-made diesels) will result in reduced engine power. The higher a fuel tank is situated in an engine room, the warmer it will get. Keep tanks down low. In any event, the top of the tank should not be above the level of the injectors because occasionally fuel may leak down through the injectors into the combustion chambers. Low tanks will help provide stability to the boat; centrally placed tanks will not affect athwartships trim.

Measure the distance from the bottom of a fuel tank to the fuel lift (feed) pump and compare it to the lifting capacity of the pump. You may need an auxiliary pump in certain exceptional cases (e.g., a deep-bilge sailboat with the tank at the bottom of the bilge).

Fuel fill hoses, vent hoses, and fuel lines must be made of a USCG- and ISO-approved, fire retardant material and adequately fastened against vibration. Avoid high spots where pockets of air can gather.

On wooden boats, you must ensure adequate air circulation between the tank and the hull against which it is mounted, or rot is likely to set into the woodwork (steel and aluminum hulls may also be susceptible to corrosion). It is often recommended that tanks be removable to permit proper hull inspections, but this is sometimes a very tall order to fill. At the least, the tank should be accessible enough to allow a close inspection of the woodwork around it.

COOLING

A raw-water seacock must be below the waterline at all angles of heel. It needs an adequate strainer (preferably see through). The seacock must be made of marine-grade materials such as bronze or Marelon and installed in such a way that it will withstand any conceivable abuse without failing. Ideally it will be UL listed, which means it has passed certain tests, including what is popularly known as the 500-pound-man test, which consists of mounting a seacock in a simulated hull section and then applying a static load of 500 pounds to its inboard end (this is done by hanging a 500-pound weight from the fitting). The fitting must be able to support the weight for 30 seconds without suffering damage or deformation that impairs its performance. The test is followed by a leak test to see if the valve still seals properly.

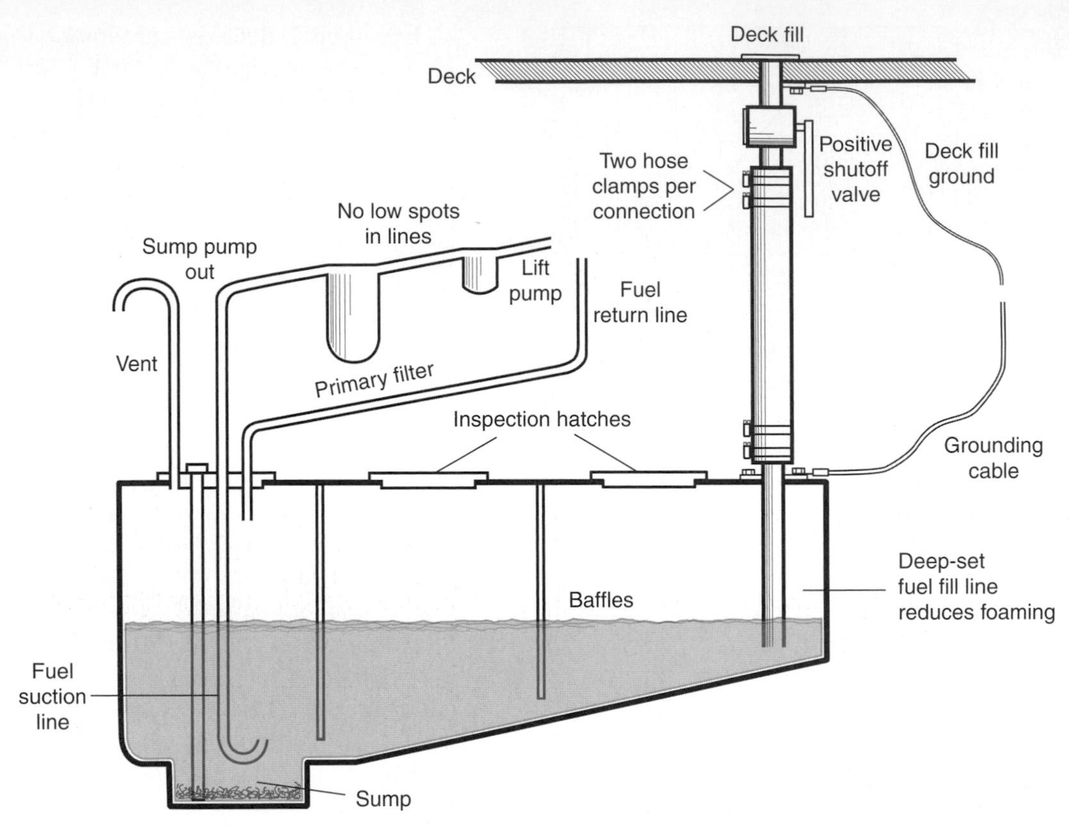

FIGURE 9-44. *A proper fuel tank installation. (Fritz Seegers)*

If a seacock is installed through a cored hull, all coring needs to be removed from the vicinity and the hull taken down to a single skin. This is to prevent any possible moisture penetration into the core.

Any kind of a gate valve is a very poor choice for a seacock (and contravenes ABYC standards) because these valves are prone to jamming and failure. A conventional seacock or a ball valve will give more reliable service.

When running raw-water hoses, avoid undrainable low spots where freezing might cause damage. Double-clamp all connections below the waterline with 300-series stainless-steel hose clamps that have 300-series stainless screws (see page 63).

EXHAUST

Every year I get a number of e-mails from people with flooded engines, mostly in sailboats. More often than not, the engine has functioned fine for years, but then a long-dreamed-of cruise was undertaken and at some point the engine flooded. The common thread is that on an offshore passage, the boat got into rougher conditions or bigger seas than it had seen before. The large waves rushing past the boat set up hydrostatic pressures that caused water to siphon into the engine. Even on new boats, way too many engines are still installed in a manner that makes this scenario possible.

A less common cause of engine flooding is the installation of a scoop-type water inlet facing forward on the outside of a sailboat hull. Anytime the boat is moving at more than a few knots, this generates pressure in the raw-water system. In normal circumstances, the vanes in the rubber impeller raw-water pump will hold this pressure at bay when the engine is not running. But if the vanes get damaged, water will be driven up through the heat exchanger and into the exhaust, where it will build up and flood the engine. Scoops, if used, should be fitted backward (i.e., with the opening facing aft), but then on a hull moving at speed (especially when planing), this may cause a vacuum that stalls the water flow. In general, it's best not to fit a scoop!

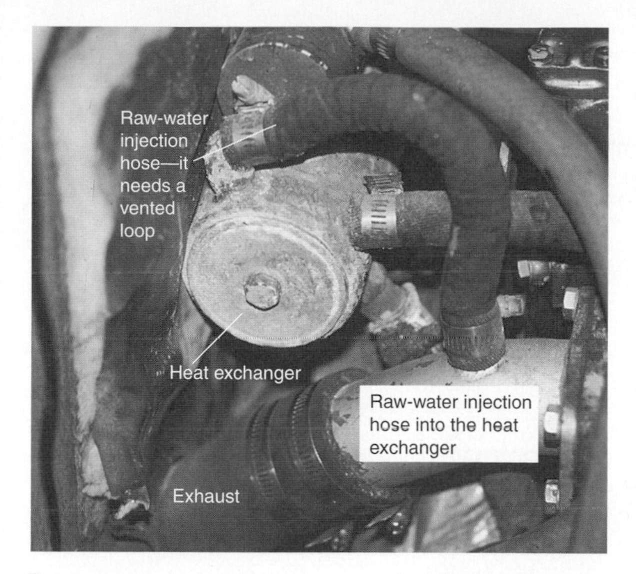

FIGURE 9-45A. *An improper raw-water installation. The water injection line goes directly from the heat exchanger to the exhaust without a vented loop.*

On any engine that is below the waterline (most sailboat engines), both the water-injection line into the exhaust and the exhaust pipe itself create the potential for water to siphon back into the exhaust, fill it, and flow into the engine via open exhaust valves (see Figures 9-45A, 9-45B, and 9-45C). The injection line must be looped at least 6 inches (15 cm) above the loaded waterline at all angles of heel, and preferably 12 inches (30 cm), with a siphon break at the top of the loop (see Figures 9-45D and 9-45E).

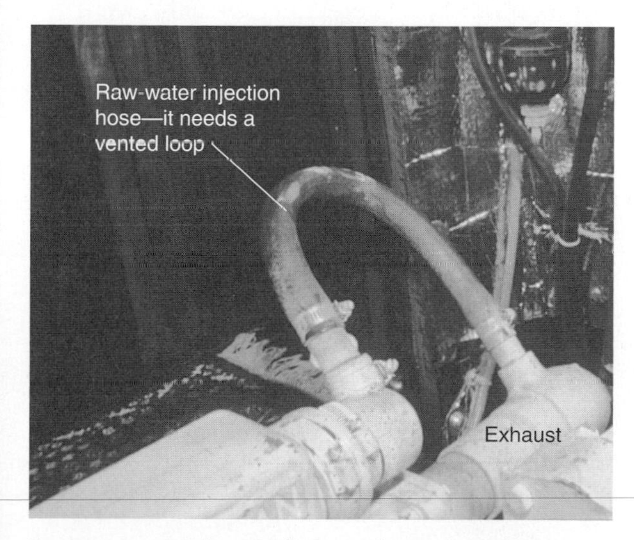

FIGURE 9-45B. *Another installation without a vented loop.*

FIGURE 9-45C. *Yet another installation without a vented loop!*

Traditional rubber-flap siphon breaks (see Figure 9-45F) tend to plug with salt. Once plugged, they are inoperative and may also spray salt water all over the running engine and its electrical systems (see Figure 9-45G), adding insult to injury. The spring-loaded types (e.g., Scot Pump—www.scotpump.com—see Figure 9-45H) are far less prone to plugging than the traditional rubber-flap type. My preference, however, is to remove the valve element altogether, add a hose to the top of the vented loop, and discharge this well above the waterline (into the cockpit works well; see Figure 9-45I). If you plan to do this, you must discharge above the highest point in the exhaust hose to prevent water weeping out of the discharge fitting when the engine is running. Installed like this, the vent hose will never plug and cannot put salt water in the engine.

The exhaust and injection water end up in a water-lift-type muffler (silencer). From here, loop the exhaust hose well above the loaded waterline (at least 12 inches/30 cm) and then maintain a slope of

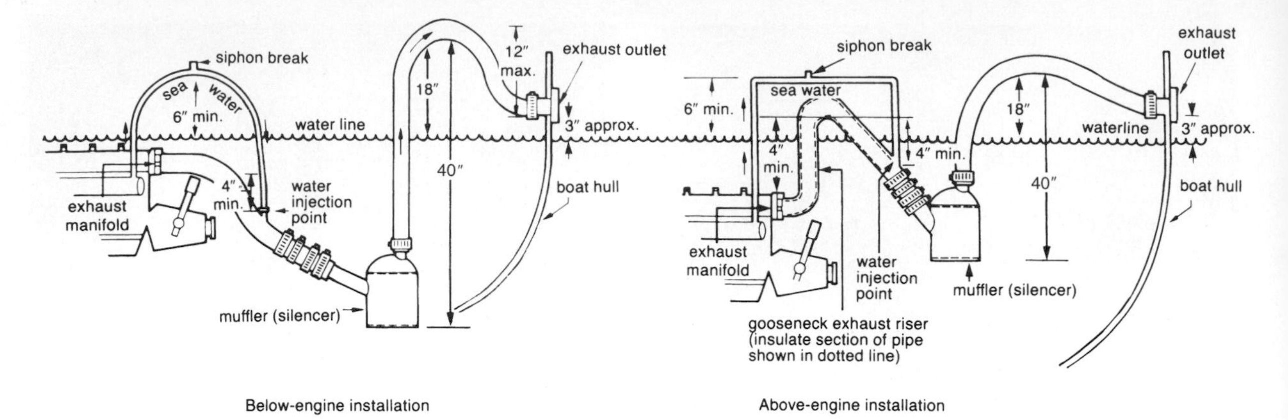

Figure 9-45D. *Correct water-lift muffler installations. Notes: 1. It would be preferable to fit a shutoff valve in the exhaust line, especially on sailboats, to prevent following seas from driving up the exhaust pipe when the engine is shut down. 2. The water-lift muffler should have a volume at least as great as the volume of the vertical section of exhaust pipe exiting the muffler. (Jim Sollers and Allcraft Corporation)*

Figure 9-45E. *The same engine as in 9-45A after installation of a vented loop in the raw-water circuit, well above the waterline.*

Siphamatic valve

Mount with rubber valve down

Open to atmosphere

Rubber valve shown in closed position

Figure 9-45F. *A Kohler siphon break. (Courtesy Kohler)*

FIGURE 9-45G. *This boat has a vented loop that was spraying water into the engine room, so the owner capped the loop and rendered it inoperative!*

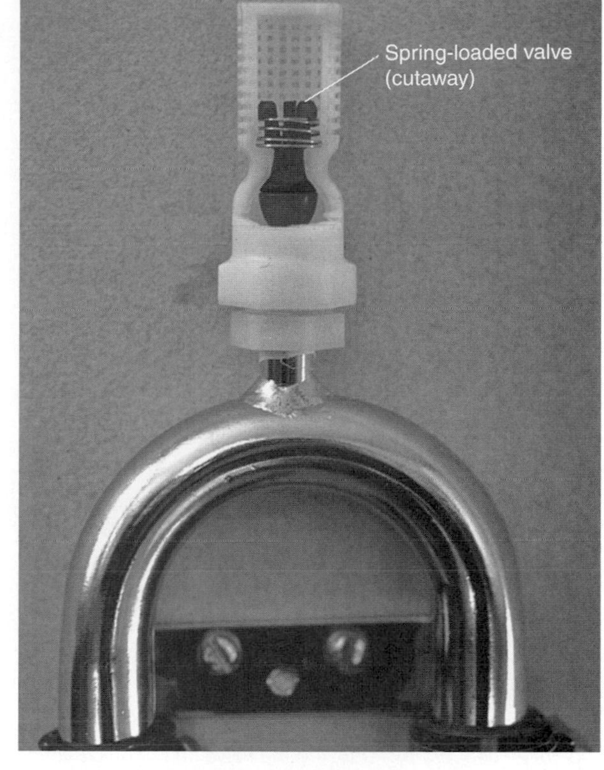

FIGURE 9-45H. *A spring-loaded siphon break.*

at least ½ inch per foot down to the through-hull. If the exhaust hose is long, place the loop close to the engine to minimize the volume of water that can flow back into the muffler when the engine is shut down. Also make sure there are no low spots where water can be trapped to flow back into the engine when the boat pounds and rolls. In sailboats, it is a good idea to install an *accessible* positive shutoff valve at the through-hull in case following seas threaten to drive up the back of the boat when the engine is not in use.

On many modern boats, a water separator is installed at the top of the exhaust loop (see, for example, Halyard, the originators of this device—www.halyard.eu.com—or Centek Industries—www.centekindustries.com). This separates the cooling water from the exhaust gases, allowing them to be discharged separately. It also, in effect, functions as a giant siphon break, adding another layer of protection against engine flooding. Another way to minimize the chance of backflooding through the exhaust is to install a plenum, or gooseneck, at the high point in the hose. This is nothing more than a chamber in which the exhaust hose from the engine comes in at the top, and the hose to the through-hull exits from the bottom. If a wave drives a slug of water up the exhaust, the sudden increase in exhaust cross section and volume in the plenum acts as a temporary reservoir that drains as soon as the hydrostatic pressure eases.

The higher an exhaust, water separator, or plenum/gooseneck is looped above the waterline, the greater the security from siphonic action, but the greater the back pressure (see the Measuring Exhaust Back Pressure sidebar) since the cooling water has to be lifted to this high point. (Note also that when the engine is shut down, the water in the vertical section of the exhaust pipe will run back into the water-lift muffler; therefore, the muffler must have a volume at least as great as that of this vertical section of pipe.) To keep back pressure within acceptable limits, the vertical lift of the cooling water

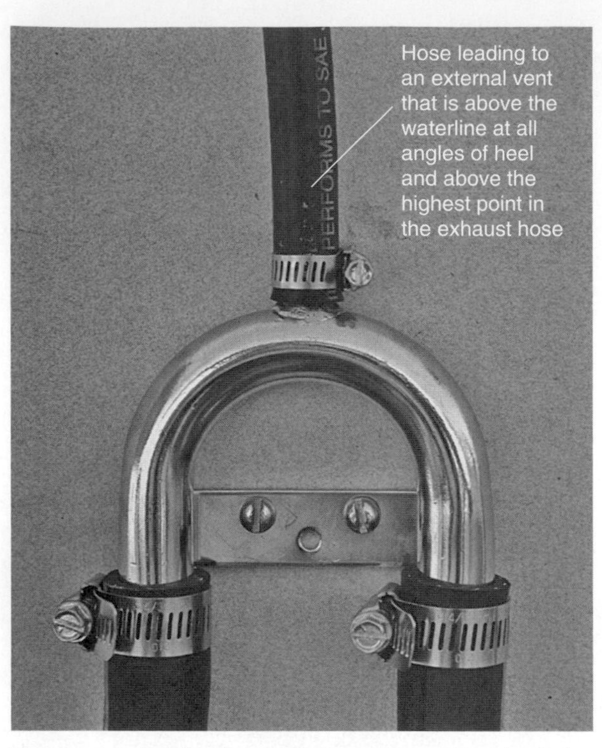

FIGURE 9-45I. *A siphon break on an engine cooling circuit that has been adapted by removing the valve and adding a length of hose, which is vented into the cockpit.*

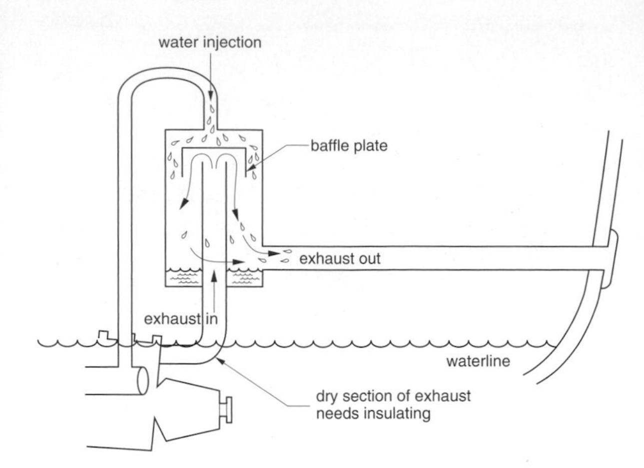

FIGURE 9-45J. *A standpipe-type muffler and exhaust installation.*

should not exceed 40 inches (1 m) on naturally aspirated engines (i.e., without turbochargers). This corresponds to a back pressure at full load of somewhere around 1.5 psi. On turbocharged engines, the water lift should be kept down to 20 inches (½ m) to give a full-load back pressure of around 0.75 psi.

In situations where a standard water-lift exhaust will create excessive back pressure, a *standpipe* exhaust may be used in which the muffler is raised as in Figure 9-45J. In general, the less the back pressure, the less the silencing effect, so an additional in-line muffler (silencer) may be required. The dry section of exhaust from the exhaust manifold to the standpipe needs effective insulation.

AUXILIARY EQUIPMENT

An increasing amount of extra equipment is driven off the engine these days, including alternators, bilge pumps, washdown pumps, hydraulic pumps, refrigeration compressors, watermakers, and AC generators. The normal method of driving this equipment is to add extra pulleys to the crankshaft or to fasten a small auxiliary stub shaft to the forward end of the crankshaft, install a couple of pulleys on it, and belt-drive the equipment from these pulleys. The following are caveats to keep in mind:

1. Equipment mounted in this fashion exerts a lateral pull on the crankshaft. There is a strict limit to how much side-loading an engine can tolerate without damaging crankshaft oil seals and bearings. Check the manufacturer's specifications to see that this limit is not exceeded. (Note: Placing loads on opposite sides of the engine partially cancels side-loading forces.)
2. If the engine is flexibly mounted (most are) but auxiliary equipment is mounted off the engine (i.e., fastened to the hull or a bulkhead), this equipment can flex the engine on its feet, pulling it out of alignment with the propeller shaft. This flexing will also alter the drive-belt tension on the auxiliary equipment, which may cause problems. All such equipment should be mounted on a common base plate with the engine.
3. The driving pulleys on an engine stub shaft must be in alignment with the driven pulleys on the auxiliary equipment. A straight rod or dowel held in the groove of one pulley should drop cleanly into the

Measuring Exhaust Back Pressure

Exhaust back pressure can be checked with a sensitive pressure gauge designed to measure in *inches of mercury* or *inches of water* rather than in pounds per square inch (psi). (See Table 9-1 for conversion of one to the other.) It can also be checked with a homemade manometer. This need be nothing more than a piece of clear plastic tubing of any diameter greater than 1/4 inch, fixed in a U-shaped loop to a board about 4 feet long. The board is marked off in inches (see Figure 9-46).

Set the board on end and half fill the tubing with water. Connect one side of the tubing to a fitting on the exhaust as close to the exhaust manifold as possible (but 6 to 12 inches *after* a turbocharger). Leave the other end of the tubing open. If the manifold has no suitable outlet to make the connection, drill an 11/32-inch hole, tap this for a 1/8-inch pipe fitting (standard pipe thread), and screw in an appropriate fitting. When finished, remove the fitting and fit a 1/8-inch pipe plug.

Note the level of the water in the tube with the engine at rest. Then crank the engine and *fully load* it. *This is important; if necessary, tie the boat off securely to a dock, put it in gear, and open the throttle.* The exhaust back pressure will push the water down one side of the tubing and up the other. The difference between the two levels, measured in inches, is the back pressure in *inches of water column.* On naturally aspirated engines, it should not exceed 40 inches of water (3 inches Hg; approximately 1.5 psi); on turbocharged engines and Detroit Diesel two-cycles, 20 inches of water (1.5 inches Hg; 0.75 psi).

All too often an engine is found to be well outside these limits, especially where long exhaust runs with numerous bends are involved. Anything that can be done to shorten the hose run, reduce bends, and *enlarge the hose size* will help reduce back pressure.

TABLE 9-1. Pressure Conversion Factors

To Convert	Into	Multiply By
Psi	Hg″	2.036
Hg″	Psi	0.4912
Psi	H_2O″	27.6776
H_2O″	Psi	0.03613
H_2O″	Hg″	0.07355
Hg″	H_2O″	13.5962

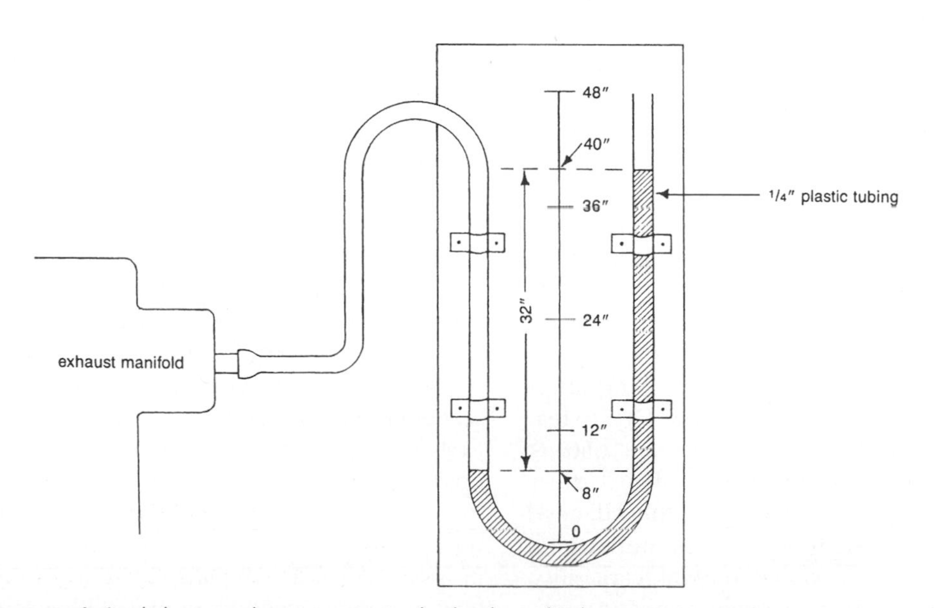

FIGURE 9-46. *A simple homemade manometer to check exhaust back pressure.*

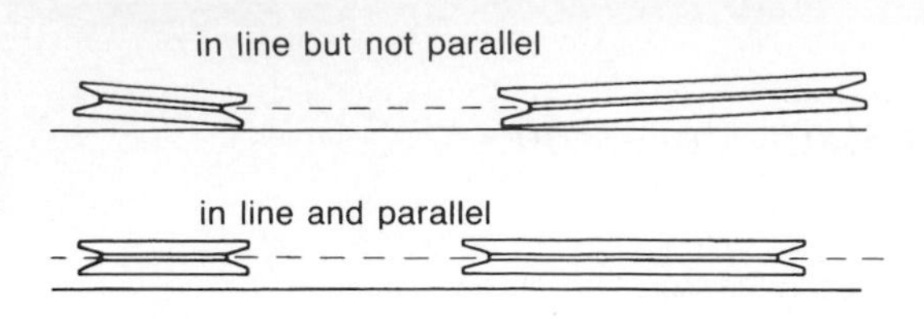

FIGURE 9-47. *Pulley alignment.*

groove on the other pulley (see Figure 9-47 for examples of misalignment).

4. The auxiliary loads may interfere with the engine's cranking speed enough to make starting difficult in cold weather. It may be necessary to have a mechanism to take the loads off-line.

SOME ELECTRICAL CONSIDERATIONS

Diesel engines have compression ratios well above those of gasoline engines (two to three times higher). They are therefore harder to turn over. In spite of this increased resistance to cranking, turning over briskly is essential for successful starting, especially in cold weather. These two things taken together mean that diesel engines require powerful batteries for effective starting. You'd be wise to install a battery at least as powerful as the engine manufacturer recommends; additional capacity will be a bonus.

If an engine-cranking battery is also used to power the ship's electrics, it will steadily be discharged when the engine (and alternator) are shut down, then later recharged when the engine runs. This is called *cycling*. Any battery designed for engine cranking that is repeatedly cycled will soon fail. A deep-cycle battery is needed for cycling applications.

Batteries are generally rated in amp-hours, a measurement of their ability to deliver a relatively low current over a 20-hour period. Engine cranking needs a very large current for a short period of time, and a rating known as cold cranking amps (CCA) is used to measure this capability. Due to features of internal construction, a deep-cycle battery with the same amp-hour rating as a cranking battery will deliver fewer cold cranking amps; you will need to buy a deep-cycle battery with a larger overall capacity to give you the same cranking capability. The key here is to check the cold cranking amps rating rather than the amp-hour rating.

An engine-driven alternator is normally the primary charging device on a boat once the boat is away from the dock. It is not uncommon to see anchored-out boats running their engines 2 hours or more a day to recharge batteries. Diesels do not like this kind of prolonged low-load running at near-idling speeds.

If you plan to go cruising, you will save handsomely on fuel and repair bills in the long run if you buy a high-output alternator and a purpose-built multistep voltage regulator with your new engine. The alternator will impose a heavier demand on the engine while keeping the low-load running hours to a minimum. If you can load up the engine some more when at anchor (e.g., with a refrigeration compressor), then so much the better. Even better still would be to take care of your charging needs altogether with a wind generator, solar panels, or both. (See my *Boatowner's Mechanical and Electrical Manual* for a detailed discussion of all these subjects.)

SERVICEABILITY

As I have repeatedly pointed out, two things are absolutely critical to the long life of a diesel engine—clean fuel and clean oil. If you keep the fuel uncontaminated and properly filtered, and you change the oil and filter at the prescribed intervals (generally every 100 to 250 running hours), most diesels will run for years without trouble.

An engine installation needs to be designed to simplify these items of basic maintenance. The fuel and oil filters must be readily accessible, and the engine oil easy to change (see Figure 9-48). Changing the oil in most boats requires an oil change pump because the drain plug in the engine pan is generally hard to get at, and even if you can remove it, there is no room to slip a suitable container under the pan to catch the oil.

The raw-water pump must be accessible, if only so you can loosen its cover to drain it when you winterize the boat. And what about any grease points (most modern engines don't have any)? The propeller shaft seal is another item to think about; so often you have to slide in headfirst through a cockpit locker and hang upside down in the bilges to get at it. What if the engine develops other problems? I once worked on a boat in which the entire engine had to be pulled off its mounts in order to change the starter motor. If the engine needs a major overhaul, will you be able to remove the cylinder head? The time to think about

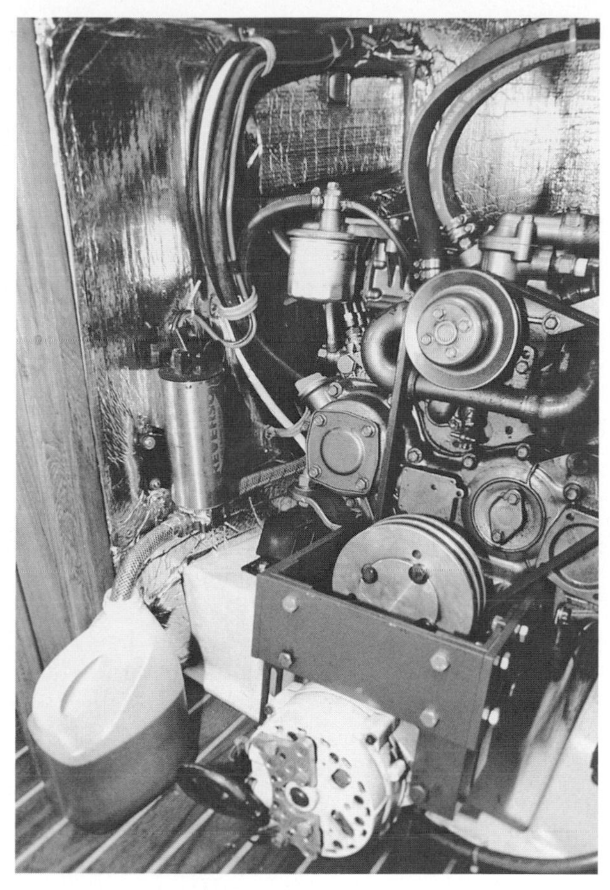

FIGURE 9-48. *Good access, and an oil change pump, made oil changes easy on our old boat.*

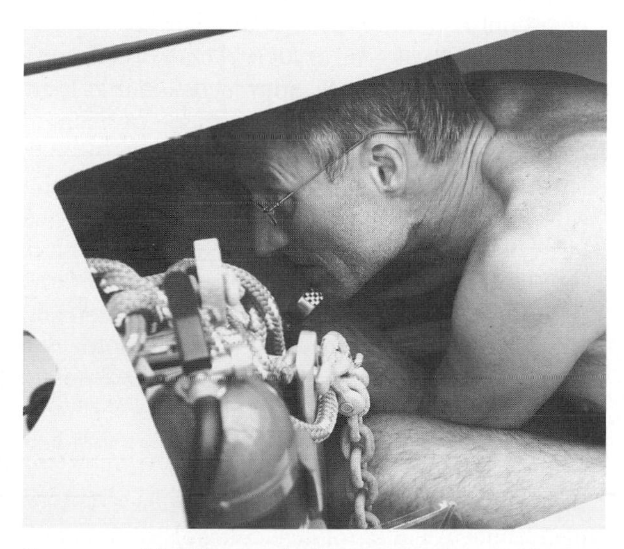

FIGURE 9-49. *"Now how am I going to get at that . . . ?"*

these things is now, not when you are faced with a breakdown out in the boonies (see Figure 9-49).

POSTSCRIPT: DIESEL-ELECTRIC PROPULSION

"Diesel-electric propulsion seems kind of dumb to me." I was involved in yet another debate sparked by our decision to sell our new Malö 45 in order to replace it with the exact same boat, but with a diesel-electric propulsion system and also a distributed power system.

My new acquaintance continued, "You take a diesel engine that is directly coupled to a propeller, and then you uncouple it and place a generator and an electric motor—two more energy conversion processes, with losses at each conversion—between it and the propeller. Where's the sense in that?"

EFFICIENCY

The added efficiency has something to do with the fact that electric motors develop full torque at 0 rpm, and thus can swing a more efficient propeller with a much larger blade area at a slower speed, and also that the generator's speed can be adjusted to match the load on the electric motor in a more efficient manner than when the engine is directly coupled to the propeller.

Remember the horsepower curves we discussed at the beginning of this chapter? Curves showing the horsepower required to spin a propeller at different speeds are concave in shape, whereas curves showing the power produced by an engine at different speeds are convex (see the Understanding Engine Curves sidebar on pages 227–28). In a properly sized boat installation, these two curves only meet at one point—maximum boat (propeller) speed and full engine speed. At anything less than full boat and engine speed, there is an increasing divergence.

Boats rarely run at full speed, however. Let's assume we are cruising at a little under hull speed. This requires us to spin our propeller at a certain speed, which in turn requires the typical diesel engine we used earlier, rated for a top speed of 3,600 rpm, to run at around 3,000 rpm. Because of the divergence between the propeller's power curve and the engine's power curve, at this particular engine speed the engine is capable of producing more power than the propeller needs, and as such it runs underloaded and inefficiently. In general, the slower we motor, the more inefficient the installation. If we uncouple the engine from the propeller and use it to power a generator that drives an electric motor, we can now slow down the engine to the minimum speed necessary to produce the power absorbed by the propeller and no more. The slower the boat (and propeller) speed, the greater the likely improvements in efficiency.

All the anecdotal data that I have seen from a number of diesel-electric installations show a significant increase in efficiency with substantial improvements in fuel economy. If the boat is large enough to have a stand-alone generator in a conventional installation (as does our Malö 45), then the generator in the diesel-electric installation can also be used to supply house loads, eliminating the stand-alone generator. In this case, the overall weight of the installation is likely to be less (plus there's one less engine to be fitted into the boat and serviced at regular intervals).

Other Benefits

Decoupling the diesel engine from the propeller has a number of other major benefits, the most important of which are:

1. The power plant can be placed anywhere on the boat, which gives the hull designer new options in terms of weight distribution, and gives the interior designer a much freer hand with the interior spaces.
2. On many boats, the power plant can be located so it provides much better servicing access than conventional installations.
3. The power plant can be soft mounted and fully sound shielded to the point that there is almost no vibration and the installation is just about inaudible—a huge improvement over most traditional installations.
4. On boats with two propellers (e.g., catamarans), a single power plant can be used to provide power to both propellers.
5. Low-speed maneuvering capabilities are improved partly because of the powerful, low-speed torque of electric motors, but also because the electric motor can be run at much lower speeds than the propeller in a conventional installation. (In a conventional installation, the propeller speed is governed by the idle speed of the engine, which is typically around 700 rpm.)

This compelling list of benefits has led to the adoption of diesel-electric propulsion systems in many large-ship applications, including cruise ships, navy vessels, tugs, and ferries. The obvious question is, why haven't we seen this technology in the recreational marketplace sooner?

Overcoming Obstacles

The efficiency gains on larger vessels are achieved by operating a series of AC generators in place of a single propulsion engine. At low loads, only one generator is run, and then as the load increases, more are brought online. In this way, the generators are always operated at the optimum (most efficient) point on their power curves, as opposed to a conventional installation in which the engine has to be powerful enough for maximum power, but then spends almost all its time running at an inefficient point on its power curve.

Obviously, on most recreational boats there are not going to be multiple propulsion generators. It has taken a fair amount of research to develop alternative approaches that optimize engine efficiency in a single engine installation. One approach (most notably pursued by Fischer Panda in Europe) is built around variable-frequency (and therefore variable-speed) AC generators feeding AC motors. The other (most notably pursued by Solomon Technologies, Glacier Bay, and Siemens in the United States) uses a DC generator to feed DC motors. Because it is not frequency sensitive, the DC generator can be run at whatever speed is appropriate to the load. In all cases, the propulsion motors are run at relatively high voltages (up to 800 volts AC or DC in larger installations).

What has made both approaches viable is the development of powerful, lightweight, permanent-magnet generators and motors (they use high-density, rare earth, neodymium magnets) together with powerful electronic power conversion devices derived from the DC-to-AC inverter marketplace. The DC motors, for example, are, in fact, AC motors that accept a DC input, which is then converted to variable frequency AC by a motor controller that employs inverter-based technology. Fischer Panda's AC motors, which are not directly driven by the AC generators, use similar technology between the generator and motor.

An outgrowth of this generator and motor development has been a range of high-voltage motors suitable for such things as windlasses, electric winches, bow thrusters, and, in the case of the Glacier Bay product line, refrigeration and air-conditioning compressors. These are typically smaller, lighter, and more efficient than the traditional motors they replace. An added benefit is that anytime motor voltage is raised, the supply cables can be reduced in size (doubling of the voltage reduces the cable size by 75% for the same rated output), which in many instances (notably windlasses and bow thrusters) gets a considerable amount of heavy cabling out of the boat.

The Breakthrough Concept

For years, the most active proponent of diesel-electric propulsion for sailboats was Solomon Technologies. Their particular focus was on running the electric propulsion motor for as long as possible off a battery bank, with the ability to recharge the batteries from a freewheeling propeller when the boat is under sail. But this approach results in a large, heavy, expensive battery bank and drives the installation cost through the roof, confining the technology to a fringe application.

The breakthrough comes when you accept that just about anytime the electric propulsion motor is in use, you'll run a generator to power it. Now the battery bank becomes essentially irrelevant and can be sized just as it would be on a normal boat. This is the approach followed by Fischer Panda, Glacier Bay, Siemens, and others. The advanced generator and motor designs they have developed complete the picture. Conceptually, it is now all there, although at the time of writing (2006), there is still a lack of real-life testing of the latest generation of products and of their increasingly sophisticated (read: complex) electronic controllers.

At the end of the day, the power plant is still a diesel engine powering the generator end of things, so the prescriptions in this book remain as relevant as ever.

APPENDIX A

CARBON MONOXIDE POISONING

All forms of heating (except electric heat) and engine combustion involve burning a fuel, which converts oxygen from the air to carbon dioxide. If the air that is consumed comes from cabin spaces, oxygen levels may be steadily depleted. In itself, this is potentially hazardous. If combustion is at all incomplete, as it almost always is, carbon monoxide will also be created, as opposed to carbon dioxide. Carbon monoxide can be deadly.

Carbon monoxide poisoning is the leading cause of death by poisoning in the United States (I don't have the equivalent figures for Europe). Boatowners are particularly at risk because (1) boats contain a small volume of air; (2) boats are necessarily tightly sealed (to keep the water out); and (3) within these small, sealed spaces, there are generally several appliances capable of producing carbon monoxide (e.g., a galley stove, the main engine, an AC generator, a nonelectric cabin heater, and maybe a gas- or diesel-powered water heater).

How It Happens

What makes carbon monoxide such an insidious killer is the fact that it is odorless, colorless, tasteless, and more or less the same weight as air (just a little lighter), so that it tends to hang around. When present in the air, it is inhaled into the lungs. From the lungs, carbon monoxide enters the bloodstream, where it binds with blood hemoglobin (red blood cells), replacing critical oxygen molecules and forming something known as *carboxyhemoglobin* (COHb). As a result, the body is deprived of oxygen.

Carbon monoxide binds to blood hemoglobin far more readily than oxygen does, even if oxygen is available. *The mere presence of carbon monoxide is dangerous, with or without plenty of fresh air.* Once attached to the hemoglobin, carbon monoxide is relatively stable. As a result, very low levels can progressively poison people.

Permissible limits, as set by various U.S. regulatory agencies, range from 35 to 50 parts per million (ppm) if sustained for an 8-hour period. (To put this in perspective, 1 ppm is the equivalent of one drop of food coloring in 13.2 gallons/50 liters of water.) As little as 0.2% carbon monoxide in the air (2,000 ppm) binds up red blood cells (i.e., forms carboxyhemoglobin) at the rate of 1% (of the body's red blood cells) per minute. If the person is doing work, the rate can double to over 2% per minute. Within 45 minutes, 75% of the red blood cells are taken up with carbon monoxide, resulting in a lethal concentration. At the other end of the scale, high doses of carbon monoxide can be lethal in a matter of minutes (Table A-1). If a victim escapes death, there may still be permanent brain damage.

THE SYMPTOMS

Typically, carbon monoxide poisoning produces a range of symptoms, beginning with watery and itchy eyes and a flushed appearance, then progressing through an inability to think coherently, headaches, drowsiness, nausea, dizziness, fatigue, vomiting, collapse, coma, and death. Note that many of the early symptoms are similar to those of seasickness, flu, or food poisoning—all too often people suffering from carbon monoxide poisoning fail to recognize the problem. Note also that the poisoning creeps up on people; it dulls the senses, causing a failure to recognize the problem, which enables the fatal punch to be delivered. BoatU.S. quotes an example of a family of three on a large sailboat with a washing machine and a propane-fueled water heater. The washing machine malfunctioned, causing the water heater to stay lit. The heater produced carbon monoxide. The son fell, tried to get up, and fell again. Hearing the loud thump, the wife got up, was overcome, and collapsed.

TABLE A-1. Carbon Monoxide Poisoning Symptoms

CO Amount (ppm)[1]	Symptoms
200	Slight headache within 2 to 3 hours.
400	Frontal (migraine) headache within 1 to 2 hours.
800	Dizziness, nausea, and convulsions within 45 minutes. Insensible within 2 hours.
1,600	Headache, dizziness, and nausea within 20 minutes. Death within 60 minutes.
3,200	Headache, dizziness, and nausea in 5 to 10 minutes. Death within 30 minutes.
6,400	Headache and dizziness in 1 to 2 minutes. Death in less than 15 minutes.
12,800	Death in less than 3 minutes.

[1]PPM = parts per million carbon monoxide in the atmosphere; 1,000 ppm = 0.1% carbon monoxide in the atmosphere.
Courtesy Fireboy-Xintex.

"My husband saw me go down and thought I had fainted because of seeing our son (whose lip was bleeding). He stepped over me to assist our son, and looking back, he noticed the cat lying beside me. It clicked—I might faint, but not the cat. He picked up the phone and pushed a preprogrammed button to call our neighbors and let them know we were in trouble. He tried to pull us out, but he, too, was going down." All three were rescued by the local fire department and regained consciousness.

What is particularly interesting about this case is that the boat's hatches were wide open with a 10-knot wind blowing outside. I have other similar accounts on file. The question is, what can be done to prevent such incidents?

MINIMIZING CARBON MONOXIDE FORMATION

Engine Installations

All engine exhausts contain carbon monoxide; *gasoline engines produce far higher levels than diesels*. (For this reason, Onan, a major generator manufacturer, stopped selling gasoline-powered AC generators in the marine market and conducted an extensive advertising campaign warning of the dangers of carbon monoxide poisoning.) Cold, poorly tuned, and overloaded engines produce more carbon monoxide than warm, properly tuned, and load-matched engines. So the first task in minimizing the potential for carbon monoxide formation is to ensure that the engine is properly matched to its task, and as far as possible, will be operated as designed. There is always, however, the cruising sailor who uses the engine more for battery charging at anchor than for propulsion, which results in the engine running long, underloaded hours below its designed temperature. There is not much that can be done about this, other than to make sure that the carbon monoxide is gotten out of the boat.

In recent decades, the drive to make engines increasingly quiet has led to ever-tighter, better-insulated engine boxes and engine rooms. On the exhaust side, the near-universal use of water-lift silencers adds a measure of back pressure to the system. Both of these design trends increase the probability of restricting the airflow through an engine, and as such, increase the likelihood of carbon monoxide formation.

To keep carbon monoxide out of accommodation spaces, it is crucial to have an exhaust system that is gas-tight to the hull. This in turn requires adequate support and strain relief built into the exhaust system (to absorb engine vibration without failure), the use of galvanically compatible materials (to lessen corrosion), proper marine exhaust hose in wet exhaust systems, and double-clamping of all hose connections with all-stainless-steel hose clamps. Each engine on a boat must have its own dedicated exhaust system, with nothing teed into this exhaust (with the sole exception of a cooling-water injection line on a water-cooled exhaust).

Regardless of the quality of the initial exhaust installation, *an exhaust system is a regular maintenance item*. Leaking gasoline exhausts on boats are far and away the leading cause of death from carbon monoxide poisoning. At least annually, inspect the entire system for any signs of corrosion or leaks. Warning signs are discoloration or stains around joints, water leaks, rusting around the screws on hose clamps, corrosion of the manifold discharge elbow on water-cooled engines, and carbon buildup within the exhaust (which will increase the back pressure, the production of carbon monoxide, and the probability of leaks). To check a discharge elbow, remove the exhaust hose, which will also enable you to check for carbon buildup.

Fuel-Burning Appliances

When it comes to fuel-burning appliances, which are almost always in the accommodation spaces themselves, certain other protective measures need to be taken. The optimum situation is one in which the appliance has its combustion air ducted in from outside the accommodation spaces, with combustion occurring inside a sealed chamber, which then exhausts through an external flue (this setup is known as a *sealed combustion system*). As long as the combustion chamber does not corrode through, such an appliance cannot cause oxygen depletion, nor can it emit carbon monoxide directly into accommodation spaces.

Some cabin heaters are built in this fashion, although commonly the inlet air is drawn from within the boat, with the exhaust plumbed outside the boat. In this case, as long as there is no back draft down the flue (a matter of proper design, although there may be situations in which the airflow off sails will cause back drafting with just about any flue), and as long as the flue is not obstructed (primarily a matter of design once again, in particular ensuring that the flue cannot trap water), even in a situation of oxygen depletion and carbon monoxide formation, the carbon monoxide will be vented outside accommodation spaces.

However, note that the nature of ventilation on boats is such that it is sometimes possible to create a *negative* pressure (with respect to the outside air pressure) inside the boat. This occurs, for example, when the hatches, ventilators, openings, and canvas structures (such as dodgers) are lined up with respect to the wind so that air is sucked out of the cabin rather than driven in (this is not hard to do). In such a situation, any combustion chamber that draws its inlet air from inside the boat has the potential to feed carbon monoxide into the boat. Over the years, this has been the cause of a number of deaths.

With or without ducted inlet air, check the combustion chamber as part of your regular maintenance schedule to ensure that there is no damage or corrosion. Also inspect inlet and exhaust ducting to make sure it is gas-tight.

Safety can be enhanced by the addition of an *oxygen depletion sensor* and wired such that it automatically cuts off the fuel supply to an appliance in the event of oxygen depletion (this is an ABYC requirement on all systems that do not have sealed combustion chambers). It should be noted, however, that in a situation where carbon monoxide is produced but there is still a good airflow through the cabin, the oxygen depletion sensor will do nothing to protect the occupants.

Unvented Appliances

Then there are all those appliances that not only draw their air from accommodation spaces but also exhaust the combustion gases into the same atmosphere. These include all nonelectric galley stoves and also some cabin heaters and water heaters. These appliances are potentially the most lethal of all; boatowners need to think long and hard before using them. It is essential to understand that these appliances:

- Should *never* be used when *unattended* or when *sleeping* (carbon monoxide is especially dangerous when sleeping since victims don't feel any side effects and may simply not wake up).
- Should *only* be used in conjunction with *adequate ventilation*. In particular, *galley stoves should not be used for cabin heat* (which is easier said than done if you have no other source of cabin heat; I have to confess that we have used our stove from time to time, but never unattended, and never when sleeping).

User Education

This leaves certain potentially lethal situations that cannot be eliminated at the equipment design and installation phases. Examples include running an engine with the boat up against a dock so that the exhaust is deflected back into accommodation spaces (see Figure A-1) or running the engine when rafted to another boat, with similar results (see Figure A-2). The operation of AC generators in such situations is of particular concern, especially if they are run at night when people are sleeping, which is commonly done to keep an air conditioner going. Boats with a lot of canvas aft (dodgers, biminis, and cockpit enclosures) can create what is known as the *station wagon effect* (see Figure A-3). This is more likely on a powerboat at speed but can also occur on sailboats.

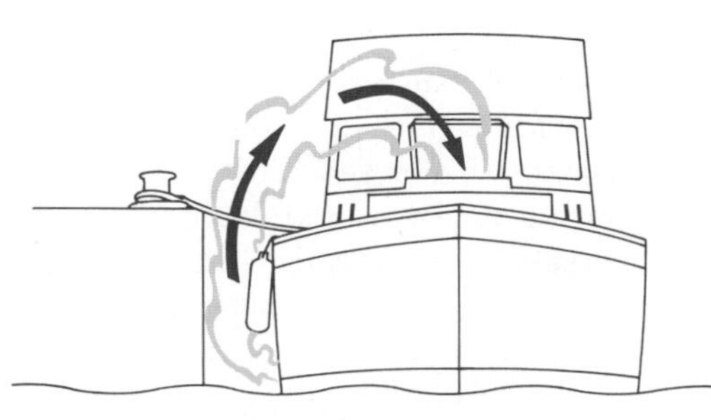

FIGURE A-1. *Running an engine alongside a dock can put carbon monoxide into the boat. (Jim Sollers)*

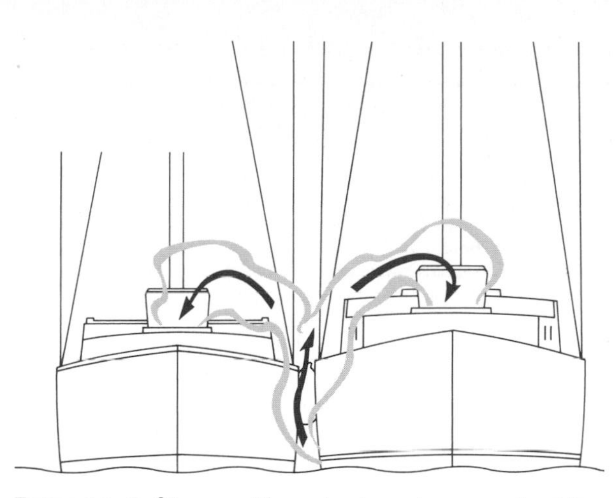

FIGURE A-2. *Rafting up, with one boat running an engine (often a generator for air-conditioning) can put carbon monoxide in the boats. (Jim Sollers)*

And then there is always the cruising sailor who decides to save the wear and tear on the main engine when battery charging at anchor by buying an inexpensive portable gasoline generator, which is placed on the foredeck for an hour a day and run with the exhaust blowing down the hatch! These kinds of hazards can only be mitigated through a process of public education and the use of effective carbon monoxide alarms.

CARBON MONOXIDE ALARMS

The design of an effective carbon monoxide alarm is a complicated business. This is because relatively low levels of carbon monoxide over an extended period of time can be just as lethal as high doses over a short period of time. Conversely, relatively high levels over a short period of time are not necessarily harmful (see Figure A-4). It is not unusual for there to be relatively high levels from time to time that rapidly disperse (for example, when an engine is first cranked at dockside, or when a boat in close proximity to other boats fires up its engine or generator, with the exhaust drifting across the other boats). If an alarm is designed simply to respond to a given threshold level of carbon monoxide, the threshold must be set at a very low level to protect against long-term, low-level contamination. However, this will cause the alarm to be triggered by any short-term rise in carbon monoxide levels, resulting in many nuisance alarms (which almost invariably results in the alarm being disconnected or bypassed by the boat operator, at which point the alarm is effectively useless). If, on the other hand, an alarm is set to respond to a higher threshold, it will provide no protection against low levels of carbon monoxide contamination sustained over long periods of time.

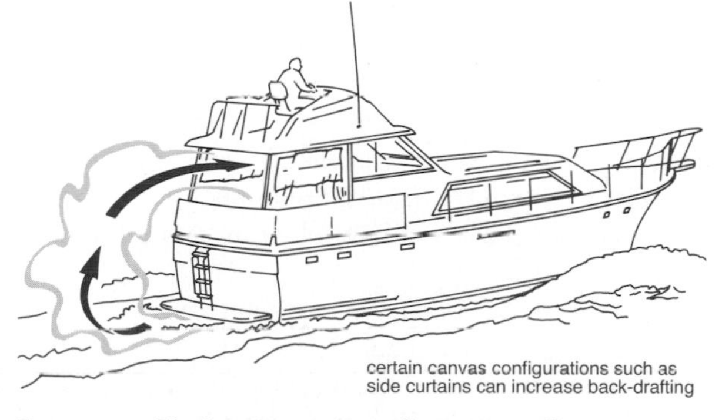

FIGURE A-3. *The "station wagon" effect. (Jim Sollers)*

Time-Weighted Average

An effective carbon monoxide alarm must have the ability to track carbon monoxide levels over time and to monitor in some fashion the likely impact on carboxyhemoglobin levels. This is a complicated process, particularly since carbon monoxide concentrations, when present, will almost certainly be constantly changing. Newer devices incorporate microprocessors that enable them to keep track of carbon monoxide concentrations over a set period of time (known as a *time-weighted average*—TWA), and to calculate the corresponding carboxyhemoglobin levels in the blood. This would filter out most nuisance alarms if it were not for the fact that it has proven extremely difficult to find affordable sensors that react solely to carbon monoxide (as opposed to styrene emissions from fiberglass and all kinds of other emissions commonly found on new boats). Cell phones and other communications devices in the 866 to 910 MHz band have also caused nuisance alarms. As a result, carbon monoxide alarms sometimes will still falsely alarm.

In the United States, Underwriters Laboratories (UL) has a standard for carbon monoxide detectors in boats (the latest version at the time of writing is UL 2034-2003; it requires the elimination of nuisance alarms caused by new-boat gases and cell phones). It is incorporated in the ABYC standard. These standards require an alarm to sound if the TWA reaches 10% carboxyhemoglobin levels, with an additional requirement that it sound after a maximum

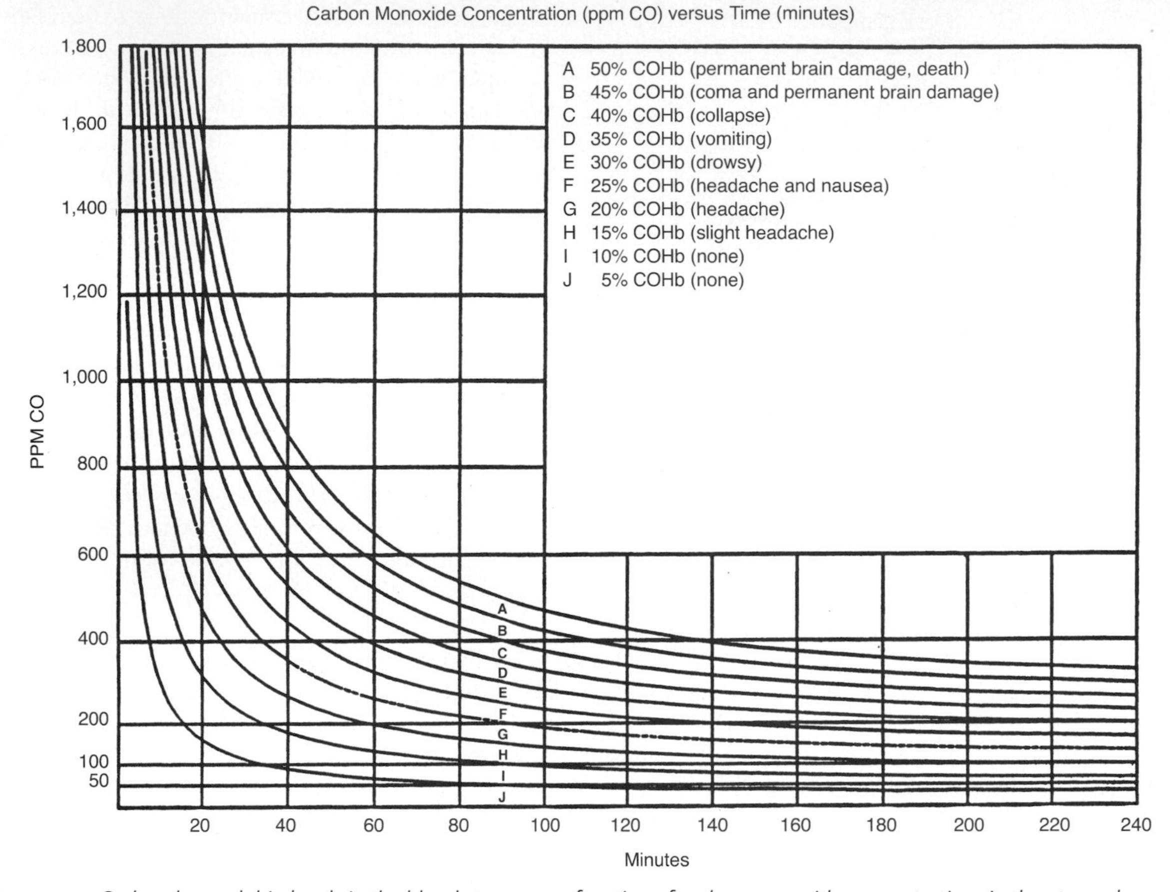

FIGURE A-4. *Carboxyhemoglobin levels in the blood stream as a function of carbon monoxide concentrations in the atmosphere and length of exposure. Note that relatively low concentrations over long periods of time can do as much damage as high concentrations over short periods of time (e.g., 200 ppm CO for 240 minutes results in the same carboxyhemoglobin level—30%—as 1,600 ppm for 10 minutes). (Courtesy ABYC)*

of 3 minutes if CO levels exceed 1,200 ppm, and that it not sound at concentrations up to 30 ppm (see Figure A-5). European standards are similar. Some sources are MTI Industries (www.mtiindustries.com), Fireboy-Xintex (www.fireboy-xintex.com), Fyrnetics (www.fyrnetics.com), and SF Detection (www.sfdetection. com).

Alarm Override

Aside from sensitivity issues, with many carbon monoxide alarms, once they go off, they may continue to sound for some time, which can be annoying to the boat operator who has taken steps to deal with the carbon monoxide problem but must still listen to the alarm. The alarm continues to sound because the internal microprocessor is simulating the carboxyhemoglobin levels in the blood. As noted previously, once attached to hemoglobin, carbon monoxide is relatively stable. Even if the atmosphere is cleared of all carbon monoxide (or other contaminants), the simulated carboxyhemoglobin levels will only reduce slowly; it may take 15 or 20 minutes for an alarm to clear (many alarms will not clear until CO levels are below 70 ppm).

Newer alarms (in the United States) include a button that can be pushed to silence an alarm for up to 6 minutes (this is the limit allowed by the UL standards), after which—if the device still records dangerous levels of

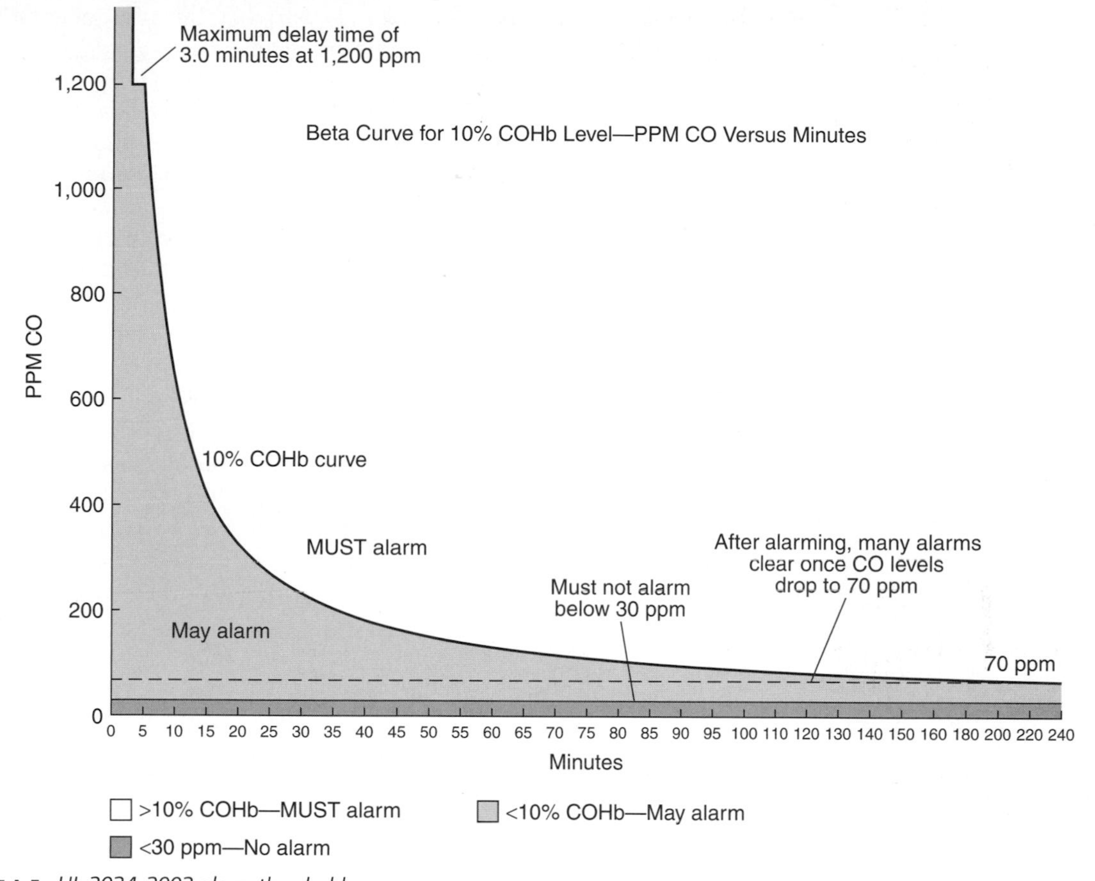

FIGURE A-5. *UL 2034-2003 alarm thresholds.*

carbon monoxide—it will start alarming again, with no additional override allowed. In combination with the kinds of adjustments to microprocessor logic already mentioned, this has resolved many past problems.

Installation Issues

Beyond the manufacturing issues with carbon monoxide detectors, there are installation issues. Ideally, there should be a detector in all sleeping areas. Detectors need to be located away from corners and other dead-air areas that do not experience the natural circulation of air through the boat; on the other hand, they should not be in the airstream from ventilators or air-conditioning ducts that may dilute any concentrations of carbon monoxide in a cabin. In other words, a detector needs to be located somewhere where it receives a representative sample of air. This location needs to be protected from spray and out of the way of likely physical abuse. To minimize nuisance alarms, it should also be at least 5 feet from any galley stove (particularly alcohol stoves, since alcohol is another substance that will trigger most alarms). The detector should be at eye level to make it easy to monitor.

The detector should be wired so that anytime the boat's batteries are in service (i.e., the battery isolation switch is on), the detector is in service. Some boat manufacturers go a step further and wire carbon monoxide detectors to the battery side of the isolation switch so that they are in continuous service. The one disadvantage to this is the fact that many detectors draw 90 to 240 mA at 12 volts. This amounts to a 2 to 6 Ah battery drain per day, which may be enough to kill a Group 27 battery (approximately 100 Ah capacity) stone dead over a period of a month if the boat is not in use and not plugged into shore power or other charging sources. However, some detectors now draw as little as 60 mA, which amounts to less than 1.5 Ah a day in continuous use, or 40 Ah per month. (The ESL SafeAir from GE Interlogix draws 8 mA, but I don't know if it complies with the marine standards; in general, the power drain of detectors is constantly being lowered). Given a moderately powerful DC system, it is practicable to leave these on permanently.

Many CO detectors for homes are relatively cheap and operate on their own battery, which is good for five years or more. Although they may not be tested for compliance with marine standards, one of these will be better than nothing!

Install an Alarm!

The bottom line is that although carbon monoxide alarms have had something of a bad reputation in the past for nuisance alarming, the latest generation has resolved most problems. *They should be standard equipment on any boat with a gasoline engine (including a portable AC generator), any boat that leaves an AC generator (diesel or gasoline) running when crew members are sleeping, and any boat that has a heating system that will operate when people are sleeping.* They are recommended on all other cruising boats.

If an alarm is triggered or carbon monoxide poisoning is suspected for any reason, *the absolute first priority is to get all affected people out into fresh air*, and then shut down all potential sources of carbon monoxide formation. With mild cases of poisoning, fresh, uncontaminated air is all that is needed for recovery. If poisoning symptoms are pronounced, seek medical help.

APPENDIX B

TOOLS

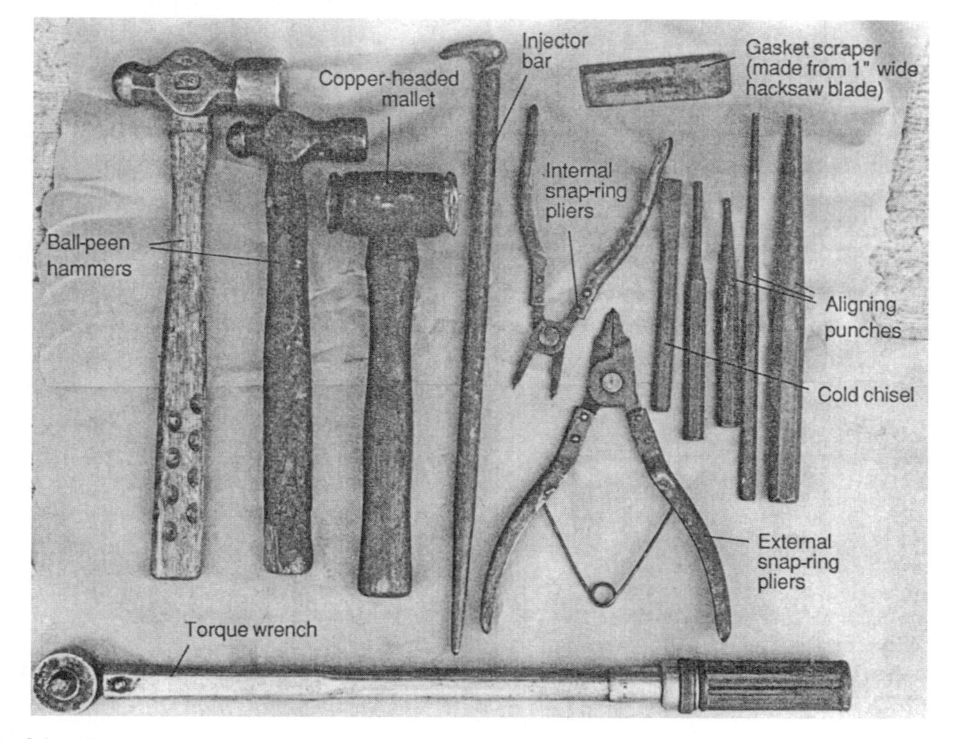

FIGURE B-1. *Useful tools.*

A basic set of mechanic's tools is required for working on engines—wrenches, a socket set (preferably ½-inch drive), screwdrivers, hacksaw, crescent wrench, Vise-Grips, pipe wrench, etc. The tool kit should also include a copy of the appropriate manufacturer's shop manual. This appendix covers one or two more specialized items.

1. Oil squirt can, preferably with a flexible tip, for putting oil into the air-inlet manifold.
2. Grease gun, preferably with a flexible hose.
3. Feeler (thickness) gauges from 0.001 to 0.025 inch (or the metric equivalent if your clearances are specified in mm).
4. Filter clamp for spin-on-type filters. Such filters are difficult to get on and off without this special-purpose tool. More than one size may be required if the oil and fuel filters are a different size.
5. Grinding paste. This is sold in three grades: coarse, medium, and fine. There is very little call for coarse, and the medium and fine very often can be bought in one container.
6. Suction cup and handle for lapping in valves.
7. Torque wrench. An indispensable tool for any serious mechanical work, and probably

the most expensive special item, although there are some perfectly serviceable and relatively inexpensive ones on the market for occasional use. The wrench fits 1/2-inch drive sockets and other sizes with suitable adaptors.

8. Ball-peen hammer. Most people have carpenter's hammers with a jaw for pulling nails, but in mechanical work, a ball-peen hammer is far more useful. Hammers are specified by the weight of the head—an 8-ounce hammer is a good all-around size.
9. Needle-nose pliers. Handy for all kinds of tasks. Side-cutting needle-nose pliers also have a wire-cutting jaw and are preferred.
10. Scrapers for cleaning up old gaskets.
11. Mallet or soft-faced hammer. A surprisingly valuable tool, especially if you have to knock an aluminum or cast-iron casting that might crack under a steel hammer.
12. Aligning punches. Invaluable from time to time, especially the long ones (8 to 10 inches). These punches are tapered—a lighter one (with a tip around 1/8 inch) and a heavier one (around 1/4 inch) will do nicely.
13. Injector bar for prying or levering. An injector bar is about 15 inches long, with a tapered point on one end and a heel on the other. It is a very useful tool.
14. Allen wrenches. You'll almost certainly require them at some point; keep an assortment on hand.
15. Hydrometer (if the boat has wet-cell batteries). Needed for testing batteries. It is best to get one of the inexpensive plastic ones, since the regular glass ones, though more accurate, break sooner or later.
16. Snap-ring pliers. We are now getting into the realm of very specialized equipment for the serious mechanic. Snap rings can almost always be removed with needle-nose pliers, regular pliers, or the judicious use of screwdrivers.
17. Valve-spring clamp.
18. Piston-ring expander. This can be dispensed with as indicated in the text (see pages 183 and 186).
19. Piston-ring clamp. This can also be dispensed with (see pages 188, 190).
20. Gear pullers. These come in all shapes and sizes. There are a number of gears and pulleys in any engine that require the use of pullers in order to remove them without damaging the engine castings or other parts. (Although it is frequently possible to improvise, as indicated in the text—see pages 165–66.) However, what you should never do is to put levers behind a gear to try and force it off—the effort generally ends in failure, and frequently in damage to some casting, the gear, the pulley, or the shaft.
21. Injector nozzle cleaning set. Includes a brass bristle brush and the appropriate nozzle hole prickers for your injectors. Might be a worthwhile investment for the long-distance cruiser. Lucas/CAV and others sell these sets with appropriate tools for their own injectors.

A good tool kit represents a considerable expense but will last a lifetime if cared for. In general, it is not worth buying cheap tools; sooner or later they break, and long before that point, they drive you crazy by slipping and bending. There are quite enough problems in engine work without creating any unnecessary ones.

APPENDIX C

SPARE PARTS

The extent of your spare parts inventory will obviously depend on your cruising plans and personal level of paranoia. The following is a fairly comprehensive list (the farther down the list you go, the more comprehensive it gets!) with an ocean-cruising sailboat in mind. If your plans are less ambitious, you can scale it down appropriately (however, focus on the top end of the list).

1. Oil filters.
2. Fuel filters—include quite a few in case you take on a dirty batch of fuel and need to do repeated filter changes.
3. An air filter (if the engine has one)—one only. This should rarely need changing in the marine environment.
4. A raw-water pump overhaul kit, or at the very least, a diaphragm or impeller (depending on the type of pump).
5. A complete set of belts (alternator, plus auxiliary equipment).
6. An alternator.
7. A starter motor solenoid and bendix unit.
8. An O-ring kit with an assortment of O-rings. May one day be worth its weight in gold.
9. A roll of high-temperature gasket paper.
10. A roll of cork-type gasket paper.
11. A tube of gasket compound.
12. Packing for the stern-tube stuffing box (if fitted).
13. Assorted hose clamps.
14. Flexible fuel line.
15. Hoses.
16. Oil and grease.
17. Penetrating oil.
18. A lift-pump overhaul kit, or at the very least, a diaphragm.
19. A fuel injection pump diaphragm (if one is fitted).
20. Exhaust elbow.
21. Heat exchanger.
22. A cylinder head overhaul gasket set.
23. An inlet valve and an exhaust valve.
24. Two sets of valve springs and keepers (exhaust and inlet, if they differ).
25. A set of piston rings.
26. A set of connecting rod bearing shells.
27. One or two injectors, or matched sets of replacement needle valves and seats and nozzles.
28. A set of high-pressure injection lines (from the fuel injection pump to the injectors).
29. A complete engine gasket set.
30. A transmission oil seal.

APPENDIX D

USEFUL TABLES

This is a rather mixed bag of information that may come in handy at some time or another.

Conversion factors

Metric to U.S. or Imperial conversion factors:
U.S. or Imperial to metric conversion factors:

	To convert from	To	Multiply by	To convert from	To	Multiply by
Length	mm	inch	0.03937	inch	mm	25.40
	cm	inch	0.3937	inch	cm	2.540
	m	foot	3.2808	foot	m	0.3048
Area	mm^2	sq.in.	0.00155	sq. in.	mm^2	645.2
	m^2	sq. ft.	10.76	sq. ft.	m^2	0.093
Volume	cm^3	cu. in.	0.06102	cu. in.	cm^3	16.388
	litre, dm^3	cu. ft.	0.03531	cu. ft.	litre, dm^3	28.320
	litre, dm^3	cu. in.	61.023	cu. in.	litre, dm^3	0.01639
	litre, dm^3	imp. gallon	0.220	imp. gallon	litre, dm^3	4.545
	litre, dm^3	U.S. gallon	0.2642	U.S. gallon	litre, dm^3	3.785
	m^3	cu. ft.	35.315	cu.ft.	m^3	0.0283
Force	N	lbf	0.2248	lbf	N	4.448
Weight	kg	lb.	2.205	lb.	kg	0.454
Power	kW	hp (metric)	1.36	hp (metric) [1)]	kW	0.735
	kW	bhp	1.341	bhp	kW	0.7457
	kW	BTU/min	56.87	BTU/min	kW	0.0176
Torque	Nm	lbf ft	0.738	lbf ft	Nm	1.356
Pressure	Bar	psi	14.5038	psi	Bar	0.06895
	MPa	psi	145.038	psi	MPa	0.006895
	Pa	mm Wc	0.102	mm Wc	Pa	9.807
	Pa	in Wc	0.004	in Wc	Pa	249.098
	kPa	in Wc	4.0	in Wc	kPa	0.24908
	mWg	in Wc	39.37	in Wc	mWg	0.0254
Energy	kJ/kWh	BTU/hph	0.697	BTU/hph	kJ/kWh	1.435
Work	kJ/kg	BTU/lb	0.430	BTU/lb	kJ/kg	2.326
	MJ/kg	BTU/lb	430	BTU/lb	MJ/kg	0.00233
	kJ/kg	kcal/kg	0.239	kcal/kg	kJ/kg	4.184
Fuel consump.	g/kWh	g/hph	0.736	g/hph	g/kWh	1.36
	g/kWh	lb/hph	0.00162	lb/hph	g/kWh	616.78
Inertia	kgm^2	lbft^2	23.734	lbft^2	kgm^2	0.042
Flow, gas	m^3/h	cu.ft./min.	0.5886	cu.ft./min.	m^3/h	1.699
Flow, liquid	m^3/h	US gal/min	4.403	US gal/min	m^3/h	0.2271
Speed	m/s	ft./s	3.281	ft./s	m/s	0.3048
	mph	knots	0.869	knots	mph	1.1508
Temp.	°F=9/5 x °C+32			°C=5/9 x (°F–32)		

FIGURE D-1. *Conversion factors. (Courtesy Volvo Penta)*

Inches	Millimeters	Inches	Millimeters	Inches	Millimeters
0.001	0.0254	0.010	0.2540	0.019	0.4826
0.002	0.0508	0.011	0.2794	0.020	0.5080
0.003	0.0762	0.012	0.3048	0.021	0.5334
0.004	0.1016	0.013	0.3302	0.022	0.5588
0.005	0.1270	0.014	0.3556	0.023	0.5842
0.006	0.1524	0.015	0.3810	0.024	0.6096
0.007	0.1778	0.016	0.4064	0.025	0.6350
0.008	0.2032	0.017	0.4318		
0.009	0.2286	0.018	0.4572		

FIGURE D-2. *Inches to millimeters conversion table.*

Pound-Feet (lb.-ft.)	Newton Metres (Nm)	Newton Metres (Nm)	Pound-Feet (lb.-ft.)
1	1.356	1	0.7376
2	2.7	2	1.5
3	4.0	3	2.2
4	5.4	4	3.0
5	6.8	5	3.7
6	8.1	6	4.4
7	9.5	7	5.2
8	10.8	8	5.9
9	12.2	9	6.6
10	13.6	10	7.4
15	20.3	15	11.1
20	27.1	20	14.8
25	33.9	25	18.4
30	40.7	30	22.1
35	47.5	35	25.8
40	54.2	40	29.5
45	61.0	50	36.9
50	67.8	60	44.3
55	74.6	70	51.6
60	81.4	80	59.0
65	88.1	90	66.4
70	94.9	100	73.8
75	101.7	110	81.1
80	108.5	120	88.5
90	122.0	130	95.9
100	135.6	140	103.3
110	149.1	150	110.6
120	162.7	160	118.0
130	176.3	170	125.4
140	189.8	180	132.8
150	203.4	190	140.1
160	216.9	200	147.5
170	230.5	225	166.0
180	244.0	250	184.4

FIGURE D-3. *Metric conversion tables.*

Fractions	Decimal In.	Metric mm.	Fractions	Decimal In.	Metric mm.
1/64	.015625	.397	33/64	.515625	13.097
1/32	.03125	.794	17/32	.53125	13.494
3/64	.046875	1.191	35/64	.546875	13.891
1/16	.0625	1.588	9/16	.5625	14.288
5/64	.078125	1.984	37/64	.578125	14.684
3/32	.09375	2.381	19/32	.59375	15.081
7/64	.109375	2.778	39/64	.609375	15.478
1/8	.125	3.175	5/8	.625	15.875
9/64	.140625	3.572	41/64	.640625	16.272
5/32	.15625	3.969	21/32	.65625	16.669
11/64	.171875	4.366	43/64	.671875	17.066
3/16	.1875	4.763	11/16	.6875	17.463
13/64	.203125	5.159	45/64	.703125	17.859
7/32	.21875	5.556	23/32	.71875	18.256
15/64	.234375	5.953	47/64	.734375	18.653
1/4	.250	6.35	3/4	.750	19.05
17/64	.265625	6.747	49/64	.765625	19.447
9/32	.28125	7.144	25/32	.78125	19.844
19/64	.296875	7.54	51/64	.796875	20.241
5/16	.3125	7.938	13/16	.8125	20.638
21/64	.328125	8.334	53/64	.828125	21.034
11/32	.34375	8.731	27/32	.84375	21.431
23/64	.359375	9.128	55/64	.859375	21.828
3/8	.375	9.525	7/8	.875	22.225
25/64	.390625	9.922	57/64	.890625	22.622
13/32	.40625	10.319	29/32	.90625	23.019
27/64	.421875	10.716	59/64	.921875	23.416
7/16	.4375	11.113	15/16	.9375	23.813
29/64	.453125	11.509	61/64	.953125	24.209
15/32	.46875	11.906	31/32	.96875	24.606
31/64	.484375	12.303	63/64	.984375	25.003
1/2	.500	12.7	1	1.00	25.4

FIGURE D-3. *(continued)*

APPENDIX E

FREEING FROZEN PARTS AND FASTENERS

Problems with frozen fasteners are inevitable on boats. One or more of the following techniques may free things up.

LUBRICATION

- Clean everything with a wire brush (preferably one with brass or stainless steel bristles), douse liberally with penetrating oil, and wait. Find something else to do for an hour or two, overnight if possible, before having another go. Be patient.
- Clevis pins: After lubricating and waiting, grip the large end of the pin with Vise-Grips (mole wrench) and turn the pin in its socket to free it. If the pin is the type with a cotter pin (also known as a cotter key or split pin) in both ends, remove one of the cotter pins, grip the clevis pin, and turn. Since the Vise-Grips will probably mar the surface of the pin, it should be knocked out from the other end.

SHOCK TREATMENT

An impact wrench is a handy tool to have around. These take a variety of end fittings (e.g., screwdriver bits and sockets) to match different fasteners. Hit the wrench hard with a hammer and hopefully it will jar the fastener loose. If an impact wrench is not available or does not work, apply other forms of shock but with an acute sense of the breaking point of the fastener and adjacent engine castings, etc. Unfortunately this is generally only acquired after a lifetime of breaking things! Depending on the problem, shock treatment may take different forms:

- A bolt stuck in an engine block: Put a sizable punch squarely on the head of the bolt and give it a good knock into the block. Now try undoing it.
- A pulley on a tapered shaft, a propeller, or an outboard motor flywheel: Back out the retaining nut until its face is flush with the end of the shaft (this is important to avoid damage to the threads on the nut or shaft). Apply pressure behind the pulley, propeller, or flywheel as if trying to pull it off, then hit the end of the retaining nut or shaft smartly. The shock will frequently break things loose without the need for a specialized puller.
- A large nut with limited room around it or one on a shaft that wants to turn (for example, a crankshaft pulley nut): Put a short-handled wrench on the nut, hold the wrench to prevent it from jumping off, and hit it hard.
- If all else fails, use a cold chisel to cut a slot in the side of the offending nut or the head of the bolt, place a punch in the slot at a tangential angle to the nut or bolt, and hit it smartly.

LEVERAGE

- Screws: With a square-bladed screwdriver, put a crescent (adjustable) wrench on the blade, bear down hard on the screw, and turn the screwdriver with the wrench. If the screwdriver has a round blade, clamp a pair of Vise-Grips to the base of the handle and do the same thing.
- Nuts and bolts: If using wrenches with one box end and one open end, put the box end of the appropriate wrench on the fastener and hook the box end of the next size up into the free open end of the wrench. This will double the length of the handle and thus the leverage.

- Cheater pipe: Slip a length of pipe over the handle of the wrench to increase its leverage.

Heat

Heat expands metal, but for this treatment to be effective, frozen fasteners must frequently be raised to cherry-red temperatures (see Figures E-1 and E-2), which will upset tempering in hardened steel. Uneven heating of surrounding castings may cause them to crack. Heat must be applied carefully.

Heat applied to a frozen nut will expand it outward, and it can then be broken loose. But equally, heat applied to the bolt will expand it within the nut, generating all kinds of pressure that helps break the grip of rust, etc. When the fixture cools, it will frequently come apart quite easily.

Broken Fasteners

- Rounded-off heads: Sometimes there is not enough head left on a fastener to grip with Vise-Grips or pipe (stillson) wrenches, but there is enough to accept a slot made by a hacksaw. A screwdriver can then be inserted and turned as above.
- If a head breaks off it is often possible to remove whatever it was holding, thus exposing part of the shaft of the fastener. It can be lubricated, gripped with Vise-Grips, and backed out (see Figures E-3 and E-4).
- Drilling out: It is very important to drill down the center of a broken fastener. Use a center punch and take some time putting an accurate "dimple" at this point before attempting to drill. Next, use a small drill to make a pilot hole to the desired depth. If Ezy-Outs or screw extractors (hardened, tapered steel screws with reversed threads—available from tool supply houses) are on hand, drill the correctly sized hole for the appropriate Ezy-Out and try extracting the stud. Otherwise drill out the stud up to the insides of its threads but no farther, or

Figures E-1 and E-2. *Using heat to free frozen parts.*

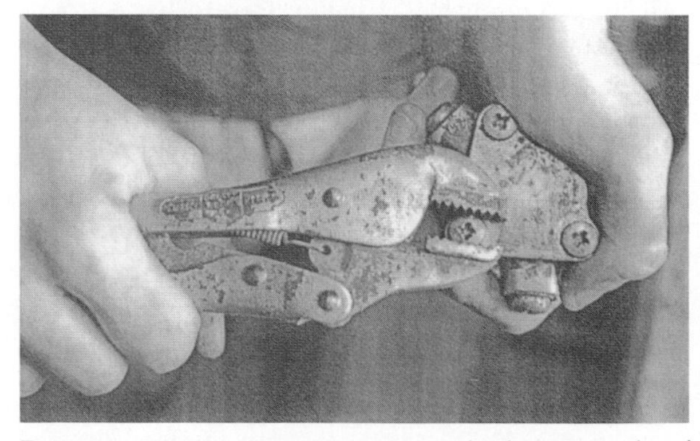

Figures E-3 and E-4. *Using Vise-Grips to back out a stud and screw.*

you will do irreparable damage to the threads in the casting. The remaining bits of fastener thread in the casting can be picked out with judicious use of a small screwdriver or some pointed instrument. If a tap is available to clean up the threads, so much the better.

- Pipe fittings: If you can get a hacksaw blade inside the relevant fittings (which can often be done using duct tape to make a handle on the blade), cut a slit in the fitting along its length, then place a punch on the outside, alongside the cut. Hit it, and collapse it inward. Do the same on the other side of the cut. The fitting should now come out easily.

Miscellaneous

- Stainless steel: Stainless-to-stainless fasteners (for example, many turnbuckles) have a bad habit of *galling* when being done up or undone, especially if there is any dirt in the threads to cause friction. Galling (also known as *cold-welding*) is a process in which molecules on the surface of one part of the fastener transfer to the other part. Everything seizes up for good. Galled stainless fastenings cannot be salvaged—they almost always end up shearing off. When doing up or undoing a stainless fastener, if any sudden or unusual friction develops, *stop immediately*, let it cool off, lubricate thoroughly, work the fastener backward and forward to spread the lubrication around, go back the other way, clean the threads, and start again.
- Aluminum: Aluminum oxidizes to form a dense white powder. Aluminum oxide is more voluminous than the original aluminum and so generates a lot of pressure around any fasteners passing through aluminum fixtures—sometimes enough pressure to shear off the heads of fasteners. Once oxidation around a stainless or bronze fastener has reached a certain point, it is virtually impossible to remove the fastener without breaking it.
- Damaged threads: If all else fails, and a fastener has to be drilled out, the threads in the casting may be damaged. There are two options:
  1. Drill and tap for the next-larger fastener.
  2. Install a Heli-Coil insert. A Heli-Coil is a new thread. An oversized hole is drilled and tapped with a special tap, and the Heli-Coil insert (the new thread) is screwed into the hole with a special tool. You end up with the original size hole and threads. Any good machine shop will have the relevant tools and inserts.

GLOSSARY

AFTERCOOLER. Also called an intercooler. A *heat exchanger* fitted between a *turbocharger* and an engine air-inlet manifold to cool the incoming air.

ALIGNMENT. Bringing together of two coupling halves in near-perfect horizontal and vertical agreement.

ALTERNATOR. A machine for generating electricity by spinning a magnet inside a series of coils. The resulting power output is alternating current. In DC systems, this output is rectified via silicon diodes.

AMBIENT CONDITIONS. The surrounding temperature or pressure, or both.

AMPERE-HOUR. A measure of the amount of electricity stored in a battery.

ANNEALING. A process of softening metals.

ATMOSPHERIC PRESSURE. The pressure of air at the surface of the earth, conventionally taken to be 14.7 pounds per square inch (psi).

ATOMIZATION. The process of breaking up diesel fuel into minute particles as it is sprayed into an engine cylinder.

BABBITT. A soft, white metal alloy frequently used to line replaceable shell-type engine bearings.

BACK PRESSURE. A buildup of pressure in an exhaust system.

BEARING. A device for supporting a rotating shaft with minimum friction. It may take the form of a metal sleeve (a bushing), a set of ball bearings (a roller bearing), or a set of pins around the shaft (a needle bearing).

BEARING RACE. See *race*.

BENDIX. The drive gear (pinion) arrangement on a starter motor.

BLEEDING. The process of purging air from a fuel system.

BLOWBY. The escape of gases past *piston rings* or closed *valves*.

BOTTOM DEAD CENTER (BDC). A term used to describe the position of a *crankshaft* when the #1 *piston* is at the very bottom of its *stroke*.

BRAKE HORSEPOWER (BHP). The actual power output of an engine at the flywheel.

BRUSH. A carbon or carbon-composite spring-loaded rod used to conduct current to or from *commutators* or *slip rings*.

BTU (BRITISH THERMAL UNIT). A unit used to measure quantities of heat.

BUSHING. See *bearing*.

BUTTERFLY VALVE. A hinged flap connected to a throttle that is used to close off the air-inlet manifold on gasoline engines and some diesel engines.

CAMS. Elliptical protrusions on a *camshaft*.

CAMSHAFT. A shaft with *cams* used to operate the *valve mechanism* on an engine.

CAVITATION. The formation and collapse of steam bubbles at the surface of a propeller blade as a result of reduced pressure. It decreases performance and can erode the surface of the propeller.

CIRCLIPS. See *snap rings*.

COLLETS. See *keepers*.

COMBUSTION CHAMBER. The space left in a cylinder (and *cylinder head*) when a *piston* is at the top of its *stroke*.

COMMON RAIL. A type of fuel injection system in which fuel circulates at full injection pressures to all the *injectors* all the time.

COMMUTATOR. The copper segments that are arranged around the end of an armature and on which the *brushes* ride.

COMPRESSION RATIO. The volume of a combustion chamber with the *piston* at the top of its *stroke* as a proportion of the total volume of the cylinder with the piston at the bottom of its stroke.

CONNECTING ROD. The rod connecting a *piston* to a *crankshaft*.

CONNECTING ROD BEARING. The bearing at the crankshaft end of a *connecting rod*.

CONNECTING ROD CAP. The housing that bolts to the end of a connecting rod, holding it to a *crankshaft*.

CRANK. An offset section of a *crankshaft* to which a *connecting rod* is attached.

CRANKING SPEED. The speed at which a starter motor turns over an engine.

CRANKSHAFT. The main rotating part in the base of an engine, transmitting power to the flywheel and *power train*.

CUTLESS BEARING. A ribbed rubber sleeve in a tube used to support the propeller shaft.

CVJ (CONSTANT VELOCITY JOINT). A type of propeller shaft coupling that tolerates substantial engine misalignment.

CYLINDER BLOCK. The housing on an engine that contains the cylinders.

CYLINDER HEAD. A casting containing the *valves* and *injectors* that bolts to the top of a *cylinder block* and seals off the cylinders.

CYLINDER LINER. A machined sleeve that is pressed into a *cylinder block* and in which a *piston* moves up and down.

Decarbonizing. The process of removing carbon from the inside surfaces of an engine and of refurbishing the *valves* and *pistons*.

Decompression levers. Levers that hold the exhaust valves open so that no compression pressure is built up, making it easy to turn over the engine.

Dial indicator. A sensitive measuring instrument used in engine and alignment work.

Diaphragm. A reinforced rubber membrane that moves in and out of certain pumps.

Displacement. The total swept volume of an engine's cylinders expressed in cubic inches or liters.

Distributor pump. A type of *fuel injection pump* using one central pumping element with a rotating distributor head that sends the fuel to each cylinder in turn.

Dribble. Drops of unatomized fuel entering a cylinder through faulty injection.

Feeler (thickness) gauges. Thin strips of metal machined to precise thicknesses and used for measuring small gaps.

Field windings. Electromagnetic coils used to create magnetic fields in alternators, generators, and electric motors.

Filter. A device for screening out impurities in fuel, air, or water.

Flexible-impeller pump. A pump with a rubber *impeller* and a *cam* on one side of the pump chamber. As the impeller passes the cam, its vanes are squeezed down, expelling fluids trapped between them. The vanes then spring back, sucking in more fluid.

Flyweight. A small pivoted weight used in mechanical *governors*.

Fuel injection pump. A pump designed to meter out precisely controlled amounts of fuel and then raise them to injection pressures at precisely controlled moments in an engine cycle.

Gasket. A piece of material placed between two engine parts to seal them against leaks. Gaskets are normally fiber but sometimes metal, cork, or rubber.

Gear ratio. The relative size of two gears. If the gears are in contact, their relative speed of rotation will be given by the gear ratio. Example: If the gear ratio is 8:1, the smaller gear will rotate eight times faster than the larger gear.

Generator. A machine for generating electricity by spinning a series of coils inside a magnet. The resulting power output is alternating current. In DC systems, this output is rectified via a *commutator* and *brushes*.

Glow plug. A heating element installed in a diesel engine precombustion chamber to aid in cold starting.

Governor. A device for maintaining an engine at a constant speed, regardless of changes in load.

Header tank. A small tank set above an engine on heat exchanger–cooled systems. The header tank serves as an expansion chamber, coolant reservoir, and pressure regulator (via a pressure cap).

Head gasket. The *gasket* between a *cylinder head* and a *cylinder block*.

Heat exchanger. A vessel containing a number of small tubes through which the engine cooling water is passed, while raw water is circulated around the outside of the tubes to carry off heat from the cooling water.

Hole-type nozzle. An *injector nozzle* with one or more very fine holes—generally used in direct (open) combustion chambers.

Horsepower. A unit of power used in rating engines. See also *brake horsepower*; *shaft horsepower*.

Hunting. A rhythmic cycling up and down in speed of a governed engine, usually caused by governor malfunction.

Hydrometer. A float-type instrument used to determine the state of charge of a battery by measuring the specific gravity of the electrolyte (i.e., the amount of sulfuric acid in the electrolyte).

Impeller. The rotating fitting that imparts motion to a fluid in a rotating pump.

Inches of mercury ("Hg). A scale for measuring small pressure changes, particularly those below *atmospheric pressure* (vacuums).

Injection pump. See *fuel injection pump*.

Injection timing. The relationship of the beginning point of injection to the rotation of the *crankshaft*.

Injector. A device for atomizing diesel fuel and spraying it into a cylinder.

Injector nozzle. The part of an *injector* containing the *needle valve* and its seat.

Injector nut. The nut that holds a fuel line to an *injector*.

In-line pump. A series of *jerk pumps* in a common housing operated by a common *camshaft*.

Intercooler. See *aftercooler*.

Inverter. A device for changing DC to AC.

Jerk pump. A type of *fuel injection pump* that uses a separate pumping element for each cylinder.

Keepers. Small, dished, metal pieces that hold a valve spring assembly on a valve stem.

Lapping. Grinding two parts together to make an exact fit.

Lift pump. A low-pressure pump in a fuel injection system supplying fuel from the tank to an injection pump.

Line contact. The machining of two mating surfaces at different angles so that they make contact only at one point.

Lip-type seal. A seal, using automotive-style oil seals, that is used in place of a *stuffing box*.

Main bearing. A bearing within which a *crankshaft* rotates, and which supports the crankshaft within an engine block.

Manifold. A pipe assembly attached to an engine block that conducts air into the engine or exhaust gases out of it.

Mechanical seal. A spring-loaded seal used in place of a *stuffing box* on a propeller shaft.

Micrometer. A tool for making precision measurements.

Naturally aspirated. Refers to an engine that draws in air solely by the action of its *pistons*, without the help of a *supercharger* or *turbocharger*.

Needle bearing. See *bearing*.

Needle valve. The valve in an *injector nozzle*.

NOZZLE BODY. The housing at the end of an *injector* that contains the *needle valve*.

NOZZLE OPENING PRESSURE. The pressure required to lift an injector *needle valve* off its seat so that injection can take place.

ORIFICE. A very fine opening in a nozzle.

PACKING. Square, grease-impregnated, natural fiber rope, usually hemp (flax), used to seal *stuffing boxes*.

PACKING GLAND. See *stuffing box*.

PINION. A small gear designed to mesh with a large gear (for example, a starter motor drive gear).

PINTLE NOZZLE. An *injector nozzle* with one central hole. Generally used in engines with precombustion chambers.

PISTON. A pumping device used to generate pressure in a cylinder.

PISTON CROWN. The top of a *piston*.

PISTON PIN (WRIST PIN). A pin connecting a *piston* to its *connecting rod*, allowing the piston to oscillate around the rod.

PISTON-RING CLAMP. A tool for holding *piston rings* tightly in their grooves to enable the piston to be slid into its cylinder.

PISTON-RING GROOVE. The slot in the circumference of a *piston* into which a piston ring fits.

PISTON RINGS. Spring-tensioned rings set in grooves in the circumference of a *piston* that push out against the walls of its cylinder to make a gas-tight seal.

PITCH. The theoretical distance a propeller will move a boat in one revolution of the propeller, assuming no losses. See also *slip*.

POINT LOADING. Uneven loading on a *bearing*, which throws all the pressure on one part of the bearing instead of distributing it evenly over the whole bearing.

PORTS. Holes in the wall of a cylinder that allow gases in and out.

POUNDS PER SQUARE INCH (PSI). Pressure measurement. *Psia* (pounds per square inch absolute) measures actual pressure with no allowance for *atmospheric pressure*. *Psig* (pounds per square inch gauge) measures pressure with the gauge set to zero at *atmospheric pressure* (14.7 psia). In other words, psig = psia – 14.7. Unless otherwise stated, psi always refers to psig.

POWER TRAIN. The components used to turn an engine's power into a propulsive force.

PUMPING LOSSES. Energy losses arising from friction in the inlet and exhaust passages of an engine.

PUSH ROD. A metal rod used to transfer the motion of a *camshaft* to a *rocker arm*.

PYROMETER. A gauge used to measure exhaust temperatures.

RACE. The inner and outer cases on a *bearing* between which the balls are trapped.

RAW WATER. The seawater side of cooling systems; also the water in which a boat is floating.

RECIPROCAL. Up-and-down motion.

RESISTANCE. The opposition an appliance or wire offers to the flow of electric current, measured in ohms.

ROCKER ARM. A pivoted arm that operates a *valve*.

ROCKER COVER. See *valve cover*.

ROD-END BEARING. The *bearing* at the *crankshaft* end of a *connecting* rod.

ROLLER BEARING. See *bearing*.

ROTOR. The name given to the rotating *field winding* arrangement in an *alternator*.

SACRIFICIAL ANODES. Anodes of a less noble metal (generally zinc) electrically connected to underwater hardware and designed to corrode, thereby protecting the rest of the hardware.

SCAVENGING. In a two-cycle diesel, the process of replacing the spent combustion gases with fresh air.

SEIZURE. The process by which excessive friction brings an engine to a halt.

SERIES-WOUND MOTORS. A DC motor in which the *field winding* is connected in series with the armature. If unloaded, series-wound motors can run away and can self-destruct.

SHAFT HORSEPOWER (SHP). The actual power output of an engine and *power train* measured at the propeller shaft.

SHIM. A specially cut piece of *shim stock* used as a spacer in specific applications.

SHIM STOCK. Very thin, accurately machined pieces of metal.

SIPHON. The ability of a liquid to flow through a hose if one end is lower than the liquid level, even if the hose is looped above the liquid level.

SLIP. The difference between the theoretical movement of a propeller through the water and its actual movement. See also *pitch*.

SLIP RINGS. Insulated metal discs on a *rotor* or armature shaft through which current is fed, via *brushes*, to or from armature or rotor windings.

SNAP-RING PLIERS. Special pliers for installing and removing *snap rings*.

SNAP RINGS. Spring-tensioned rings that fit into a groove on the inside of a hollow shaft or around the outside of a shaft.

SOLENOID. An electrically operated *valve* or switch.

SPECIFIC GRAVITY (SG). A measure of the density of a liquid as compared to that of water.

SPEEDER SPRING. The spring in a *governor* that counterbalances the centrifugal force of the *flyweights*.

STROKE. The movement of a *piston* from the bottom to the top of its cylinder.

STUFFING BOX. A device for making a watertight seal around a propeller shaft at the point where it exits a boat.

SUPERCHARGER. A mechanically driven blower used to pressurize the inlet air.

SWEPT VOLUME. The volume of a cylinder displaced by a *piston* in one complete *stroke* (i.e., from the bottom to the top of its cylinder).

THERMOSTAT. A heat-sensitive device used to control the flow of coolant through an engine.

THICKNESS GAUGES. See *feeler gauges*.

THRUST BEARING. A *bearing* designed to take a load along the length of a shaft (as opposed to perpendicular to it).

TIMING. The relationship of valve and fuel pump operation to the rotation of the *crankshaft* and to each other.

TOP DEAD CENTER (TDC). A term used to describe the position of a *crankshaft* when the #1 *piston* is at the very top of its *stroke*.

Torque. A twisting force applied to a shaft.

Torque wrench. A special wrench that measures the force applied to a nut or bolt.

Turbocharger. A blower driven by engine exhaust gas used to pressurize the inlet air.

Vacuum. Pressure below *atmospheric pressure*.

Valve clearance. The gap between a valve stem and its *rocker arm* when the valve is fully closed.

Valve cover. The housing of an engine bolted over the valve mechanism.

Valve guide. A replaceable sleeve in which the valve stem fits and slides up and down.

Valve keepers. See *keepers*.

Valve overlap. The period of time in which an exhaust valve and inlet valve are both open.

Valves. Devices for allowing gases in and out of a cylinder at precise moments.

Valve seat. The area in a *cylinder head* on which a valve sits to seal that head.

Valve spring. The spring used to hold a valve in the closed position when it is not actuated by its *rocker arm*.

Valve-spring clamp. A special tool to assist in removing valves from their *cylinder heads*.

Viscosity. The resistance to flow of a liquid (its thickness).

Volatility. The tendency of a liquid to evaporate (vaporize).

Volumetric efficiency. The efficiency with which a four-cycle diesel engine replaces spent combustion gases with fresh air.

Wrist pin. See *piston pin*.

Yoke. The hinged and forked lever arm that couples a *governor flyweight* to its driveshaft.

INDEX

Numbers in **bold** refer to pages with illustrations

G

H

I

J

K

L

M

N

O

P

R

S

T

U

V